Anachronisms in the History of Mathematics

The controversial matters surrounding the notion of anachronism are difficult ones: they have been broached by literary and art critics, by philosophers, as well as by historians of science. This book adopts a bottom-up approach to the many problems concerning anachronism in the history of mathematics. Some of the leading scholars in the field of history of mathematics reflect on the applicability of present-day mathematical language, concepts, standards, disciplinary boundaries, indeed notions of mathematics itself, to well-chosen historical case studies belonging to the mathematics of the past, in European and non-European cultures. A detailed introduction describes the key themes and binds the various chapters together.

The interdisciplinary and transcultural approach adopted allows this volume to cover topics important for history of mathematics, history of the physical sciences, history of science, philosophy of mathematics, history of philosophy, methodology of history, non-European science, and the transmission of mathematical knowledge across cultures.

NICCOLÒ GUICCIARDINI is Professor in History of Science at the State University of Milan. He holds degrees in physics and philosophy. He has been awarded the Gil Prize (Gulbenkian Foundation), the Bacon Prize (Caltech and the Francis Bacon Foundation), and the Sarton Medal (Ghent University). He is the author of two monographs on Newton's mathematics published by Cambridge University Press.

Anachronisms in the History of Mathematics

Essays on the Historical Interpretation of Mathematical Texts

Edited by

NICCOLÒ GUICCIARDINI
Universitá degli Studi di Milano

CAMBRIDGE
UNIVERSITY PRESS

CAMBRIDGE
UNIVERSITY PRESS

University Printing House, Cambridge CB2 8BS, United Kingdom

One Liberty Plaza, 20th Floor, New York, NY 10006, USA

477 Williamstown Road, Port Melbourne, VIC 3207, Australia

314–321, 3rd Floor, Plot 3, Splendor Forum, Jasola District Centre, New Delhi – 110025, India

103 Penang Road, #05–06/07, Visioncrest Commercial, Singapore 238467

Cambridge University Press is part of the University of Cambridge.

It furthers the University's mission by disseminating knowledge in the pursuit of education, learning, and research at the highest international levels of excellence.

www.cambridge.org
Information on this title: www.cambridge.org/9781108834964
DOI: 10.1017/9781108874564

First published 2021

A catalogue record for this publication is available from the British Library.

ISBN 978-1-108-83496-4 Hardback

Contents

Contributors *page* xi

Figures xii

Preface xvii

1 **Introduction: The historical interpretation of mathematical texts and the problem of anachronism**
Niccolò Guicciardini 1

1.1 Purpose of this book 2
1.2 Defining anachronism 3
1.3 On mathematical anachronism: the problem 5
1.4 Deceptive familiarity 14
1.5 Tasks and criteria of quality control 19
1.6 Changing present viewpoints 26
1.7 Reading the text 29
1.8 Summing up 32
References 34

2 **From reading rules to reading algorithms: textual anachronisms in the history of mathematics and their effects on interpretation**
Karine Chemla 42

2.1 Introduction 42
2.2 Edouard Biot's approach to Chinese mathematical texts: a literal interpretation 46
2.3 Mikami on rules and problems: a mathematical and contextual interpretation 57

2.4 How a text refers to the computation meant: Joseph Needham, Wang Ling, and Donald Knuth 69
2.5 Conclusion 78
References 79

3 Anachronism and anachorism in the study of mathematics in India
Kim Plofker 83
3.1 Introduction 83
3.2 The "error" of division by zero in medieval Sanskrit algebra 87
3.3 Mādhava's infinitesimal "calculus" in fourteenth-century Kerala 92
3.4 Conclusion 101
References 102

4 On the need to re-examine the relationship between the mathematical sciences and philosophy in Greek antiquity
Jacqueline Feke 105
4.1 Introduction 105
4.2 Claudius Ptolemy 117
4.3 Hero of Alexandria 121
4.4 Archytas of Tarentum 125
4.5 Conclusion 126
References: Primary Sources 127
References: Secondary Sources 129

5 Productive anachronism: on mathematical reconstruction as a historiographical method
Martina R. Schneider 131
5.1 Introduction 132
5.2 Reconstructing a Chinese approach to indeterminate analysis 134
5.2.1 Background: Sources and their reception 134
5.2.2 On Ludwig Matthiessen's biography 136
5.2.3 Matthiessen's reconstructions of the dayan rule 138
5.3 Discussion 149

5.3.1 Different types of mathematical reconstruction 149
5.3.2 A historiographic success or failure? 153
5.3.3 The dayan rule and anachronism 156
5.3.4 On the example's complex time structure 157
5.4 Concluding remarks 158
References 161

6 Anachronism in the Renaissance historiography of mathematics
Robert Goulding 167
6.1 Introduction 167
6.2 Ramus on the early Academy 169
6.3 Mathematics in the early Academy 175
6.4 Proclus unstuck in time 178
6.5 Conclusion 191
References 193

7 Deceptive familiarity: differential equations in Leibniz and the Leibnizian school (1689–1736)
Niccolò Guicciardini 196
7.1 The "birth of analytical mechanics:" qualifying a historiographical category 196
7.2 Deceptive familiarity 198
7.3 Bernoulli's equation for central force motion 200
7.3.1 The equation and its domestication 200
7.3.2 Reduction to quadrature 201
7.3.3 Solution 203
7.4 Equations and constants 205
7.5 Solutions as constructions 209
7.6 The slow transition from geometry to algebra 212
7.7 Some lessons 217
References 221

8 Euler and analysis: case studies and historiographical perspectives
Craig Fraser and Andrew Schroter 223
8.1 Introduction 223
8.2 Euler and the invariance of the variational equations 224
8.2.1 Euler's variational equation 224

8.2.2 Invariance in calculus of variations and analytical dynamics 227
8.2.3 Some critical reflections 231
8.2.4 Euler and the foundations of analysis 233
8.3 Euler and divergent series 234
8.3.1 Convergence and rigor 234
8.3.2 Infinite series in the eighteenth century 234
8.3.3 Cauchy's new definitions 236
8.3.4 Summability theory 238
8.3.5 Different kinds of definition 241
8.3.6 Euler's definitions 242
8.4 Conclusion 246
References 247

9 Measuring past geometers: a history of non-metric projective anachronism
Jemma Lorenat 251
9.1 Introduction 252
9.2 The non-metric projective anachronism in the historiography of geometry 255
9.3 The distinction between metric and the non-metric among early-nineteenth-century geometers 256
9.4 *Geometrie der Lage* as an introduction to the geometry of modern times 260
9.5 The beginning of "projective geometry" 263
9.6 Klein's historiography of projective geometry 267
9.7 The historiography of projective geometry in the *Encyklopädie der mathematischen Wissenschaften* 270
9.8 Conclusion 274
References 275

10 Anachronism: Bonola and non-Euclidean geometry
Jeremy Gray 281
10.1 Introduction 281
10.2 Anachronism 282
10.3 Bonola (1906) 284
10.3.1 Bonola's sources 285
10.3.2 Roberto Bonola (1874–1911) 286
10.4 Elementary mathematics 287

10.5 Bonola's "elementary" account 292
10.5.1 Lobachevskii 293
10.5.2 Distance 297
10.6 The acceptance and rejection of non-Euclidean geometry 300
10.7 Conclusion 303
References 304

11 Anachronism and incommensurability: words, concepts, contexts, and intentions
Joseph W. Dauben 307
11.1 Introduction 308
11.2 Transfinite set theory, nonstandard analysis, and Charles Sanders Peirce 308
11.3 Anachronisms and ancient Chinese mathematics 328
11.3.1 Ancient Chinese surveying methods: the double-distance method 328
11.3.2 Chinese algorithms for calculating square roots 337
11.3.3 The Chinese 勾股 *Gou-Gu* (Pythagorean) theorem: proofs and diagrams 344
11.4 Conclusion: hedgehogs and foxes 348
References 352

Index 358

Contributors

Karine Chemla *SPHERE (CNRS–Université Paris), Paris, France*

Joseph W. Dauben *Department of History, Herbert H. Lehman College, and The Graduate Center, City University of New York, New York NY, USA*

Jacqueline Feke *Department of Philosophy, University of Waterloo, Waterloo ON, Canada*

Craig Fraser *Institute for the History & Philosophy of Science & Technology, University of Toronto, Toronto ON, Canada*

Robert Goulding *Department of History, University of Notre Dame, Notre Dame IN, USA*

Jeremy Gray *School of Mathematics and Statistics, The Open University, Milton Keynes, and University of Warwick, Warwick, UK*

Niccolò Guicciardini *Dipartimento di Filosofia "Piero Martinetti", Università degli Studi di Milano, Milan, Italy*

Jemma Lorenat *Department of Mathematics, Pitzer College, Claremont CA, USA*

Kim Plofker *Department of Mathematics, Union College, Schenectady NY, USA*

Martina R. Schneider *Institut für Mathematik FB 08 – Physik, Mathematik und Informatik, Johannes Gutenberg-Universität, Mainz, Germany*

Andrew Schroter *Branksome Hall, Toronto, ON, Canada*

Figures

2.1 Mikami transforms the "rule" into a formula. 60

3.1 The circle quadrant of radius R inscribed in the southeast corner of a square of side $2R$, on which the derivation of Mādhava's result for the circumference is based. 93

5.1 Overview of texts related to Wylie (1852) and published in Europe between 1856 and 1882: "translations" of Wylie (1852) or Biernatzki (1856) in light grey, interpretations and reconstructions of Biernatzki (1856) in black. 136

5.2 Excerpt from Ludwig Matthiessen's *Zur Algebra der Chinesen* (1874). 140

5.3 Excerpt from Ludwig Matthiessen's *Ueber eine antike Auflösung des sogenannten Restproblemes in moderner Darstellung* (1882). 146

5.4 Resources of Matthiessen's study of the dayan rule. 148

7.1 Bernoulli's differential equation for central force motion (1710). 200

7.2 Substitutions of variables in Johann Bernoulli's paper on central force motion (1710). 202

7.3 ignorespaces Equation of the "curve of speeds" in Johann Bernoulli's paper on central force motion (1710). 206

7.4 ignorespaces Pierre Varignon's graphs representing kinematical magnitudes (1700). 207

7.5 Construction of conic trajectories for an inverse-square force according to Johann Bernoulli (1710). 210

8.1 Tabula I, Fig. 4, from Leonhard Euler's *Methodus inve-*

niendi lineas curvas maximi minimive proprietate gaudentes (1744). 225

8.2 Tabula I, Fig. 7, from Leonhard Euler's *Methodus inveniendi lineas curvas maximi minimive proprietate gaudentes* (1744). 226

11.1 (a) Robinson at the Hebrew University, Jerusalem, 1957–1958. (b) Robinson's MAA lecture, *Nonstandard Analysis*, filmed in 1970. (c) Robinson celebrating his Brouwer medal with Arend Heyting, Leiden, April 26, 1973. 309

11.2 (a) Georg Cantor, 1870s. (b) Richard Dedekind, 1870s. (c) First page of Cantor's paper on the denumerability of all algebraic numbers (1874). 311

11.3 (a) C.S. Peirce, ca. 1867. (b) Georg Cantor, ca. 1895. 314

11.4 First page of Peirce's manuscript "Multitude and Number" (1897), 1. 316

11.5 "Multitude and Number" (1897), 5. 317

11.6 Detail of "Multitude and Number" (1897), 32. 318

11.7 "Multitude and Number" (1897), 32. 319

11.8 Detail from the published version of "Multitude and Number" (1931), 4.196. 319

11.9 Detail of "Multitude and Number" (1897), 32. 319

11.10 Page and detail from Jerome Keisler's textbook, *Elementary Calculus: An Infinitesimal Approach* (2012), 25. 325

11.11 (a) Title page from Robinson's *Non-Standard Analysis* (1966). (b) Abraham Robinson, Sterling Professor, Yale University (1971). (c) Frontispiece from Euler's *Introductio in Analysin Infinitorum* (1748). 326

11.12 (a) Illustration of the first problem in Liu Hui's *Sea Island Mathematical Manual*, from the 古今圖書集成 *Gujin tushu jicheng* (Encyclopedia of Illustrations and Writings from Antiquity to the Present (1726). (b) First page of the 海算經 *Haidao suanjing* (Sea Island Mathematical Manual). 328

11.13 (a) Title page of the journal, 通報 *Tong Pao*. (b) Title page of Louis van Hee's article on the *Sea Island Mathematical Manual*, in *Tong Pao* (1920), 51. (c) Diagram from van Hee's article on the *Sea Island Mathematical Manual*, in *Tong Pao* (1920), 57. 330

11.14 Diagram illustrating the first problem in the *Sea Island*

Mathematical Manual, from Li Yan and Du Shiran's *Chinese Mathematics. A Concise History* (1987), 76. 330
11.15 Diagrams for proving the equality of areas inscribed in a rectangle. 331
11.16 "Hypotenuse" diagram for proving the *gou-gu* (Pythagorean) theorem. 331
11.17 "Hypotenuse" diagram for proving the *gou-gu* (Pythagorean) theorem. 333
11.18 "Hypotenuse" diagram for proving the *gou-gu* (Pythagorean) theorem. 333
11.19 (a) Title page of the 周髀算經 *Zhoubi suanjing* (Mathematical Classic of the Zhou Gnomon). (b) 日高圖 *ri gao tu* (height of the Sun diagram) from the 周髀算經 *Zhoubi suanjing*. 335
11.20 Wu Wenjun's diagram illustrating a proof of the "double-difference" method. 335
11.21 Autograph page from Li Yan's copy of the 楊輝算 *Yang Hui suanfa* (Arithmetical Methods of Yang Hui). 336
11.22 Wu Wenjun's illustration of the "Western" approach to proving the double-difference rule. 337
11.23 Cover showing nine of the bamboo slips published by Peng Hao in his edition of the 算數書 *Suan shu shu* (A book on numbers and computations) (2001). 338
11.24 Cover of the 永樂大典 *Yongle dadian* and the oldest surviving diagram illustrating the square root extraction method (1403–1408). 340
11.25 Diagram showing the first step of the square root extraction method. 340
11.26 Diagrams showing the next steps of the square root extraction method. 341
11.27 (a) Wang Ling and Arnold Koslow at the VIIIth International Congress of the History of Science, Excursion to Vinci, Italy, September, 1956. (b) Wang Ling and Joseph Needham, Gonville and Caius College, Cambridge University. 345
11.28 (a) Page discussing the *gou-gu* (Pythagorean) theorem, *Science and Civilisation in China* (1959), 22. (b) The "hypotenuse diagram" (enlarged) for proving the *gou-gu* theorem. 346

11.29 (a) Arnold Koslow's original sketches for his "proof" of the *gou-gu* theorem. (b) and (c) Reconstructed diagrams for ancient proof of the *gou-gu* theorem. 347

Preface

My awareness that there are several historians of mathematics who have asked themselves questions about the legitimacy and limitations of anachronistic readings of past texts led me to the project of organizing a workshop on this theme. The workshop took place in Pasadena in April 2018 in the context of the Francis Bacon Award in the History and Philosophy of Science and Technology, which I had the honor of receiving from the Francis Bacon Foundation and the California Institute of Technology.

The scholars that convened in Pasadena contributed excellent talks and engaged in a wonderful discussion. The present volume gathers much improved and expanded versions of most of the talks, while one chapter (Goulding's) has been commissioned for the production of this volume. I express my gratitude to all the authors for their wonderful commitment: this volume was made possible by their enthusiasm, competence, and impeccable scholarship.

Plan of the book

The book opens with a chapter in which I have tried to provide a definition of "anachronism" as well as an overview of the different positions concerning anachronism taken by historians of mathematics. This chapter originally consisted of just a few pages, but the invitation of two anonymous referees and questions from the Cambridge editor, David Tranah, somewhat "compelled" me to expand it into a rather long essay.

In the introductory Chapter 1 I ask some of the questions that, in my opinion, deserve our attention and that indeed find different answers by the authors of this volume. The issue of anachronism is a vexed one in historical interpretation, but is particularly thorny when one attempts to historicize a discipline that is, perhaps naively, celebrated for being independent of the context. The reader will notice the central importance I give to the work of Henk Bos, a historian of Descartes's and Leibniz's mathematics who has, I believe, defined a historical methodology, based on the notions of "tasks and criteria of quality control," which answers in a very original and promising way the many issues raised by the use of anachronisms in the historical interpretation of past mathematical texts. One of the most fascinating features of this book, which I highlight in Chapter 1, is that for the historian of mathematics anachronism is not *per se* a sin we should try our best to avoid. There are indeed inevitable, even virtuous, ways in which our present knowledge of mathematics plays a positive heuristic role in the interpretation of past mathematical texts. But the projection of our present knowledge onto the past can lead to erroneous readings too. The authors of the following chapters broach the theme of the book by considering case studies taken from their own fields of expertise.

Karine Chemla identifies a kind of anachronism which she terms "textual anachronism," and which is most widespread in the history of mathematics. It consists in taking the forms of texts and inscriptions that we read in ancient sources (e.g., a mathematical problem, an algorithm, a proof, a diagram) as being the same as their modern counterparts. This kind of anachronism has caused misinterpretations of several kinds. Chemla examines how some ancient Chinese procedures were read successively by Edouard Biot (1803–1850), Mikami Yoshio (1875–1950), Joseph Needham (1900–1995), and Wang Ling (1917–1994). Thanks to a close analysis of the texts, she shows how the assumptions about these procedures that each of these authors made, and the material conditions of their work, influenced the views they formulated about the mathematics of ancient China. In the last part of her chapter, Chemla argues that the development of algorithms by mathematicians such as Donald Knuth since the 1950s has introduced new mathematical ideas that can be deployed to interpret ancient texts less anachronistically. What is fascinating about Chemla's chapter is that it demonstrates how a very recent viewpoint brought about by a research program on numerical algorithms can lead to a reading of ancient Chinese mathematics that is valuable for the historian of mathematics. Our present knowledge of mathematics, can thus, in this case, be a heuristically fruitful tool for the historian.

The lineal descent of fundamental ideas and methods in modern mathematics from classical antiquity makes the concept of "anachronism" a natural concern when studying pre-modern successors to Hellenistic mathematics: are we projecting our own awareness of modern versions and developments of these ideas onto earlier thinkers? Kim Plofker introduces another fundamental concept. In studying the history of other mathematical traditions such as those of India and China, we should be aware of a form of projection that is more geographical in nature than temporal: what she calls "anachorism." For instance, when we label/equate an ancient non-Greek mathematical development with its European "equivalent," our interpretation is not so much "out of time" as "out of place." Important aspects of this combined phenomenon of anachron/chorism include its potential for both insights and misunderstandings in the reading of texts (both of which arise in the historiographical controversy about whether the infinitesimal mathematics of the late medieval Kerala school of south India amounts to "calculus"). Thus, Plofker throws into relief the tension, so characteristic of the history of mathematics, between the mathematical search for the equivalence between results expressed in texts separated by time and distance, and the recognition of their situatedness. As she shows, in trying a trade-off between these two aims one risks falling into either Eurocentric teleology (what is praiseworthy about fourteenth-century Mādhava's results is their proximity to Newton's) or orientalism (Mādhava's results are incommensurably "Indian").

Jacqueline Feke examines the relationship between mathematics and philosophy in Greek antiquity, and the perspectives brought to the study of ancient Greek thought by modern historians. For the ancient Greeks, mathematics was part of philosophy, along with physics and theology. Modern disciplinary differences between philosophy and mathematics are reflected in the approach that historians of philosophy take to classical Greek texts today. The implicit modern separation of mathematics and philosophy is imposed anachronistically on the study of ancient Greek thought, where no such separation existed. The examination of some ancient Greek mathematical texts, moreover, reveals that some mathematicians were aiming to solve the traditional, most fundamental, problems of philosophy, understood, as Pierre Hadot has argued, as a way to attain a "good life" and a just society. Feke's chapter thus deals with the form of anachronism concerning disciplinary boundaries that I discuss in Chapter 1 in relation to Geoffrey Lloyd's work.

Martina Schneider examines a nineteenth-century European example of mathematical reconstruction pertaining to the history of Chinese mathematics. In 1852 the English missionary Alexander Wylie published "Jottings on the Science of Chinese Arithmetics" in the North-China Herald. This article was "translated" into German by Karl Biernatzki in 1856. Biernatzki rearranged and changed Wylie's text, thereby adding mistakes. This led to an unfavorable evaluation of Chinese arithmetic by Moritz Cantor and Hermann Hankel. Ludwig Matthiessen, professor of physics in Rostock, spotted Biernatzki's mistakes regarding the dayan rule (the Chinese remainder theorem). On the basis of the translated paper on Chinese mathematics and a close study of its examples, Matthiessen was able to reconstruct the dayan rule for arbitrary moduli and discussed conditions of solvability. He stressed that the Chinese had "exactly the same" method as Gauss, for moduli that were relatively prime, and pointed out that they even had one for not relatively prime moduli, thus surpassing Gauss. This finding led to a more favorable reception of Chinese mathematics. Schneider points to the complexity of the time structure (pluritemporality, in Achim Landwehr's terminology) at work in the example. Indeed, Landwehr, rejecting a linear conception of historical time and elaborating in its stead a more complex time scheme (pluritemporality), has rethought the concept of anachronism. Schneider, following Landwehr and Reinhart Koselleck, argues that the case of Matthiessen's historical reconstruction of the dayan rule can be seen as productive of new knowledge about the mathematical past. The final part of this philosophically deep chapter is devoted to study the conditions under which mathematical reconstructions can become a heuristic tool in the history of mathematics generating new hypotheses and interpretations. She concludes that in the dynamic process of historical interpretation, historiographically sensitive mathematical reconstructions (such as Matthiessen's), even when based on seemingly erroneous historiographical methods (no use of original sources, use of contemporary mathematics), might lead to a deeper understanding of mathematical traces of the past.

Robert Goulding devotes his chapter to Petrus Ramus's *Prooemium*

Mathematicum (1567), one of the most influential Renaissance histories of mathematics. Goulding carefully analyses three, for us obvious, mistakes in Ramus's narrative: his account of the early Platonic Academy as dominated by anti-metaphysical materialists, his evaluation of the work of Eudoxus of Cnidus, and the significance for his philosophy of mathematics of a gross misdating of the neo-Platonic commentator of Euclid, Proclus. Goulding's chapter is the most telling proof that the history of anachronism can generate historical insights that make us aware both of the distance of past authors from our own present viewpoint – a viewpoint that would make our predecessors' anachronism impossible – and of the contingency of our own present narrative of the mathematical past, which – for sure – will be seen as anachronistic by our successors. As Goulding observes in his conclusion, when we criticize Renaissance historiography for being "anachronistic," we should neither forget how recently modern understandings of the history of mathematics and of philosophy emerged, nor miss the fact that the Renaissance history of mathematics answered desiderata situated in contemporary conceptions of university pedagogy and the philosophy of mathematics.

Niccolò Guicciardini discusses the dangers and advantages we encounter when we translate past mathematics into our modern language. An apparently innocent (and ever so slight!) change in notation in rewriting a differential equation by Johann Bernoulli in such a way that it can be received in our textbooks leads to a way of writing that would have been disowned by its author – and for reasons that were explicitly discussed in his correspondence with Leibniz. By rewriting the equation according to our conventions, we miss how this equation should have been written in order to fulfill what Bos calls the "self-imposed tasks" Bernoulli had in mind. On the other hand, a modernizing (domesticating in Lawrence Venuti's terms) translation delivers a number of advantages that Guicciardini discusses in concluding his chapter.

Leonhard Euler is celebrated for his contributions to the calculus of variations and the theory of infinite series. According to Craig Fraser and Andrew Schroter, certain concepts from these subjects occupy a fundamental place in modern analysis, but do not appear in the work of either Euler or his contemporaries. In the case of variational calculus there is the concept of the invariance of the variational equations which several historians read back into Euler's work. Further, modern commentators such as Morris Kline believe that the theory of summability – developed at the end of the nineteenth century by Ferdinand Georg Frobenius, Ernesto Cesàro, and others – may be found in Euler's work on divergent series. This thesis is in line with the view of some historians, such as Adolph-Andrei Pavlovich Yushkevich, that Euler possessed a kind of visionary intuition of mathematical concepts and lines of development that would only develop much later. Fraser and Schroter examine the anachronism implicit in this point of view. Euler understood divergent series as things that were given as part of objective reality, and not simply defined in such a way as to make them whatever the investigator wishes them to be, as they are in summability theory. The subject of summability grew out of researches in complex analysis within a mathematical framework that was foreign to Euler's mathematics. In their conclusion, Fraser and Schroter comment on some parallels between claims for the historiographical relevance of summability theory and non-Archimedean analysis. Fraser and Schroter's chapter illustrates what is perhaps the most typical kind of anachronism in the history of mathematics, the kind which occurs when an object at first assumed as given becomes in a later period the result of contested deliberations. Definitions are not fixed and unchangeable, but often have to be fine-tuned in the light of later understanding. In his classic *Proofs and Refutations* Imre Lakatos has shown how this phenomenon can occur in the case of polyhedra.

With Jemma Lorenat's chapter we move into the nineteenth and twentieth centuries. She observes that by the early twentieth century projective geometry had come to be seen as an approach to geometry that was essentially non-metric. This conception is apparent, for instance, in Oswald Veblen and John W. Young's 1910 book on the subject. While historians and mathematicians anachronistically attributed the conception to Jean-Victor Poncelet, the French geometer in fact employed metric notions in his definition of cross ratio in a fundamental way. The same holds true for other early-nineteenth-century geometers, such as Michel Chasles and Jakob Steiner, who, according to a recurrent historical narrative, are depicted as striving and failing to create a non-metric projective geometry. Not until Karl Georg von Staudt's *Geometrie der Lage* in 1847 were projective properties defined as properly non-metric. Lorenat challenges this historical narrative, which does not capture the aims of mathematicians such as Poncelet, whose work is read under the assumption that they shared the same agenda of mathematicians belonging to later generations. Non-metric projective geometry developed alongside an increased focus on the axiomatic foundations of geometry in late-nineteenth and early-twentieth-century research. Authors writing at this time tended to read a foundational interest that was not there into the work of early-nineteenth-century authors.

Jeremy Gray examines Roberto Bonola's influential history of non-Euclidean geometry *La Geometria non Euclidea: Esposizione Storico-Critica del suo Sviluppo*, published in 1906, focusing on Bonola's account of Lobachevskii's work. Bonola was writing within an educational tradition prevalent in early-twentieth-century Italy that drew a distinction between elementary and advanced mathematics. Elementary geometry was geometry in the style of Euclid, perhaps updated to include axiomatic ideas in the manner of Moritz Pasch and David Hilbert. Advanced geometry was the differential geometry of Bernhard Riemann and Eugenio Beltrami. Bonola anachronistically placed Lobachevskii within the tradition of elementary geometry, and in doing so missed the central importance for Lobachevskii's pioneering work of basing geometry on a primitive concept of distance and of establishing non-Euclidean geometry through the formulae of hyperbolic trigonometry.

Both Lorenat's and Gray's chapters are closely related in their focus on nineteenth-century geometry and in that they deal with the situatedness of disciplinary boundaries. Yet, their discussions bring out different aspects of anachronism. In her chapter, Lorenat discusses a two-part anachronism: on the one hand, the identification of a body of mathematical work as work concerned with projective geometry, and, on the other hand, the assumption that anything called projective geometry had to have a program for developing a non-metric geometry. Gray's chapter looks at the portrayal of the two central figures in early non-Euclidean geometry, Lobachevskii and János Bolyai. Gray examines the misunderstanding of their work by Bonola. This misunderstanding partly arises from an anachronistic view of what is elementary and what is not. He shows how this then affects how one sees the development of a mathematical subject, in particular the degree to which Riemann's work represented a leap or step forward in the development of non-Euclidean geometry. Lorenat's anachronism, accordingly, concerns the validity of characterizing an area of mathematics in terms of its content and presumed purpose, while Gray focuses on the mathematical methods employed and their characterization as either elementary or advanced.

In his broad-ranging concluding chapter, Joseph Dauben presents examples of anachronism in the history of mathematics, two that involve non-standard analysis and several concerning historical accounts of ancient Chinese mathematics. A case study that is perhaps not widely known concerns the 1989 attempt by philosopher Hilary Putnam to use ideas from non-standard analysis (in a suitably qualified form) to illuminate Charles Peirce's conception of continuity, set forth by Peirce in one of the Cambridge Conferences Lectures (1898). Dauben argues that Putnam's study provides an instance of the successful application of anachronistic "rational reconstructions" in the history of mathematics. On the other hand, using Cantorian alephs to examine Peirce's conception of the infinite is obfuscating.

In the second part of his chapter, Dauben examines examples from ancient Chinese mathematics that have been read in misleading anachronistic ways. Dauben shows that the translation of technical terms from ancient Chinese into modern English, or the use of diagrams and geometrical arguments foreign to ancient Chinese mathematical practice, can lead to serious misinterpretations. Dauben's learned chapter thus concludes the book by warning us about the risks in looking for the "predecessors who 'got it right' " with the aim of discovering the "path to our current truths" or the "royal road to me," as Naomi Oreskes and Ivor Grattan-Guinness wrote in seminal papers cited in the closing lines of the chapter.

It is fitting to conclude the volume with a chapter written by a historian who has been so authoritative in defending the idea that mathematical thought should be seen as punctuated with discontinuities. Avoiding anachronism, from Dauben's viewpoint, is crucial for a historical narrative that recognizes the existence of "conceptual revolutions" in mathematics.

Some of the themes explored by the authors concern the cultural situatedness of disciplinary boundaries, styles, and genres of writing in mathematics; the inevitable use of translations and the importance of being aware of the hermeneutic possibilities and possible dangers of the renderings of mathematical formulas in different mathematical languages; the opportunities and the difficulties implied in attempts to conceive the interpretation of past mathematical texts as dependent upon the understanding of authorial tasks and criteria of adequacy; the study of past mathematical texts through the history of their reception; the contrasting readings and evaluations of past mathematics achieved through a change in the historian's perspective. All these are daunting and well-studied topics, which have been dealt with by an intimidating array of philosophically minded scholars. These vexed problems, though, acquire a different status because of the peculiar character of mathematical thought. As historians, we hope to have contributed to this debate by adopting a bottom-up approach, which is to say by offering the reader a rich palette of historical cases.

Acknowledgements

The Francis Bacon Award implies an appointment as Visiting Professor, a position that gave me the chance to experience the exciting life of Caltech. I very much enjoyed teaching such bright students and the conversations with many wonderful colleagues, most notably Jed Buchwald, Tracy Dennison, Frederick Eberhardt, Moti Feingold, Stefano Gattei, Kristine Haugen, Christopher Hitchcock, Diana Kormos-Buchwald, Jean-Laurent Rosenthal, Tilman Sauer, Noel Swerdlow, and Nico Wey Gomes. The secretarial help of Emily de Araujo, Sinikka Elvington, and Francine Tise made everything possible and enjoyable. I thank the Francis Bacon Foundation and the California Institute of Technology for providing generous financial support during 2018. Later on, my research was funded by the Department of Philosophy "Piero Martinetti" of the University of Milan under the Project "Departments of Excellence 2018–2022" awarded by the Ministry of Education, University and Research (MIUR).

Many people contributed to the appearance of the book. I wish to thank Cristina Chimisso, Michael N. Fried, Nick Jardine, Sébastien Maronne, David Rowe, and Michalis Sialaros for their comments on a draft of the Introduction (Chapter 1), Craig Fraser for his help in writing this Preface and Sergio Knipe for his careful revision of the English. My former student Sebastián Molina (Bergamo) helped me prepare the name and subject index. Marco Cassé, a professional art director and illustrator, improved the quality of the images.

I thank David Tranah, Anna Scriven, and all the staff at Cambridge University Press, for their professional and kind assistance, and the two anonymous referees, for their useful reports.

Niccolò Guicciardini

1

Introduction: The historical interpretation of mathematical texts and the problem of anachronism

Niccolò Guicciardini
University of Milan

The worst of all sins, the sin that cannot be forgiven: anachronism[1] (Febvre, 1982, p. 5).

To be sure, unchecked anachronism is the stuff of popular myth [. . .] . Nevertheless, historical thinking should be concerned above all else with understanding temporality, how the present arises out of the past. So it will do no good to treat the past as its own country subject to no other law but its own and severed from moments in time that precede or succeed it (Lynch, 2004, pp. 241–2).

Abstract: I provide a definition of "anachronism" as well as an overview of the different positions concerning anachronism taken by historians of mathematics, since the important debate on the "algebra" of the Babylonians and the "geometrical algebra" of Euclid and Apollonius. Should we stress the continuity of past mathematics with the mathematics practiced today, or should we emphasize its difference, namely what makes it a product of a distant, Babylonian or Greek, mathematical culture? The issue of anachronism is a vexed one in historical interpretation, but is particularly thorny when one attempts to historicize a discipline that is, perhaps naively, celebrated for being independent of the context. The overview made available in this chapter provides the reader with information on the historiographical background against which the following chapters have been written.

[b] This research was funded by the Department of Philosophy "Piero Martinetti" of the University of Milan under the Project "Departments of Excellence 2018–2022" awarded by the Ministry of Education, University and Research (MIUR).

[1] To quote Febvre in full: "le problème est d'arrêter avec exactitude la série des précautions à prendre, des prescriptions à observer pour éviter le péché des péchés – le péché entre tous irrémissible: l'anachronisme" (Febvre, 1947, p. 5).

1.1 Purpose of this book

The debate on anachronism has been a long and vexed one in the field of historical interpretation. Forms of anachronism are often declared the greatest failure a historian can commit – almost a moral sin, as exemplified by Lucien Febvre's words quoted in the epigraph. Yet, many have spoken in favor of anachronism, considering it an inevitable, if not desirable, feature of historical works (Lynch, 2004; Jardine, 2000a). Historians of mathematics have also discussed this issue, sometimes in polemical terms. The example of the quarrel concerning the notion of "geometrical algebra" in ancient Greek mathematics begun by Sabetai Unguru on which we will have more to say below comes to mind. The purpose of this volume is to reflect on the "uses and abuses" of anachronism in the historical study of the mathematical sciences.[2] The debate among historians of mathematics is polarized in relation to two key issues: the *legitimacy of translations* of past mathematics into modern language, and the *legitimacy of evaluations* of past mathematics according to standards accepted today.

The first question emerges when, for example, we employ algebraic equations in order to analyze Galileo's kinematics, which was expressed in terms of proportions.[3] Thus, Galileo's statement that the distances of fall are proportional to the square of the times might be translated as $s = (g/2)\,t^2$. Yet, Galileo did not have algebraic equations at his disposal and even expressed ideas concerning the applicability of mathematics to the science of motion at odds with the application of algebra to continuous magnitudes, such as distance and time (Giusti, 1993; Blay, 1998).

The second question emerges when, to give another well-known example, one claims that the handling of infinitesimal magnitudes by Gottfried Wilhelm Leibniz, and by eighteenth-century mathematicians in general, lacks rigor, and that a logically tight version of the calculus was provided in the nineteenth century by the likes of Augustin-Louis Cauchy and Karl Weierstrass. Yet, Leibniz discussed the nature of differentials at length, and many eighteenth-century mathematicians had their own way

[2] I avail myself of Nicholas Jardine's turn of phrase in Jardine (2000a).

[3] A similar well-studied problem in translation emerges when one interprets the so-called Renaissance "abacus" or "Coss" problem-solving techniques in terms of symbolic algebra, that is, in terms of an algebraic language that was developed in the seventeenth century. See Corry (2015, pp. 125–53).

of making sense of the definition and use of infinitesimal magnitudes (Malet, 1996; Mancosu, 1996).

The controversial matters surrounding the notion of anachronism are difficult ones: they have been often broached by literary and art critics, by philosophers, as well as by historians of science divided, most notably, over the issue of Whiggism.[4] This book has not been written to attempt to provide an answer, even though its authors found the above debates illuminating. Rather, we would like to propose a bottom-up approach to the many problems concerning anachronisms in the history of mathematics. We are historians, and what we wish to do is to open our toolboxes and see what happens when we apply present mathematical language, concepts, standards, disciplinary boundaries, and indeed notions of mathematics itself, to well-chosen historical case studies belonging to the mathematics of the past, in both European and non-European cultures.

1.2 Defining anachronism

> The sense of anachronism, or "obsolescence," is a characteristic feature of modern historical thinking. ... An awareness that the past differs in fundamental respects from the present (Ritter, 1986, p. 9).

The notion of anachronism is an elusive one. The term has been used to refer to several mistakes in historical interpretation, and indeed the history of this term is an interesting topic for the intellectual historian. Thus, I have used a plural in the title of this book, and in the following chapters we shall encounter several forms of anachronism. *Anachronismos* was defined by chronologists such as Joseph Justus Scaliger (1540–1609). It meant misplacing an event against a correct temporal ordering. In this sense it was used by the Jesuits Denys Petau and Sforza Pallavicino, and by bishops Jacques Bossuet and Pierre-Daniel Huet (Grafton, 1991, pp. 104–44; Burke, 2006, p. 291).

Such temporal mistakes are not what we are interested in. One of the first statements concerning the historical interpretation of past texts that may be seen as relevant for the theme of this book was made by Jean Le Clerc (1657–1736) in his work on textual erudition entitled *Ars Critica* (1697):

[4] A philosophically innovative study is Syrjämäki (2011).

> So we must beware of lending our notions to the Ancients and then judging their discourse on the basis of these notions, as often happens. If we wish their thought to be understood, our opinions should be as if forgotten. [. . .] We should not compare their sayings with the nature of the things about which they speak, so as to be able to say that their knowledge of them is greater or less than ours, but should as far as possible interpret them from their very words.[5]

Le Clerc puts it beautifully and his words, made possible by the Renaissance discovery of the alterity of the past (Burke, 1969), can still be taken as an acceptable first definition of what we mean by anachronism in textual interpretation.[6]

Note that Le Clerc is not worried about using modern notions to describe past deeds and works. Rather, in his view, the mistake in historical interpretation consists in "lending" our notions to past authors and "judging their discourse on the basis of these notions." One might claim that using present knowledge and concepts in order to study the past is not only inevitable, but also profitable for historical interpretation. For example, when we study the decline of the Etruscan civilization (to use its Latin name), we might use categories and data drawn from disciplines such as demography or economy that would have been totally inconceivable for the Rasenna (to use the local name), who tended to believe, after a study of the livers of sheeps, that their misfortunes were due to the wrath of the Gods (Pallottino et al., 1986). A historian of Neolithic culture who would set himself the task of giving a talk using only the actors' categories and language, as one might try to reconstruct them, would surprise his audience! Furthermore, analogies between past events and present-day situations are routinely deployed by historians in order to describe the past. It is not only our present-day knowledge that helps us assess past events: what often informs historical narratives is the assumption of there being certain similarity between past and present.

When it comes to the interpretation of texts, rather than events, though, things become more complicated. Can we "use" our present-day concepts for descriptive purposes without "lending" them to the texts handed

[5] "Igitur cavere debemus ne notiones nostras mutuas Veteribus demus, deinde ex notionibus illis de eorum sermone judicemus, ut passim fieri solet. Si velimus eorum mentem adsequi, oportet veluti nostrarum opinionum oblivisci. [. . .] Non debemus eorum dicta cum ipsa rerum, de quibus loquuntur, natura conferre, quatenus nobis nota est, nam ut diximus eorum cognitio amplior, aut tenuior esse potuit, quam nostra; sed eos ex ipsis eorum verbis, quantum licet, interpretari" (Le Clerc, 1697, pp. 534–5).

[6] Le Clerc asked himself how ancient pagan works should be translated and evaluated by modern Christian readers. See Collis (2016).

down by past authors? One might claim that we are allowed – or even compelled – to use our own concepts, provided that we do so with an awareness of the present-day elements shaping our historical reading and choice of the texts we order at the desk of the Rare Books Room. It would be too simplistic, however, to assume, with Le Clerc, that we can easily sanitize our historical discourse from our present "notions" by a forced act of forgetfulness. This point can be illustrated by a metaphor used by Jonathan Clark, who writes:

> The sin of anachronism in historical method is a mortal one because it re-arranges the ideas and values of the past in ways which make past actions inexplicable except as attempted anticipations of the present. The historian is always condemned to see the past through a glass, darkly; the introduction of anachronistic categories turns that glass into a mirror (Clark, 2000, p. 13).

To elaborate on the metaphor Clark employs in his provocative book on English society in the long eighteenth century, one might say that the difficulty we encounter in our daily work as historians consists in the fact that we look at past authors through the glass of our anachronism, thus rendering their texts anticipations of the present. Our glass can only be soiled by the impurities of our own preconceptions. We see the shadowy figures of our predecessors, we turn and tilt the glass, we try to eliminate the impurities distorting the image. The glass of our doctrines and prolepses will be always there: we are twenty-first-century actors ourselves, and we cannot but use the optical instruments provided by our own times and by the hermeneutic tradition we belong to. The important thing is to make sure that, by manipulating the glass of historical research too clumsily, we do not "turn that glass into a mirror," and end up seeing our face reflected in it.

1.3 On mathematical anachronism: the problem

> The reality and reliability of the human world rest primarily on the fact that we are surrounded by things more permanent than the activity by which they were produced (Arendt, 1998, pp. 95–6).

Is there a special problem with the history of mathematics and anachronisms?[7] Perhaps one might put it like this: the special problem is that an anachronistic take on the past of mathematics is oftentimes deemed

[7] In the following two pages, I draw upon Guicciardini (2018).

quite successful for historical interpretation. The stability of mathematics (which might be deemed an empirical fact) or even, according to some philosophers, its universality (which is a controversial issue) invites its historians to adopt a cumulative and teleological narrative. Indeed, German mathematician Hermann Hankel (1839–1873) once wrote:

> In most sciences one generation tears down what another has built, and what one has established another undoes. In mathematics alone each generation builds a new storey on the old structure.[8]

Yet, many practitioners of the history of mathematics – especially those active after the turn against such teleological narratives promoted by scholars such as Helène Metzger and Thomas Kuhn – think that the study of past mathematics should be devoted to placing it into a "context" (a notion variously understood and hard to pin down) revealing otherness rather than kinship. The problem is that the history of mathematics thus contextualizes a discipline encoded in texts that apparently travel across time and cultures without much loss of their original meaning and cogency. Indeed, according to some philosophers endorsing forms of Platonism, these texts transcend their historical situatedness: they are independent of the author, his language, the intentions he might have had, and the context to an extent unknown in other areas of human culture.[9] However, the stability of mathematics is a fact that does not necessarily imply Platonism: one might wish to say, rather, that some mathematical techniques have endured for long (not for "eternity") and have been transmitted across different cultures.

The history of mathematics has a character of its own: it is not the history of dead mentalities or forms of life. It is the history of ideas and methods that are still in good health, thriving, perhaps eternal. Indeed, the works of past mathematicians can still be inspiring for many practicing mathematicians: solving a problem formulated centuries ago can

[8] "In den meisten Wissenschaften pflegt eine Generation das niederzureissen, was die andere gebaut, und was jene gesetzt, hebt diese auf. In der Mathematik allein setzt jede Generation ein neues Stockwerk auf den alten Unterbau" (Hankel, 1869, p. 34; translation in Dauben, 1984, p. 81).

[9] The contrast between the two approaches to the past of mathematics has been discussed at length by Ivor Grattan-Guinness in his classic papers (Grattan-Guinness, 2004a,b). It goes without saying that I owe a great deal to my former supervisor. I have not only profited from his papers, but I still remember the many discussions we entertained on the historiography of mathematics. A recent monograph is Wardhaugh (2010). Another book I have found very useful is Hodgkin (2005). A paper that refines Grattan-Guinness's viewpoint is Fried (2018).

lead to the Abel Prize.[10] As Godfrey Harold Hardy famously recalled:[11] "Littlewood said to me once, '[the Greeks] are not clever schoolboys, or scholarship candidates, but Fellows of another college'." So Greek mathematics is "permanent, more permanent even than Greek literature." Perhaps, historians of philosophy experience a similar familiarity with the works of Plato and Aristotle.[12] Mathematicians' engagement with past mathematical texts is often remote from antiquarianism. Some mathematicians can find inspiration for their own research in past texts, as I have just noted. At the level of teaching, the mathematics of the past often enters our classrooms. And this also marks a difference compared to other scientific disciplines. We would not be happy if our children were taught Paracelsian alchemy or Galenic medicine at school. Instead, we might be interested, and maybe even enthusiastic, to know that their mathematics teacher is using Apollonius[13] or that their physics teacher is using a textbook, such as French's *Newtonian Mechanics* (1971), which is replete with mathematical problems taken from Galileo's *Discorsi* and Newton's *Principia*.

The historian's view of the past is often at odds with those of Hankel, John Edensor Littlewood, or Anthony French. Some historians of mathematics, when attempting to interpret a mathematical text from the past, encounter distance and differences; they encounter the past as a "foreign country," as all historians do, and as David Lowenthal has recounted in his fascinating study (Lowenthal, 2015). Rather than Hankel's pronouncement, many historians of mathematics would subscribe to an equally famous passage from Morris Kline (1973, p. 69):

It is safe to say that no proof given at least up to 1800 in any area of mathematics,

[10] As was the case in 2016 with Andrew Wiles and his proof of Fermat's Last Theorem. It should be added, however, that lately the ability of practicing mathematicians to read older works is somewhat declining, partly because of lack of interest, but also because knowledge of the languages used in old works is getting rare.

[11] Hardy (1940, p. 12).

[12] Such familiarity is often invoked by historians of philosophy. Richard Rorty, to cite a classic, defends the opportunities disclosed by entertaining "conversations" with past philosophers: "we also want to imagine conversations between ourselves (whose contingent arrangements include general agreements that, e.g., there are no real essences, no God, etc.) and the mighty dead. We want this not simply because it is nice to feel one up on one's betters, but because we would like to be able to see the history of our race as a long conversational interchange. We want to be able to see it that way in order to assure ourselves that there has been rational progress in the course of recorded history that we differ from our ancestors on grounds which our ancestors could be led to accept" (Rorty, 1985, p. 51; see also the more recent Antognazza, 2015).

[13] This is what Fabio Acerbi did with his high-school students at the Liceo Scientifico Magrini in Gemona (Italy), where I had the pleasure of lecturing in 2003. On the use of the history of mathematics in the teaching of mathematics, see Fried (2007).

except possibly in the theory of numbers, would be regarded as satisfactory by the standards of 1900. The standards of 1900 are not acceptable today.

Several historians and philosophers of mathematics have underlined the fact that the criteria of proof validity have changed in history, making the development of mathematics subject to discontinuities, or even revolutions.[14] For example, Leonhard Euler's handling of infinite series or Augustin-Louis Cauchy's use of limits might be rejected today as not sufficiently rigorous. Kline highlights the discontinuities encountered by the historian of mathematics, and his position is often contrasted to Hankel's.

The historian of mathematics Henk Bos, in his inaugural lecture as Extraordinary Professor in the History of Mathematics at the University of Utrecht, entitled "Recognition and Wonder," argued that we both recognize the mathematics of the past as part of our canon and wonder at its diversity. It is worth quoting from Bos's lecture:

> Recognition makes it possible to distinguish historical events and thus initiates the link of past to present. If recognition or affinity is absent, earlier events can hardly, if at all, be historically described. Wonder, on the other hand, is indispensable too. The unexpected, the essentially different nature of occurrences in the past excites the interest and raises the expectation that something can be discovered and learned. History studied without wonder reduces itself to a mere listing of recognizable past events, which differ from what is familiar only by having another date (Bos, 1989, p. 3).

I find Bos's lesson totally convincing: it captures most of what I have to say in this Introduction. Indeed, it seems that as historians of mathematics we have to strike a balance between recognition of familiarity with the past and wonder for its foreign remoteness from our present. The essence of our discipline consists in talking sense into the mathematician, trained in searching for equivalences, and in writing a narrative that meets the standards of the historian sensitive to the otherness of the past. Like all historians, when looking back to temporal developments, we have to identify permanence and similarities, and distinguish changes and idiosyncrasies as well.

A picture of the permanence of mathematics through the ages, and

[14] There has been an extensive debate on the applicability of Kuhn's ideas to the development of mathematics. This debate, incepted by the exchange between Michael Crowe and Joe Dauben, has given rise to a very interesting literature of absolute relevance for the theme to which this volume is devoted. Two collections of essays that are essential reading are the classic (Gillies, 1992) and the more recent (Sialaros, 2018).

therefore of its accessibility from a modern standpoint, is vividly expressed in the naive incipit of a textbook on ancient mathematics written by a fine historian, Asger Aaboe. It reads:

> A modern schoolboy transposed to Babylonia or ancient Greece would find the physics of classical Antiquity utterly unfamiliar. Mathematics however would look familiar to our schoolboy: he could solve quadratic equations with his Babylonian fellows and perform geometrical constructions with the Greeks. This is not to say that he would see no differences, but they would be in form only, and not in content; the Babylonian number system was not the same as ours, but the Babylonian formula for solving quadratic equations is still in use (Aaboe, 1964, p. 1).

Aaboe, a historian inspired by the work of Otto Neugebauer, read the mathematics of ancient Mesopotamia and Greece through an interpretative lens aimed at recognizing similarities with modern high-school algebra.[15] What he claimed in these opening lines is that, by reading closely the cuneiform tablets from ancient Mesopotamia, one could talk with some justification of a 4000-year-old method equivalent to our algebra. One should emphasize that Neugebauer was keenly aware that the ancient Mesopotamian scribes did not have our algebra, and the above quote does not do any justice to Neugebauer's nuanced methodology. One might perhaps say that Neugebauer used our algebra as a means to verify the results of Babylonian mathematics, and as a means to cautiously interpret its methods as well. By a close analysis of the texts, he showed that this exercise in translation is possible, and occurs quite naturally, once the code used by the authors of the cuneiform tablets is broken, a feat in decipherment prowess that is the knack of the accomplished mathematician (Høyrup, 1991; Swerdlow, 1993; Rowe, 2012, 2016).

As already noted, we cannot perform the same exercise in translation, as successfully and seamlessly, with Galenic medicine or Aristotelian physics. And this truth, or truism, is perhaps what Aaboe had in mind. It is also interesting to note that Aaboe distinguishes a "content" from a "form" in past mathematics: it is the form that might change, not the content. Recent research has called Aaboe's and Neugebauer's continuist historiography into question by highlighting the distance between ancient

15 Note that Donald Knuth, who is famous for his contributions to the mathematics of computer programming, read some ancient Babylonian texts not in terms of algebra, but rather in terms of the theory of algorithms. The unexpected advantages of Knuth's anachronism are discussed in Karine Chemla's chapter in this volume.

Babylonian mathematics and the algebra of "our schoolboys" (Høyrup, 2002; Robson, 2008). Christine Proust, one of the great experts in the field, puts it beautifully:

> The mathematics of Mesopotamia is the most ancient which has been transmitted to us. These texts, written on clay tablets in cuneiform symbols, deal with mathematical objects familiar to us, such as numbers, units of measurement, areas, volumes, arithmetical operations, linear and quadratic problems, or algorithms. However, when we look more closely, these familiar objects, reveal strange features on the clay tablets.[16]

When we turn our attention to Aaboe's other imaginary classroom, the one attended by the schoolboys of ancient Greece, we encounter the celebrated – or perhaps infamous, one should say – debate on what was termed "geometrical algebra." The whole issue, in this case, concerned the similarity of demonstrations occurring in some propositions in Books II and VI of Euclid's *Elements* and in Apollonius's *Conics* with those of present-day algebra and analytic geometry. According to this viewpoint, the demonstrations of the ancient Greeks are undoubtedly geometrical, but they reveal knowledge of a form of proto-algebraic thinking, a "geometrical algebra," so to speak. The proponents of this interpretation, Neugebauer being one of them, actually defended different theses, and it is quite improper to group them together without paying attention to subtle, yet decisive, distinctions between their views (Høyrup, 2017). One might say that several scholars, following Hieronymus G. Zeuthen's pioneering works, agreed that a mathematically trained eye discerns methods in Euclid's and Apollonius's geometrical works that can be seamlessly rendered into the language of high-school algebra.[17] Thus, the story goes, while the ancient Greeks developed geometrical demonstrations concerning lines, surfaces, angles, and so forth, they possessed knowledge of rules applied to magnitudes that translate into our elementary algebra (most notably, the rule to solve second-degree algebraic equations) and into analytic geometry (especially, the definitions of

[16] "Les mathématiques de Mésopotamie sont les plus anciennes qui soient parvenues jusqu'à nous. Ces textes, écrits en écriture cunéiforme sur des tablettes d'argile, traitent d'objets mathématiques qui nous sont familiers, tels que des nombres, des unités de mesure, des aires, des volumes, des opérations arithmétiques, des problèmes linéaires et quadratiques, ou encore des algorithmes. Cependant, à y regarder de plus près, ces objets familiers se présentent dans les tablettes d'argile sous des aspects étranges" (Proust, 2015, p. 17).

[17] For example, Zeuthen (1886); Heath (1908); Neugebauer (1936); Van der Waerden (1954). Zeuthen distinguishes two phases, "geometrische Arithmetik" und "geometrische Algebra" in Zeuthen (1896, pp. 40–53).

conic sections in Cartesian coordinates). Neugebauer developed the bold idea that the analogies between the geometrical algebra of Euclid and Apollonius and the algebra inscribed in the cuneiform tablets was evidence of a transmission from Semitic to Greek cultures, an interpretation that, as is evident from the final pages of Neugebauer (1936), pleased an anti-Nazi German, but which he most often left implicit and somewhat hidden behind his non-ostentatious and austere style of writing.

Several scholars, most notably philosopher Jacob Klein and philologist Árpád Szabó, considered the renderings of Euclid's and Apollonius's demonstrations in terms of geometrical algebra as obfuscating the original, essentially geometric, character of ancient Greek mathematics.[18] This criticism was expounded in the most systematic way in the middle of the 1970s by Sabetai Unguru.[19] The ensuing debate, unfortunately a heated one, took the form of a disciplinary clash between Unguru, a professional historian well trained in philology and philosophy as well, and a number of eminent mathematicians, such as Bartel Leendert Van der Waerden, Hans Freudenthal, and André Weil.[20] The crucial issue was to determine who had the competence to write about the history of Greek mathematics: the classical scholar who has full command of the ecdotic criteria needed to understand a text and contextualize its transmission, or the practicing mathematician who has the technical ability to produce a conjectural reconstruction of the text (Weil, 1980)? The answer to this dilemma, one might think, would be cooperation between the two, but, unfortunately, this did not occur at the time.[21]

The debate on "geometrical algebra" clearly illustrates the problem of anachronism that historians of mathematics often encounter when translating the mathematics of the past into modern language. No doubt, these translations shed light on past mathematical practices, but the light

18 See, especially, Klein (1968) and Szabó (1978).

19 The two important papers that instigated the controversy are Unguru (1975, 1979). See also Unguru and Rowe (1981, 1982). David Rowe, it should be noted, was a very young scholar when he had the exciting experience of co-authoring these wonderful papers. As a mature scholar he left Greek mathematics for modern and contemporary mathematics and developed a nuanced and original historiographical position. In this context, the important contributions by Michael Mahoney should be also considered. See Mahoney (1970).

20 Van der Waerden (1975); Freudenthal (1977); Weil (1978). For the context of this debate, see Schneider (2016). A new interpretation based on the notion of "pre-modern algebra" is Sialaros and Christianidis (2016) in which the reader can find a complete, up to 2016, bibliography on "geometrical algebra."

21 Fried and Unguru (2001) is a detailed edition and commentary of Apollonius's *Conics*. This is the most elaborate and sustained criticisms of the notion of "geometrical algebra." For a revisionist viewpoint, see Blåsjö (2016). An innovative position is found in Netz (2002a,b).

– so to speak – can be too strong and make subtle differences fade away. These nuances are often considered insignificant by mathematicians interested in finding equivalences rather than distinctions between past and present.[22] Contrariwise, historians sensitive to the cultural world in which the mathematicians of the past operated will find these distinctions of vital importance for their discourse.

Many historians of mathematics nowadays present the development of mathematics by taking minute differences between texts into account: differences between texts that are mathematically equivalent but significantly distant for the historian. Moreover, as also shown by Fraser and Schroter's chapter in this volume, there is increasing awareness of the distinction between what is explicitly defined and what is implicit in the mathematical practice of authors who might even be unaware of what can be extracted from their texts a posteriori. In the case at hand, one might think that a preliminary definition of "algebra" would clear the field of divergencies and polemics, so that one might say that algebra was "implicit" in the works of Euclid and Apollonius. Yet, a more promising approach consists in resisting the temptation to start from a definition imposed from the outside, and in working instead "from within," by studying how algebra was gradually defined and extracted from Euclidean and Apollonian texts through a process that implied a slow and conceptually problematic transition from a geometrical to an algebraic paradigm.[23]

The two opposing views taken in the debate we have been reviewing in this section can be illustrated by the following quotations. The first comes from Unguru's pen:

> Different ways of thinking imply different ways of expression. It is, therefore, impossible for a system of mathematical thought (like Greek mathematics) to display such a discrepancy between its alleged underlying *algebraic* character and its *purely geometric* mode of expression (Unguru, 1975, p. 80).

Here, Unguru goes so far as to claim that one cannot separate form from content. This position thus offers no room for compromise with those who might wish to draw out the "essential" mathematical substance in

22 An interesting comparison between the two viewpoints can be found in Fried (2014) and Blåsjö and Hogendijk (2018).

23 It is interesting to note that two scholars who have masterfully dealt with the historical reconstruction of such paradigm transitions in mathematics are two Israelis deeply influenced by Unguru. See Corry (2013, 2015), and Fisch (2017).

a text and compare it with that in another. And here is Weil's pointed rejoinder:

> To berate Heath and others for betraying Euclid when all they do is to use a certain amount of notation to clarify the contents of his writings does not merely indicate a lack of mathematical sense; it argues a deficiency in logic. As everyone knows, words, too, are symbols. The content of a theorem does not change greatly, whether it is expressed in words or in formulas; the choice, as we all know, is mostly a matter of taste and of style (Weil, 1978, p. 92).

A statement that, when accepted with no qualification, might lead to a negation of the historian's aim to capture how even minute variances between symbolic expressions may reveal important differences of conceptions, that is, differences that must be taken into account in order to provide a narrative of the chronological development of, or of the geographical distance between, mathematical texts. To Weil's statement, Unguru would reply:

> In mathematics (like in anything else) form and content are not independent variables. On the contrary, they mutually condition one another and neither is immune to change. A certain form permits only a certain content, and a new content requires a new form. This is why the methodological approach which casts indiscriminately the algebraic shadow over the garden of Greek mathematics obscures precisely those features which make it *Greek* mathematics (Unguru, 1979, p. 563).

A statement that, when accepted with no qualification, might lead to an incommensurability between languages that would make any attempt of translation an impossible task.

Interestingly, both Unguru and his critics added qualifications to their most vehement statements, such as the ones quoted above, making thus their confrontation methodologically nuanced and deep. It is clear to me that they showed us that we have to strike a balance between familiarizing and foreignizing past texts, between recognition and wonder. The chapters of this volume try to achieve this end. The debates surrounding the notions of the "algebra" of Ancient Mesopotamia and the "geometrical algebra" of Euclid and Apollonius have been foundational for the discipline of history of mathematics. In a way, the discipline grew to maturity thanks to this confrontation. It would be unwise for the editor of this volume, given his ignorance on ancient mathematics, to expand further on these vexed issues. I am a historian of a less distant and exotic past: I study early-modern European mathematics. I will turn to my field of competence, therefore, with the purpose of offering some general observations that might be useful to exemplify the problem with anachronism in the history of mathematics that I have set out in this section.

1.4 Deceptive familiarity

The authors are near and far. At one and the same time, they are paradoxically both near and far [. . .]. All that is needed is to choose aptly among well-known texts. In this case, it is not hard to find passages that present themselves as curiously modern, so that they could be easily integrated into ongoing debates. And then again, it is easy to set them alongside texts by the same authors that awaken the feeling of an unbridgeable distance or a nearness that is merely apparent.[24]

In order to make sense of past texts, we have to translate the language of their authors into our own language, and this "domesticating" translation achieves – to use Bos's terminology – a "recognition" of the texts as familiar, as "easily integrated into ongoing debates." But according to Rossi's words cited in the epigraph, as historians we must also appreciate the otherness, the "unbridgeable distance" of past authors: we must achieve a sense of "wonder." In order to attain this goal, it is more appropriate to adopt a "foreignizing" strategy in translation.[25]

Rossi, a historian of magic and of conceptions of time and memory in the early-modern period, was looking for the "other present" of our predecessors. He aimed to recover, or re-enact – as he wrote re-enacting Robin Collingwood, if I may put it so – their voice and to interpret the texts by bearing the intentions of the historical actors in mind.[26] Referring frequently to Clifford Geertz, Rossi fashioned himself

[24] "Gli autori sono vicine e lontani. Contemporaneamente e paradossalmente vicini e lontani [. . .] Basta sçegliere bene entro testi ben conosciuti. Allora non è difficile trovare pagine che appaino singolarmente moderne, tali da essere facilmente integrabili nel dibattito contemporaneo. Ed è poi facile contrapporre ad esse testi dello stesso autore che danno il senso di una irrimediabile distanza o di una vicinanza che era solo apparente" (Rossi, 1999, p. 28).

[25] The classic text discussing the two translation strategies of domestication and foreignization is Venuti (1995). Domestication is the strategy of making text closely conform to the culture of the language being translated to, which may involve the loss of information from the source text. Foreignization is the strategy of retaining information from the source text, and involves deliberately breaking the conventions of the target language to preserve its meaning (Gile, 2009, pp. 251–2).

[26] Robin G. Collingwood (1939, pp. 111–2) claimed that in order to interpret a text historically, a historian must experience the thoughts of the author. He called this process "re-enactment": the historian "must be able to think over again for himself the thought whose expression he is trying to interpret."

as an anthropologist of the past capable of inhabiting bygone cultures in which metals could vegetate, planets could have moving souls, and natural philosophers "were encouraged to carve talismans and images, and surround themselves with precise colors or herbs connected to planetary influences in order to exploit the conjunction of the whole [the macrocosm], a living being of which man [the microcosm] is both part and lord" (Del Soldato, 2016).

A consequence of Rossi's viewpoint is the rehabilitation of past worldviews as foreign to our own but not reducible to errors, different but nonetheless reasonable, robust, explanatory, and even rational in their own terms. Unfortunately, as Rossi was well aware, neither time travel allowing the cultural anthropologist to inhabit the other present of past actors, nor an act of historical necromancy allowing him to resurrect the intentional illocutions of past authors seem like viable options. Most past events, texts, and constructs experienced and produced by our predecessors are irretrievable, not to speak of their intentional stance. The parallel between the historian and the anthropologist, fascinating as it might be, is misleading: as historians we are confronted not with the "thick" description detailed by Geertz. Rather, in the absence of sufficient information on the context and the actors, we have to cope with a "thin" description.[27]

Early-modern European mathematics has been transmitted to us and has been seamlessly embedded into our canon: think of Descartes's analytic geometry, Leibniz's calculus, or Newton's laws of motion. Thus, early-modern European mathematics in a way poses a challenge to historians precisely because it is so similar, so "near," to present-day mathematics: it has seeped into our textbooks. We encounter it in the genealogies constructed in the narratives imagined again and again, throughout the centuries, by those scientists who have received and further developed the works of Descartes, Leibniz, and Newton. One might say that this similarity is deceptive: for example, we tend to read the algebraic equations of Descartes, or the differential equations of Johann Bernoulli and Leonhard Euler, as if back then they meant what they mean for us today. There is, however, a difference, a distance, also in these cases; a distance that we, as historians, should recognize, as Rossi urges us to do,

27 In the preceding two paragraphs I have drawn on Guicciardini (2018).

when we seek to capture the idiosyncrasies and contingencies of past mathematics.[28]

As I show in Chapter 7, some early-modern mathematical techniques (the differential equations of Leibniz and his immediate followers) look very similar to the ones in use nowadays. Indeed, in order to make them fully compatible with the conventions adopted in our textbooks, we only need to change a few symbols. This cosmetic massaging, however, generates a remarkable shift in meaning that makes those very equations – when so reformulated – unacceptable for their early-modern authors. They would have considered our equations, as we write them, wrong and would have corrected them back, for reasons that they explicitly stated as being important. An apparently innocent change of symbols and notation can project the present into the past, it can turn Clark's dirty "lens" into a mirror: we end up seeing our own face, not the shadowy image of our predecessors. It is this experience of meaning-shifts (significant ones for early-modern authors) due to a slight change of notation that induces me to recommend caution when translating ancient geometry into present-day algebraic terms. If apparently trifling translations give rise to obfuscations of early-modern texts, one might reasonably suspect that even worse results ensue when we are overhasty in rendering cuneiform or ancient Greek texts in our algebraic terms.

An interesting case of "domesticating translation," in Venuti's terms, of early-modern mathematics into present-day notation can be found in the rightly acclaimed commentary provided by Tom Whiteside in his eight-volume edition of Newton's mathematical papers (Newton, 1967–1981). Very much as in the case of another great anachronistic historian of last century, Neugebauer, nobody denies that Whiteside obtained fundamental results in the history of mathematics. His commentary is considered indispensable reading for any scholar of early-modern mathematics. Whiteside had no qualms in commenting on Newton's mathematical manuscripts and works by deploying our contemporary calculus notation, a notation which is indebted, but not equivalent, to the one used by Leibniz. Whiteside has been a master for all those interested in studying Newton's mathematical work, yet his commentaries sometimes fail to capture important features that require a consciously foreignized reading of the text. To take a telling example, Whiteside commented on some of Newton's geometrical proportionalities as formulated in the

[28] On continuity versus revolutions, see Dauben (1984).

Principia in terms of differential equations. In some of these equations, proposed by Whiteside as renderings of Newton's text, the differential of time dt occurs. Yet, as Michel Blay has shown in his seminal monograph, the algebrization of time can be viewed as a major advance in the birth of eighteenth-century analytic mechanics; an innovation mostly due to the brothers Jacob and Johann Bernoulli, to Jacob Hermann, and to Pierre Varignon (Blay, 1992). Certainly, in Blay's narrative, it would be unjustified to use dt when analyzing Newton's mathematics. From Blay's standpoint, a translation of Newton's mathematics such as the domesticating one proposed by Whiteside, enlightening as it certainly is, risks erasing from view a major achievement in the mathematization of motion and force due to Continental mathematicians.

As in the case of ancient mathematics, what is at stake are two conceptions of history: one (Whiteside's) aimed at discovering some kinship between past and present, the other (Blay's) aimed at appreciating the distance of the past. Several historians of mathematics, especially those active at the end of the nineteenth and the first half of the twentieth century, adopted the first viewpoint and routinely produced translations of early-modern mathematics into present-day language. A Whitesidean domesticating anachronism has its own justifications, though, which at this juncture it is appropriate to enumerate.

(i) The fact that translations from early-modern mathematical language to our own are possible is in itself interesting: it is of some consequence for the historian (as all facts should be!).

(ii) By embedding past mathematics into present-day mathematics, these translations help the historian make use of modern knowledge in order to assess the possibilities and limitations encountered by past actors (and which were most often opaque to them: think of the case of the Etruscans we toyed with above).

(iii) Modernizing translations might allow us to appreciate the reception of past mathematics: it is often the case that our modernizing translations are those of readers of past texts who, active – say – some generations after the author, read the text beyond the author's conceptual horizon.

(iv) Good historical practice is often based upon the drawing of analogies between past and present (Burke, 2006; Leutzsch, 2019). If we assume that the past is completely alien to the present, it becomes totally opaque to our understanding.

(v) We learn by noting differences, not only analogies, between past and present: a modernizing translation can be a useful springboard to analyze in what respects the languages of past actors differs from ours.

(vi) As historians of mathematics, we address readers who are often mathematicians (or students of mathematics): if using present-day language may be tricky for our historical interpretations, it can, on the other hand, assist readers who are not professional historians.

Anachronistic renderings of past mathematics are, therefore, not to be demonized as intrinsically sinful. To be sure, if our aim as historians is to study the "foreign country" of past mathematicians, we will be inclined to avoid modernizing renderings, or to use them with great caution, since our efforts will be aimed at stressing the disanalogies and the otherness of past mathematics. However, to expand on point (vi) above, since our audience is constituted by mathematicians whose profession consists in searching for equivalences, an excessive bias towards the search for disanalogies might isolate us from the practitioners of the discipline whose history we are writing about. And this, for obvious reasons, would be a bad policy.

The reader will, I am sure, forgive me if in the next section I expand somewhat upon Bos's influential methodology, as put into practice in his work on Descartes's *Géométrie*. I wish to do this for two reasons. I need to put my hands on a concrete example – and what a wonderful example of scholarship this is! Further, Bos's methodology offers a credible way out of the dilemma facing contextualist historians of early-modern mathematics which we have considered in this section. As we have seen, early-modern mathematics is in some cases very similar to present-day mathematics, and successful historical interpretations have been provided by assuming this similarity is unproblematic. But how, then, can we contextualize and foreignize mathematical practices of the past that are apparently – I would say deceptively – so close to our present?

1.5 Tasks and criteria of quality control

The historical perception of mathematics is through the actions of mathematicians [. . .] the actions of mathematicians are to be understood as performing self-imposed tasks according to self-created criteria of quality control (Bos, 2004, p. 65).

René Descartes's *Géométrie* (1637) is one of the great classics in the history of mathematics. It is, for good reasons, viewed as a turning point in the history of geometry and algebra. With this small treatise – one might be tempted to say – modern mathematics was born, since in it Descartes gave birth to "analytic geometry." Indeed, Descartes's text sounds familiar to the modern reader. Points on plane curves are identified by pairs of variables z and y, the equations of conics, cubics, and higher-order algebraic curves are expressed in a notation that is very close to the one standard nowadays, and geometrical problems are translated into systems of equations, very much as we would do in our classrooms.

Descartes's text can be interpreted in a more contextual and less presentist way, as Bos has shown in a series of essays that were brought together in his monograph *Redefining Geometrical Exactness* (2001). Bos began his research by emphasizing the difficulties that he had encountered in understanding the structure, genre, and content of Descartes's mathematical essay, which did not appear familiar to him, but rather unclear, alien and enigmatic. Bos writes (and one could not find a more perspicuous expression of the sense of "wonder" at the diversity of the foreign past of mathematics):

the reader who opens the *Géométrie* expecting to read a book on analytic geometry is likely to have a confusing experience [. . .] the equivalence of curve and equation, which is the core of analytic geometry, appears to be rather a side issue in the *Géométrie*. Had that equivalence been a central one, one would have expected Descartes to deal with the curves according to their degrees, starting with the straight line [. . .] In fact, an equation of the straight line occurs only once in the text, more or less in passing, and Descartes discussed several curves without giving the equation at all. Furthermore, the book contains much algebraic theory about equations in one unknown, which at first sight seem unrelated to the theme of expressing curves by equations. [. . .] These enigmatic aspects have to do with the structure of Descartes' theory and with the resulting structure of his book (Bos, 1990, p. 351).

The plain fact is that – as Henk Bos has demonstrated – the *Géométrie* is not a treatise on analytic geometry, as we would understand this discipline nowadays. Indeed, Descartes devoted many pages of the *Géométrie* to a description not so much of the equations that define curves, but rather of several instruments, consisting of linked sliding rulers and strings, tracing curves by continuous motion. What would be the role of these tracing mechanisms in a work on analytic geometry? According to Bos, the standard reading of the *Géométrie* as the foundational text of analytic geometry is a myth, based on a proleptic reading of Descartes's mathematical essay. We read analytic geometry where its author had something else in mind.[29] As Bos has shown, in order to avoid projecting our discipline upon Descartes's text, we have to recognize that the study of plane curves is not the only issue for its author. Or, better, curves are not only the focus of study in the *Géométrie*: they are also the tools employed for geometrical constructions. Indeed, geometrical constructions are the task Descartes sets himself to. Curves are the instruments to achieve such a task. Just as in the *Elements* problems are constructed by intersections of circles and straight lines traced by compass and straightedge, so in the *Géométrie* problems are constructed by intersections of plane curves, such as ellipses, hyperbolas, and higher-order algebraic curves, traced by instruments of Descartes's conception.

One should bear in mind that the point of departure of the *Géométrie*, namely the translation of geometric problems into algebraic equations, which makes Descartes's essay recognizable as the founding text in analytic geometry, is not the end result of Descartes's canon for problem solving. In Descartes's mind, the algebraic calculation of the roots of a system of equations does not yield a solution to a geometric problem, as might be the case nowadays. In the Euclidean tradition, the solution of a geometric problem must end with a Q.E.F. ("quod erat faciendum," what was to be done), a geometrical construction: this was Descartes's "task." In other words, Descartes's agenda is dominated by the Greek tradition of geometrical problem-solving, which Pappus related to the methods of classical analysis.

In the very first pages of the *Géométrie*, Descartes explained how geometrical problems could be reduced to algebraic equations. This was a resolution, or analysis, of the problem. But algebra does not

[29] Similarly, in her chapter in this volume, Lorenat shows that early nineteenth-century geometers, when working on projective geometry, had in mind tasks different from those of mathematicians of later generations.

provide a solution: a solution is attained via a geometrical construction. For Descartes, as in classic Greek geometry, in order to produce a geometric construction, points must be determined via intersection of curves traced by continuous motion. One, therefore, needs instruments that trace curves by continuous motion so that an intersection can be found. In the *Géométrie*, curves, which were later seen as the main objects studied in analytic geometry, are tools used to construct points.

But how should one proceed? In the *Géométrie* Descartes developed a sophisticated theory aimed at constructing, by the intersection of curves, segments whose length, once a unit segment is arbitrarily chosen, is equal to the roots of polynomial equations. In his theory for the "construction of equations," Descartes determined a series of "self-imposed criteria of adequacy or quality control." First, he established that only a class of curves (what we call "algebraic" curves) were admissible as "exact." Second, he also indicated how one can choose between two algebraic curves the one which is "simpler." By addressing these issues concerning the adequacy and simplicity of construction tools, Descartes was answering desiderata about the solution of geometrical problems that were discussed at length in ancient Greek texts such as Pappus's *Mathematical Collection* (1588).

Descartes's theory of the "construction of equations" constituted an established topic of mathematical research for more than a century, and it attracted the attention of many mathematicians. For example, René-François de Sluse, Philippe de La Hire, John Wallis, Guillaume François de L'Hôpital, Isaac Newton, Christian Wolff, Charles-René Reyneau, Colin Maclaurin, Gabriel Cramer, and Leonhard Euler all wrote about this theory. The theory of the construction of equations remained an important mathematical topic for over a century. However, it disappeared around the middle of the eighteenth century. As Bos (1984, pp. 372–3) observes:

> the construction of equations did not disappear because some problems in the field were unsolvable [. . .] neither did the theory lose interest because all problems were already solved. Neither, again, can the loss of interest be attributed to a rival theory which produced better results. [. . .] The causes that made the construction of equations disappear lay in the sphere of motivation and method, rather than in the sphere of the mathematical problems and techniques. The causes were connected with the reasons why mathematicians considered certain problems as meaningful, and with the criteria of adequacy which mathematicians set for the solution of these problems. Such reasons and

criteria are very essential in the development of mathematics. They guide the research in a field, and as such they are necessary for its development.

At the middle of the eighteenth century, European mathematicians did not seek solutions as geometrical contractions, rather the equation $p(x) = 0$ itself was a solution. If a geometrical representation of the roots was needed, it was given by the intersection of the x-axis with the graph of $y = p(x)$. Yet, constructing the roots via the intersection of the graph and the horizontal axis was not a viable option for the Cartesian theory, because in this case one would construct the roots by intersection of a higher-degree curve (a quartic, say) with a straight line, rather than by the intersection of two lower degree curves (two conics). Pappus's precept, according to which problems should be constructed by the simplest means available, was important for Descartes and his contemporaries, who developed a sophisticated theory concerning the tools admissible in geometric construction. They discussed the properties of simplicity, elegance and exactness that curves should satisfy in order to be accepted as tools for constructing the roots of equations.

What is important here from a historiographical point of view – which is why I wish to expand on the issue – is Bos's attention for the motivations (the "task," as he aptly says) and the criteria of adequacy mathematicians set themselves. It is this viewpoint that allows him to shed light not only on Descartes's text, but also on the tasks Descartes wished to achieve (e.g., exactness, finitism, the visual intuition of mathematical constructions, simplicity). It is worth quoting Bos at length:

> Rather than – or perhaps in addition to – reconstructing the [mathematical] concepts, we might attempt to understand the tasks which led mathematicians to form their ideas and to write the texts we study. These tasks were self-imposed by the mathematicians in response to available knowledge, open questions and more diffuse challenges. They were therefore time-dependent – indeed it is difficult to imagine a Platonic eternal universe of mathematical tasks. Moreover, the tasks were governed by criteria of adequacy whose formulation was itself part of the general mathematical endeavour (Bos, 2004, p. 64).

Indeed, "self-imposed tasks" (constructing the roots of equations) and "self-created criteria of adequacy" or "of quality control" (using algebraic curves of the lowest possible degree) do not belong to the eternal realm of Platonic truths, but to the contingent, culturally situated agenda shared by the network of mathematicians whom Descartes corresponded with and upon whom he exerted his influence. Thus, Bos focuses on the practices and linguistic conventions[30] adopted in a community of mathematicians intent on addressing issues that are no longer important for us, that do not belong to the canon practiced in contemporary analytic geometry. It is such insightful attention to the culturally situated tasks and criteria of this community that allows Bos to ascribe the *Géométrie* to a genre practiced by Descartes's contemporaries and immediate successors, a genre that eventually faded away.

Bos's historical methodology helps us to address the problematic nature of the interpretation of past mathematical texts, which are both stable over time and produced in order to meet exigencies emerging in geographically and chronologically limited contexts. When Bos speaks of different "tasks," different "tools," and different "criteria of quality control, " as elements defining the wondrous difference of past mathematical practices, he sets a credible aim for the contextualist historian of mathematics. Tasks (such as constructing problems via the intersection of a determined class of curves), tools (such as the compasses of Descartes's inventions), and criteria of quality control (using the tools according to given postulates, using curves of the lowest possible degree) change over time and across cultures. The historian's aim will be to capture the mathematical practices shared by networks of mathematicians active in contexts that made those now forgotten tasks and criteria meaningful and important.

Indeed, it is difficult to imagine tasks and criteria as "given." They were self-imposed by the historical actors themselves and for reasons that are situated in time and place. The issues of geometrical problem-solving that polarized the attention of Descartes's contemporaries were related to the dream of reviving ancient knowledge pervading Renaissance culture. The curve-tracing mechanisms that they studied turned useful in many technological enterprises that were vital for their society: an example being the machines for lense-grinding, indeed a topic that connects the *Géométrie* to the *Dioptrique*. The lesson Bos imparts us is that by avoiding a proleptic reading of the *Géométrie* – that is, by avoiding to read it as a text that answers our problems in analytic geometry, but rather as a text that addressed the mathematical challenges open in Descartes's times – we can situate it in its cultural context in a more meaningful way.

The only qualifications that I would like to add to what Bos has to

30 I am thinking here of the use of terms such as *resolution*, *construction*, *analysis*, *exactness*, *mechanical*, etc.

say about tasks and criteria is that the same author, in the same work, might pursue different tasks and might wish to adopt different criteria of quality control. Such a plurality of tasks and criteria can generate conflict, tension, and anxiety. Indeed, the fulfillment of one task, for example the generality of methods, can be in conflict with another, for example visualization. I have found such tensions in the work of Newton, when, in concluding a masterpiece of algebra such as the *Arithmetica Universalis* (1707), he reverts to geometrical methods and praises them as superior (Guicciardini, 2009, pp. 106–7). I would suggest that in the *Géométrie* too Descartes might have been pursuing projects other than the one identified by Bos, as evidenced in the method to determine the normal of an algebraic curve, a method that seems to be addressed to the study of curves rather than the use of curves as construction tools. This method was, of course, very important for Descartes, both because of its mathematical interest and because of its application to optics in the *Dioptrique*. It is, therefore, important to realize – and I believe Bos is fully aware of this – that the agenda pursued in the *Géométrie* cannot be entirely captured by focussing on the theory of the construction of equations.

Bos's methodology somewhat concurs with the program in intellectual history put forward by the Cambridge School, especially in the version propounded by Quentin Skinner. Starting with his classic (1969) paper, Skinner distanced himself from a series of "mythologies" plaguing the history of political thought as practiced by historians of ideas working in the wake of Arthur O. Lovejoy. He chastised historians of political thought who were guilty of a "mythology of prolepsis," i.e. the risk of reading into past texts an anticipation of present-day doctrines. In a way, Bos answers Skinner's desideratum by distancing Descartes's theory from present-day analytic geometry. Bos also addresses the question of authorial intent, which is so vital for Skinner and many other intellectual historians. According to this view, we should interpret a text in view of the author's "intentions," a somewhat vaguer concept compared to Bos's more solid authorial "tasks" and "criteria of quality control." Skinner (1969, pp. 48–9) writes:

> The essential question which we therefore confront, in studying any given text, is what its author, in writing at the time he did write for the audience he intended to address, could in practice have been intending to communicate by the utterance of this given utterance. It follows that the essential aim, in

any attempt to understand the utterances themselves, must be to recover this complex intention on the part of the author.

The category of intentionality, even when fleshed out in philosophical terms inspired by John Langshaw Austin – as it is in Skinner's classic works of the 1960–70s – is marred by many difficulties, as philosophers and literary critics have been quick to point out.[31] In a crystal-clear paper, Denis Dutton has shown why "intentionalism won't go away." Dutton (1987) sets out by considering the interpretative work of anthropologists. He claims that the anthropological interpretation of the meaning of a work belonging to a culture different from ours can fail when the assignment of that work to a genre incurs in some sorts of categorical mistake, for it is only "once a categorical framework for understanding has been established (using intentionalist criteria)" that criticism "can be undertaken (using non-intentionalist criteria)" (Dutton, 1987, p. 199). Let us think about the conventions that allow the attribution of a text to categorical genres such as parody, eulogy, or erudite amusement. According to Dutton, the historian has to pursue the difficult task of recapturing the rhetorical rules and the linguistic conventions – maybe even tacit conventions not explicitly evident to the historical actors – which endow texts of culturally situated meanings. And this is, one might claim, what Bos achieves through his identification of the lost genre of the Cartesian theory of the construction of equations.

Skinner's pronouncements on historical method, problematic as they might be, indicate values that have directed the research of many intellectual historians during these last decades. Nowadays, for many practitioners of the history of mathematics too, recognition of the familiarity of past texts is only the beginning of a process of interpretation that must lead one to wonder about these texts, to develop a "feeling of unbridgeable distance or nearness that is merely apparent," as Rossi would put it. In a way, one might say that the Whitesidean "domesticating" renderings of early-modern mathematical texts – on whose utility I expanded in the last section – are viewed nowadays as the first step in a process of

[31] Among the philosophers, one might cite here the polemics between Roland Barthes and Raymond Picard, or between Hans-Georg Gadamer and Eric D. Hirsch. Among the literary critics, one might still profitably consider the opposite views of Monroe C. Beardsley and William K. Wimsatt, as expressed in their classic (1946) paper on the "intentional fallacy" and the sharp essay by Elizabeth Anscombe (1963). More recently, both psychologists and literary and art critics have underlined how a presumption of intentionality is essential to what we experience when we assign a particular meaning to an utterance, a work of art, or a literary text (Gibbs, 1999).

interpretation that must lead to a "foreignization" of the text. Bos shows historians of mathematics a viable way to achieve this aim. The fact that Bos's historical study of Cartesian mathematics can be considered to be in accord with the desiderata of intellectual historians is an extremely exciting feature of the recent historiography of mathematics. It brings our discipline into conversation with the history of science and the history of philosophy, and, indeed, with general history. Last but not least, the advantage of Bos's methodology is that it avoids both the difficulties implied by the anthropological model proposed by Rossi and the subjectivity of the Wittgensteinian pragmatics of intentionality evoked by Skinner. Indeed, Bos bases his interpretation on firm and accessible evidence: a study of the tasks and criteria explicitly stated in the texts by mathematicians working on the Cartesian theory of the construction of equations.

1.6 Changing present viewpoints

> We are bound to see the Second World War differently in 2015 compared to 1945, not merely because new evidence has come to light, but also because the ensuing decades unfolded further consequences: the Bomb, decolonization, the Cold War, and much more (Lowenthal, 2015, p. 340).

So far, we have dealt with the difficulties we encounter when we translate and evaluate the mathematics of the past from a present viewpoint. As a matter of fact, as many readers will have already noticed, there are a plurality of present viewpoints. Past texts are read (and transformed in this process of reception) by different readers, so their transmission is ramified: certainly, it is not linear.[32] Furthermore, present-day mathematics, the chronological end-point at which we are temporally situated, comes in different forms, prospers in different contexts, and is promoted by different schools that can differ, and even fiercely disagree, as regards tasks, values, methods, disciplinary boundaries, and criteria of quality control.

A change in the present viewpoint, as Karine Chemla details in her chapter, can lead to considerably different evaluations of the past, to state a truism. To give an example drawn from my field of expertise: let us consider the changing appreciation of Robert Hooke as a mathematician.

[32] See Schneider's chapter on this issue.

The nature of Hooke's contributions to mathematics was contested during his own lifetime and gave rise to diverging evaluations until the last century. Twentieth-century science historians often set Hooke in contrast to Newton and Christiaan Huygens, unfavorably comparing him with these two theorists (Vickers, 1987, pp. 99–100). Famously, Rupert Hall dismissed him as a "mechanic of genius" rather than "a scientist."[33] Yet, from the point of view of some of his contemporaries, Hooke, because of his dexterity in trading and moving between the diverse locales of London – between watchmakers' shops and the laboratory in Gresham College – was better suited to the enterprise of mathematized natural philosophy as envisaged by the Fellows of the Royal Society than the lofty Newton (Shapin, 1989, pp. 235–85). One should note that Hall's statement is based upon categories of mathematics and science (when he excludes that Hooke was a "scientist") that are anachronistic. Much ink has been spilled on the need to view Hooke and his contemporaries as "natural philosophers," rather than as "scientists," in order to capture the broad cultural texture of the discipline they pursued. In what follows, I will show how the category of "mechanics" was understood, in such a way that it makes sense to view Hooke as contributing to mathematics as well.

It is only recently that Hooke's reputation has been "restored," as Lisa Jardine (2006, pp. 247–58) puts it. The tercentenary of his death in 2003 has been graced with a proliferation of studies, and it is fitting to ask why Hooke's fortune has changed recently, so much so that a commemorative plaque to him has been unveiled in Westminster Abbey, just in front of the funerary monument of his great enemy, Sir Isaac Newton. There is no easy answer to this question. The fact that Hooke is given pride of place in recent accounts of seventeenth-century natural philosophy might be a consequence of the re-evaluation of the role played by engineers and mechanicians in the development of early-modern science. In recent literature one witnesses a shift of interest from mathematics and astronomy (the two disciplines that informed the Koyrean narratives of the scientific revolution) to mixed mathematics,

33 "It is worth noting too that whereas Huygens had approached horological invention through his studies in pure mechanics, and left the work of construction to professional clock-makers, Hooke's attitude is that of a mechanic of genius, rather than that of a scientist" (Hall, 1951, p. 175).

hydrology, pneumatics, biology, medicine, microscopy, geology, and alchemy.[34]

But perhaps, at an even deeper level, our perspective has changed because of the recent advent of new scientific paradigms dominated by computer simulations and performative technologies (Pickering, 1995; Callon, 1998). Hooke, the practitioner who tested his new theory of planetary motion by letting spheres roll on conical surfaces or observing pendular motion, the "mechanician" who graphically reconstructed central force motion before the eyes of the Royal Society Fellows, might be viewed as a precursor of computer modeling and numerical simulations of planetary orbits. His social status and scientific practice somewhat resemble those of twenty-first-century practitioners of nanotechnology and genetic engineering. Many scientists of our times are active in extramural science – in scientific enterprises located outside academia – driven by a know-how nurtured via manipulations performed in the laboratory and shared with technicians and entrepreneurs, rather than via theoretical conversations with post-docs and tenured colleagues at department seminars. Our present directs our view of the past, and orients our historical lens towards authors and texts discarded by our predecessors, such as Alexandre Koyré, determining a different evaluation of past mathematical practices.

In the case of Hooke, we experience a change in our understanding of the disciplinary boundaries of "mathematics" itself. For the historians of Hall and Whiteside's generation, Hooke was neither a scientist nor a mathematician.[35] Nowadays, we see things differently. Geoffrey Lloyd has warned us to be extremely careful when writing about disciplines as historians of science. He stressed that we should use terms such as "physics," "anatomy," or "mathematics," with great care when it comes to writing the history of the ancient Chinese or Greek worlds. The Chinese category for the study of numbers ("suan shu") and the Greek one ("mathēmatikē") do not tally exactly with modern conceptions. These disciplines were different from our "mathematics": furthermore, they differed between themselves, and within the same culture, Greek or Chinese, there were debates on how to define them. Lloyd observes that these problematic issues in translation were not considered by historians

[34] This new trend is best represented in Bennett (1986) and Bertoloni Meli (2006, pp. 218–23; 242–7).

[35] Whiteside did not mention Hooke in his broad overview devoted to the "Patterns of Mathematical Thought in the Later Seventeenth Century" (1961).

such as Joseph Needham, another anachronist giant of the mid-twentieth century. Lloyd's suggestion is either to use the original terms or to use less theory-laden expressions such as the "study of the heavens" for "astronomy" (Lloyd, 2004, 2009). What Lloyd's work teaches us – a lesson that is confirmed by Plofker's chapter in this volume – is that historians are translators who have to make choices about how to relate an old language to a new one, and that such a linguistic enterprise runs up against a crucial difficulty when it deals with disciplines and categorizations that are deeply related to local social structures (in the spheres of teaching, administration, religion, etc.).[36]

It is interesting to read what Syrjämäki has to say on this issue in his unpublished doctoral thesis. He writes (Syrjämäki, 2011, pp. 48–9):

> Can the historian translate the past into a modern language? This metaphorical question is the same as asking whether the historian may ascribe modern terms to the past. I have to say that I am not very keen on the notion of the historian as a translator, and I am more attracted by the idea that the historian is essentially a language teacher whose job is to teach new – or, in the historian's case, old – languages to his audience. With regard to the danger of anachronism the use of modern terms is not so problematic when teaching older languages as it is when translating them.

Feke's chapter in this volume can be taken as exemplary of how a historian can teach twenty-first-century readers the meaning of terms such as mathematics, physics, and theology, as used in ancient Greece. As Cristina Chimisso (2019, p. 160) argues in her new book on Hélène Metzger, the French historian aimed to make herself a "contemporary" of her own sources in order to translate them for current readers. Yet, Chimisso notes that "complete and unencumbered access to past sources would not only be impossible, but it would erase the work of the historian as a constructor of narratives." She continues by observing that the historian as "constructor of narratives," has very different aims from those of the author. This difference, I claim, might justify the use of modern terms in that painstaking setting out of a dictionary that helps to relate our language to the languages of the past. As Lloyd teaches us, the work of the historian consists in providing such a dictionary and in critically discussing what are inevitably provisional linguistic correspondences.

[36] I, once again, refer the reader to the study by Venuti (1995) on the strategies of foreignization and domestication.

1.7 Reading the text

One would never reach the book, but always a mind reacting to the book and mingling with it, ours, or that of another reader.[37]

Another way to make good use of the plurality of receptions a text encounters in its *Wirkungsgeschichte* is to interpret it through the eyes of its readers. The focus of historical investigation, in this case, is the fragmentation of meanings a text acquires through its reception: and the historian will do well in not siding with one reading as superior to another, if this is reasonable enough. This historical methodology, the "receptionist" approach to textual interpretation, has been defended by a number of historians of science, including Metzger and Jardine.[38] Their positions have been lately brought into comparison by Chimisso (2019, p. 164) who claims that Jardine's approach bears similarities with Metzger's since "he shares with her a reception-based view of meaning, with a focus on the receptions contemporary to the sources." From this viewpoint, textual meaning is the result of the reader's experience, not something that preexists the reading.

In the case of mathematical texts, this process of reception is one that soon leads to a profound transformation, consciously or tacitly put into place.[39] Art historians and literary critics have taught us that a work does not entirely belong to its author. Meaning is conferred to the text by its readers too.[40] In a way, this is true of mathematics as well, even though, of course, mathematical texts have a theoretical counterpart, contrary to literary texts, and hence largely escape this kind of active reading. Even if it is true that *reading* mathematics requires *doing* mathematics, and that therefore the reading of mathematical texts is always performative (the reader is invited to write new formulas and draw fresh diagrams of his own invention), we can nonetheless assume a minimum degree of faithfulness on the reader's part to the author's meaning.

Once a text has been published or brought into circulation, the mathematician who wrote it has little control over how it might be read and interpreted. Readers will likely approach it for their own purposes, per-

[37] "On n'atteindrait jamais le livre, mais toujours un esprit réagissant au livre et s'y mêlant, le nôtre, ou celui d'un autre lecteur" (Lanson, 1919, p. 41).

[38] See Jardine (2000b) for his receptionist theory of interpretation.

[39] Gray, in his chapter in this volume, shows how the meaning of "elementary" referred to "geometry," tacitly changed at the turn of the twentieth century.

[40] As proponents of this historiographical approach, one might, of course, refer to the Constance School of Wolfgang Iser and Hans Robert Jauss, or to Stanley Fish's Reader-Response Theory.

haps developing new ideas from it, or drawing conclusions not envisaged by the author. This reception process, which gradually effaces the voice of the author, constitutes an essential part of any long-term mathematical development. In a way, the greater the mathematical reception is, the sooner the author's intentions disappear. As Charles Leedham-Green observes: "All great mathematics dies, and the greater the mathematics the sooner it dies. It is studied, improved, generalised, re-written in different ways, and is absorbed into monographs and textbooks."[41] Newton's mathematics was soon transformed into something quite different. The mathematics of Bernard Nieuwentijt, who inconclusively attempted to eliminate the use of higher-order infinitesimals from the differential calculus, is still intact and available for historical inspection.

Recently, Sébastien Maronne has been interpreting Descartes's *Géométrie* from a receptionist viewpoint that complements the "inceptionist" one promoted by Bos. For Bos, the meaning of the Cartesian essay is accessible through an understanding of the author's tasks in writing the work. Maronne searches the meaning assigned to it by the translations and commentaries carried out by Descartes's Dutch followers.[42] How did they read, translate, use, and ultimately transform Descartes's essay into a stratified, multifaceted, even contradictory text? As Maronne shows, in the commentaries on Frans van Schooten's Latin translation one finds a reinterpretation of the *Géométrie* that goes beyond Descartes's intentions, so much so that one finds the presence of material (most notably, Pierre de Fermat's method of tangents) that Descartes would have disavowed.[43] The interesting feature of Maronne's receptionist approach is that it views the meaning of Descartes's essay as the product of a collective enterprise.

This aspect has been brought into relief in one of the best receptionist studies in the history of mathematics, Catherine Goldstein's account of the reception of an elementary theorem formulated by Fermat and by Bernard Frénicle de Bessy. As she shows, the meaning and equivalence of different presentations of this theorem cannot be given a priori by

41 Charles Leedham-Green, "Appendix F, On Newton's Style and Translating the *Principia*," to appear in Newton (2021). I thank Professor Leedham-Green for sending me a preliminary version of his translation and notes of Newton's *Principia*.

42 Maronne's research is still ongoing. In this context, I might cite Maronne (2017).

43 I hope the reader will allow me to gesture to a book (Guicciardini, 1999) on the reception of Newton's *Principia* in which I adopted a receptionist approach: I did not devote that book to Newton's magnum opus, but to the ways in which it was read by competent mathematicians active in the period 1687 to 1736.

the historian, who must instead regard the network of relations between readers as the condition which confers meaning to the text and makes "equivalence" the result of a contingent historical development (Goldstein, 1995). Thus, the different receptions of a text from the past become an object of historical study – and this in a way that is anachronistic, in so far as they do not belong to the author. Receptionist studies reveal the conceptual worlds, values, and agendas of the readers of a text, and – as already noted – the historian will do her best not to favor one reading as superior, thereby applying a "principle of symmetry" that has gained currency among sociologists influenced by the so-called strong program of the Edinburgh School.[44]

1.8 Summing up

It is high time to put an end to my philosophically low-tech introductory words with a few concluding comments and an invitation for the reader to move on to the best part of this book, namely to the wonderful chapters that follow. The reader will be aware by now that the daunting philosophical problems implied by any attempt to theoretically understand anachronisms in historical interpretation remain, sadly, intact after these opening pages. These problems become even more challenging when dealing with mathematical texts, as I have argued all along in this Introduction.

As argued above, we should not be too hostile toward translations and evaluations based on present-day notations and standards. Relating the past to the present should be viewed as one of the aims we set ourselves as historians of mathematics. This aim is particularly important for the historian of mathematics, given the stability of mathematical thought and the disciplinary position of historians of mathematics as part of the community of practicing mathematicians. It is difficult to discern cracks and gaps in the development of mathematics to an extent comparable to physics or medicine; and this fact justifies a presentist approach to

[44] In this context, it is appropriate to refer the reader to the important sociological researches by Donald MacKenzie and J. Bennet Shank. Among their many wonderful works one might cite MacKenzie (2001) and Shank (2018) as particularly relevant for the themes broached in this Introduction. I should add that in this Introduction, and in general in this book, scant attention is paid to STS studies, since the methodological gap between the approaches taken in this volume, their diversity notwithstanding, and social constructivism is too wide. It would be very interesting, however, to devote attention – maybe in a new book – to this issue avoiding the animosity that often surrounds these discussions. It is a pleasure to underline my admiration for the work carried out by the historians working in this tradition.

its history. Furthermore, some of our readers are mathematicians whose expertise is designed to allow them to search for equivalences. Playing a totally different game would place us at odds with the practitioners of the discipline we have elected to study as historians. This move would be detrimental: it would alienate us from an important audience and deprive us, as Martina Schneider demonstrates in her chapter in this volume, of the expert judgement (Chang's "connoisseurship" (2008)) of practicing mathematicians interested in reading (and using) texts conceived in the past.[45]

As I have argued above, this can be only one of the aims we set ourselves. In a way, we might say that translations and evaluations based on present-day notations and standards are the first approach to the texts we seek to interpret historically. As historians, we share Rossi's ethos: we search for the features distancing the past from the present. Thus, we focus on the very human choices that made mathematics a different discipline from the one we practice today. In the past, mathematicians set themselves tasks, adopted criteria of adequacy, and understood the disciplinary boundaries in ways that differ from ours, and they did so for reasons situated in the broad cultural contexts they lived in.[46] Their lives, their flesh and blood, the quarry Marc Bloch's ogre is after, are intertwined with the abstraction of their mathematical formulas and diagrams.[47] In hunting for Bloch's big game we often devote attention

[45] Of course, it is often the case that practicing mathematicians – in some cases eminent ones – are skilled practitioners of history. The names of Weil, Van der Waerden, and Truesdell – already cited above – come to mind. On the genre of connoisseur historians see Chang (2008). One of the aims of this genre is to re-examine classical or now forgotten scientific texts in order to assess and to explain what is valuable in them and why. Chang's paper is a comment on Truesdell (1973). Before raising the charge of Whiggism against this genre of writing, one should consider that few would dismiss the value of connoisseurs' assessments in scholarly works on the history of music or painting. Should we not listen with interest and respect to what Maurizio Pollini – certainly not a historian of music – has to say about the use of chromatic harmony in Chopin's *Ballades*? Would we not do the same, if a famous conservator such as Vittorio Granchi spoke about Titian's techniques for combining pigments? Surely the opinions and interpretations of connoisseurs contribute greatly to our appreciation of past works in music and art, making them seem alive and meaningful for us. Are we, as historians of mathematics, really so different from historians of music or art that we can simply treat the views of connoisseurs as irrelevant for our own work?

[46] I find what Sabina Leonelli writes on contexts vs situations interesting and profitable. She urges us to avoid the static notion of "context," and rather to deploy John Dewey's notion of "situation." As Leonelli (2016, p. 184) observes, "Dewey's account accommodates the idea that any one research context may have flexible, dynamic boundaries, since what belongs in a situation is subject to constant change in response to the shifting goals of the inquirer(s) and the ever-evolving state of the material and social world and is thus never fixed in time or space."

[47] In *Apologie pour l'histoire*, Bloch (1949, p. 51) writes that the historian is like the ogre in fairy tales: when he smells human blood, he knows his prey is near. "Le bon historien, lui, ressemble à l'ogre de la légende. Là où il flaire la chair humane, il sait que là est son gibier."

to mathematical practices that are assessed by the mathematical connoisseur as trivial, boring, and unimportant. Yet, these features, even dead-ends such as the theory of the construction of equations, can provide valuable information on the culture in which past mathematics was embedded.

As this book demonstrates, anachronism in the history of mathematics can be both enlightening and misleading.[48] We need translations and evaluations of past mathematical texts framed in present-day language and it is instructive to read these texts according to present-day standards. But we should be aware, and consequently warn our readers, that these anachronistic approaches can be both helpful and misleading. The instruments we use, "our toolbox," are replete with the most powerful lenses ever made available to historians, but they can have an unwanted distorting effect. This book is devoted to studying how these tools have been used to produce both good and poor results.

References

Aaboe, Asger (1964). *Episodes from the Early History of Mathematics*. New York: Random House.

Anscombe, Elizabeth (1963). *Intention*, second edition. Oxford: Blackwell.

Antognazza, Maria Rosa (2015). The benefit to philosophy of the study of its history. *British Journal for the History of Philosophy* **23**, 161–84.

Arendt, Hannah (1998). *The Human Condition*, second edition. Chicago: The University of Chicago Press. (First edition, 1958.)

Beardsley, Monroe C. and William K. Wimsatt (1946). The intentional fallacy. *Sewanee Review* **54**, 468–88. Revised and republished in *The Verbal Icon: Studies in the Meaning of Poetry*. Lexington KY: University of Kentucky Press, 1954, 3–18.

Bennett, Jim A. (1986). The mechanics' philosophy and the mechanical philosophy. *History of Science* **24**, 1–28.

Bertoloni Meli, Domenico (2006). *Thinking with Objects: The Transformation of Mechanics in the Seventeenth Century*. Baltimore MD: The Johns Hopkins University Press.

Blåsjö, Viktor (2016). In defence of geometrical algebra. *Archive for History of Exact Sciences* **70**, 325–59.

Blåsjö, Viktor and Hogendijk, Jan P. (2018). On translating mathematics. *Isis* **109**, 774–81.

Blay, Michel (1992). *La Naissance de la Mécanique Analytique: La Science*

48 Most notably, see Dauben's chapter.

du Mouvement au Tournant des XVIIe et XVIIIe Siècles. Paris: Presses Universitaires de France.

Blay, Michel (1998). *Reasoning with the Infinite: From the Closed World to the Mathematical Universe*. Chicago: The University of Chicago Press.

Bloch, Marc (1949). *Apologie pour l'histoire: ou métier d'historien*. Paris: Armand Colin (edition cited 2013).

Bos, Henk J.M. (1984). Arguments on motivation in the rise and decline of a mathematical theory: the 'construction of equations,' 1637–ca. 1750. *Archive for History of Exact Sciences* **30**, 331–80.

Bos, Henk J.M. (1989). Recognition and wonder: Huygens, tractional motion and some thoughts on history of mathematics. *Tractrix, Yearbook for the History of Science, Medicine, Technology and Mathematics* **1**, 3–20. (Reprinted in *Lectures in the History of Mathematics*. Providence, RI: American Mathematical Society (1993), 1–21).

Bos, Henk J.M. (1990). The structure of Descartes' Géométrie. In: *Descartes, il Metodo e i Saggi: Atti per il 350o Anniversario della Pubblicazione del Discours de la Méthode e degli Essais*, Giulia Belgioioso, Guido Cimino, Pierre Costabel, and Giovanni Papuli (eds). Rome: Istituto dell'Enciclopedia Italiana, 349–69.

Bos, Henk J.M. (2001). *Redefining Geometrical Exactness: Descartes' Transformation of the Early Modern Concept of Construction*. New York: Springer.

Bos, Henk J.M. (2004). Philosophical challenges from history of mathematics. In: *New Trends in the History and Philosophy of Mathematics*, Tinne Hoff Kjeldsen, Stig Arthur Pedersen, and Lise Mariane Sonne-Hansen (eds). Odense: University Press of Southern Denmark, 51–66.

Burke, Peter (1969). *The Renaissance Sense of the Past*. London: Edward Arnold.

Burke, Peter (2006). Triumphs and poverties of anachronism. *Scientia Poetica* **10**, 291–98.

Callon, Michel (1998). Introduction: the embeddedness of economic markets in economics. *The Sociological Review* **46**, 1–57.

Chang, Hasok (2008). The historian of science: Painter, guide, or connoisseur. *Centaurus* **50**, 37–42.

Chimisso, Cristina (2019). *Hélène Metzger: Historian and Historiographer of the Sciences*. Abingdon, Oxon and New York: Routledge.

Clark, Jonathan C.D. (2000). *English Society 1660–1832: Religion, Ideology and Politics During the Ancien Regime*. Cambridge: Cambridge University Press.

Collingwood, Robin G. (1939). *An Autobiography*. London and New York: Oxford University Press.

Collis, Karen (2016). Reading the Ancients at the turn of the century: the third Earl of Shaftesbury (1671–1713) and Jean Le Clerc (1657–1736).

In: *Ancients and Moderns in Europe: Comparative Perspectives*, Paddy Bullard and Alexis Tadié (eds). Oxford: Voltaire Foundation, 113–30.

Corry, Leo (2013). Geometry and arithmetic in the medieval traditions of Euclid's Elements: a view from Book II. *Archive for History of Exact Sciences* **67**, 637–705.

Corry, Leo (2015). *A Brief History of Numbers*. Oxford: Oxford University Press.

Dauben, Joe (1984). Conceptual revolutions and the history of mathematics: two studies in the growth of knowledge. In: *Transformation and Tradition in the Sciences: Essays in Honour of I. Bernard Cohen*, E. Mendelsohn (ed). Cambridge: Cambridge University Press, 81–103. (Reprinted in *Revolutions in Mathematics*, Donald Gillies (ed), Oxford: Clarendon Press (1992), 49–71 with an appendix 'Revolutions Reconsidered', 72–82.)

Del Soldato, Eva (2016). Natural philosophy in the Renaissance. In: *Stanford Encyclopedia of Philosophy*, Edward N. Zalta (ed). `https://plato.stanford.edu/` (accessed October 6, 2019).

Descartes, René (1637). La Géométrie. In *René Descartes, Discours de la Méthode, pour Bien Conduire sa Raison et Chercher la Verité dans les Sciences: Plus la Dioptrique, Les Meteores, et la Géométrie, Qui Sont des Essais de Cete Methode*, 297–413. Leiden: Maire. Edition cited: *The Geometry of René Descartes with a Facsimile of the First Edition*. Edited and translated by D.E. Smith and M.L. Latham. New York: Dover (1954).

Dutton, Denis (1987). Why Intentionalism won't go away. In: *Literature and the Question of Philosophy*, Anthony J. Cascardi (ed). Baltimore: Johns Hopkins University Press, 192–209.

Febvre, Lucien (1947). *Le Problème de l'Incroyance au XVIe Siècle: La Religion de Rabelais*. Paris: Albin Michel.

Febvre, Lucien (1982). *The Problem of Unbelief in the Sixteenth Century: The Religion of Rabelais*, translated by Beatrice Gottlieb. Cambridge, MA: Harvard University Press.

Fisch, Menachem (2017). *Creatively Undecided: Toward a History and Philosophy of Scientific Agency*. Chicago: The University of Chicago Press.

French, Anthony P. (1971). *Newtonian Mechanics*. New York and London: W.W. Norton.

Freudenthal, Hans (1977). What is algebra and what has it been in history? *Archive for History of Exact Sciences* **16** (3), 189–200.

Fried, Michael N. (2007). Didactics and history of mathematics: knowledge and self-knowledge. *Educational Studies in Mathematics* **66**, 203–23.

Fried, Michael N. (2014). The discipline of history and the 'modern consensus' in the historiography of mathematics. *Journal of Humanistic Mathematics* **4** (2), 124–36.

Fried, Michael N. (2018). Ways of relating to mathematics of the past. *Journal of Humanistic Mathematics* **8** (1), 3–23.

Fried, Michael N. and Sabetai Unguru (2001). *Apollonius of Perga's Conica: Text, Context, Subtext*. Leiden: Brill.

Geertz, Clifford (1973). *The Interpretation of Cultures: Selected Essays*. New York: Basic Books.

Gibbs, Raymond W. (1999). *Intentions in the Experience of Meaning*. Cambridge: Cambridge University Press.

Gile, Daniel (2009). *Basic Concepts and Models for Interpreter and Translator Training*. Amsterdam and Philadelphia: John Benjamins Pub. Co.

Gillies, Donald (ed) (1992). *Revolutions in Mathematics*. Oxford: Clarendon Press.

Giusti, Enrico (1993). *Euclides reformatus: la teoria delle proporzioni nella scuola galileiana*. Milano: Bollati Boringhieri.

Goldstein, Catherine (1995). *Un Théorème de Fermat et ses Lecteurs*. St-Denis: Presses Universitaires de Vincennes.

Grafton, Anthony (1991). *Defenders of the Text: The Traditions of Scholarship in an Age of Science, 1450–1800*. Cambridge, MA, and London: Harvard University Press.

Grattan-Guinness, Ivor (2004a). History or heritage? An important distinction in mathematics and for mathematics education. *The American Mathematical Monthly* **111** (1), 1–12.

Grattan-Guinness, Ivor (2004b). The mathematics of the past: distinguishing its history from our heritage. *Historia Mathematica* **31**, 163–185.

Guicciardini, Niccolò (1999). *Reading the Principia: The Debate on Newton's Mathematical Methods for Natural Philosophy from 1687 to 1736*. Cambridge: Cambridge University Press.

Guicciardini, Niccolò (2009). *Isaac Newton on Mathematical Certainty and Method*. Cambridge MA: MIT Press.

Guicciardini, Niccolò (2018). Un Altro Presente: on the historical interpretation of mathematical texts. *BSHM Bulletin: Journal of the British Society for the History of Mathematics*, **33** (3), 148–165.

Hall, A. Rupert (1951). Robert Hooke and horology. *Notes and Records of the Royal Society* **8** (2), 167–77.

Hankel, Hermann (1869). *Die Entwickelung der Mathematik in den letzten Jahrhunderten*. Antrittsvorlesungen. Tübingen: Fues'sche Sortimentsbuchhandlung.

Hardy, Godfrey Harold (1940). *A Mathematician's Apology*. Cambridge: Cambridge University Press.

Heath, T.L. (1908). *The Thirteen Books of Euclid's Elements*. Cambridge: Cambridge University Press.

Hodgkin, Luke (2005). *A History of Mathematics: from Mesopotamia to Modernity*. Oxford: Oxford University Press.

Høyrup, Jens (1991). Changing trends in the historiography of Mesopotamian mathematics: An insider's view. *Filosofi og Videnskabsteori pa Roskilde*

Universiteitscenter 3 (3) (published in *History of Science* **34** (1) (1996), 1–32).

Høyrup, Jens (2002). *Lengths, Widths, Surfaces: A Portrait of Old Babylonian Algebra and Its Kin*. New York: Springer.

Høyrup, Jens (2017). What is "geometric algebra," and what has it been in historiography? *AIMS Mathematics* **2** (1), 128–60.

Jardine, Lisa (2006). Robert Hooke: A reputation restored. In: *Robert Hooke: Tercentennial Studies*, M. Cooper and M. Hunter (eds). Aldershot, Hants: Ashgate, 247–58.

Jardine, Nicholas (2000a). Uses and abuses of anachronism in the history of the sciences. *History of Science* **38**, 251–70.

Jardine, Nicholas (2000b). Original meanings and historical interpretation. In *The Scenes of Inquiry: On the Reality of Questions in the Sciences*, second edition. Oxford: Clarendon Press, 243–58.

Klein, Jacob (1968). *Greek Mathematical Thought and the Origins of Algebra*. Cambridge MA: MIT Press. Reprinted New York: Dover, 1992. [German original (1934–1936).]

Kline, Morris (1973). *Why Johnny Can't Add: The Failure of the New Math*. New York: St Martin's Press.

Lakatos, Imre (1976). *Proofs and Refutations: The Logic of Mathematical Discovery*. Cambridge: Cambridge University Press.

Lanson, Gustave (1919). Quelques mots sur l'explication de textes. In: *Méthodes de l'histoire littéraire*, cited from the reprint Paris et Genève: Slatkine, 1979.

Le Clerc, Jean (1697). *Ars critica in qua ad studia linguarum latinae, graecae & hebraicae via munitur: veterumque emendandorum, & spuriorum scriptorum à genuinis dignoscendorum ratio traditur volumen primum*. Amsterdam: Apud Georgium Gallet.

Leonelli, Sabina (2016). *Data-centric Biology: A Philosophical Study*. Chicago and London: The University of Chicago Press.

Leutzsch, Andreas (ed) (2019). *Historical Parallels, Commemoration and Icons*. London: Routledge.

Lloyd, Geoffrey E.R. (2004). *Ancient Worlds, Modern Reflections: Philosophical Perspectives on Greek and Chinese Ancient Science and Culture*. Oxford: Oxford University Press.

Lloyd, Geoffrey E.R. (2009). *Disciplines in the Making: Cross-cultural Perspectives on Elites, Learning and Innovation*. Oxford: Oxford University Press.

Lowenthal, David (2015). *The Past is a Foreign Country*, second edition. Cambridge: Cambridge University Press. (First edition, 1985.)

Lynch, William T. (2004). The utility of the present in reconstructing science's past: historical counterfactuals and contemporary possibilities. *Scientia Poetica* **8**, 241–50.

MacKenzie, Donald (2001). *Mechanizing Proof: Computing, Risk, and Trust*. Cambridge MA: MIT Press.

Mahoney, Michael S. (1970). Babylonian algebra: Form vs. content (Essay review of O. Neugebauer, *Über vorgriechische Mathematik*). *Studies in History and Philosophy of Science* **1** (4), 369–80.

Malet, Antoni (1996). *From Indivisibles to Infinitesimals: Studies on Seventeenth-Century Mathematizations of Infinitely Small Quantities*. Bellaterra, Spain: Universitat Autònoma de Barcelona.

Mancosu, Paolo (1996). *Philosophy of Mathematics and Mathematical Practice in the Seventeenth Century*. Oxford: Oxford University Press.

Maronne, Sébastien (2017). Les commentaires sur la *Géométrie de M. Descartes* (1730) de Claude Rabuel. In: *Autour de Descartes et Newton: Le Paysage Scientifique Lyonnais dans le Premier XVIIIe Siècle*, Pierre Crépel and Christophe Schmit (eds). Paris: Hermann, 111–61 and 349–55 (Annexe).

Netz, Reviel (2002a). Review of Michael N. Fried and Sabetai Unguru (2001) *Apollonius of Perga's Conica: Text, Context, Subtext*. Leiden: Brill. *Byrn Mawr Classic Review* 2002.09.34 (online).

Netz, Reviel (2002b). It's Not That They Couldn't. *Revue d'histoire des mathématiques* **8**, 263–89.

Neugebauer, Otto (1936). Zur geometrischen Algebra (Studien zur Geschichte der antiken Algebra III). *Quellen und Studien zur Geschichte der Mathematik, Astronomie und Physik, B* **3**, 245–59.

Newton, Isaac (1707) *Arithmetica Universalis: Sive de Compositione et Resolutione Arithmetica Liber*, William Whiston (ed). Cambridge: Cambridge University Press.

Newton, Isaac (1967–1981). *The Mathematical Papers of Isaac Newton*, 8 volumes, Derek T. Whiteside (ed). Cambridge: Cambridge University Press.

Newton, Isaac (2021). *The Mathematical Principles of Natural Philosophy*. Translated and Annotated by C.R. Leedham-Green. Cambridge: Cambridge University Press.

Pallottino, Massimo et al. (1986). *Rasenna: Storia e Civiltà degli Etruschi*. Milano: Scheiwiller.

Pappus (1588). *Mathematicae Collectiones à Federico Commandino Urbinate in Latinum Conversae, et Commentariis Illustratae*. Pesaro: Girolamo Concordia.

Pickering, Andrew (1995). *The Mangle of Practice: Time, Agency and Science*. Chicago: The University of Chicago Press.

Proust, Christine (2015). Mathématiques en Mésopotamie: étranges ou familières? In: *Pluralités Culturelles et Universalité des Mathématiques: Enjeux et Perspectives pour leur Enseignement et leur Apprentissage – Actes du Colloque EMF2015 Plénières*, L. Theis (ed). Alger: Université des Sciences et de la Technologie Houari Boumediene, Société Mathèmatique d'Algérie, 17–39.

Ritter, Harry (1986). *Dictionary of Concepts in History*. Westport CT: Greenwood Press.

Robson, Eleanor (2008). *Mathematics in Ancient Iraq: A Social History*. Princeton: Princeton University Press.

Rorty, Richard (1985). The historiography of philosophy: four genres. In: *Philosophy in History*, R. Rorty, J.B. Schneewind and Q. Skinner (eds). Cambridge: Cambridge University Press, 49–75.

Rossi, Paolo (1999). *Un Altro Presente: Saggi sulla Storia della Filosofia*. Bologna: Il Mulino.

Rowe, David E. (2012). Otto Neugebauer and Richard Courant: on exporting the Göttingen approach to the history of mathematics. *Mathematical Intelligencer* **34**(2), 29–37.

Rowe, David E. (2016). Otto Neugebauer's vision for rewriting the history of ancient mathematics. In: *Historiography of Mathematics in the 19th and 20th Centuries*, V.R. Remmert, M. Schneider and H. Kragh Sørensen (eds). Basel: Birkhäuser, 123–41.

Schneider, Martina (2016). Contextualizing Unguru's 1975 attack on the historiography of ancient Greek mathematics. In: *Historiography of Mathematics in the 19th and 20th Centuries*, V.R. Remmert, M. Schneider and H. Kragh Sørensen (eds). Basel: Birkhäuser, 245–67.

Shank, John Bennett (2018). *Before Voltaire: The French Origins of "Newtonian" Mechanics, 1680–1715*. Chicago: The University of Chicago Press.

Shapin, Steven (1989). Who was Robert Hooke? In: *Robert Hooke, New Studies*, M. Hunter and S. Schaffer (eds). Woodbridge: The Boydell Press, 253–285.

Sialaros, Michalis (2018). *Revolutions and Continuity in Greek Mathematics*. Berlin and Boston: Walter de Gruyter.

Sialaros, Michalis and Christianidis Jean (2016). Situating the debate on "geometrical algebra" within the framework of premodern algebra. *Science in Context* **29** (2), 129–50.

Skinner, Quentin (1969). Meaning and understanding in the history of ideas. *History and Theory* **8**, 3–53. (Reprinted in *Visions of Politics, Volume 1: Regarding Method*. Cambridge: Cambridge University Press, (2002), 57–89).

Syrjämäki, Sami (2011). *Sins of a Historian: Perspectives on the Problem of Anachronism*. Doctoral Dissertation, University of Tampere.

Swerdlow, Noel (1993). Otto E. Neugebauer (26 May 1899–19 February 1990). *Proceedings of the American Philosophical Society* **137** (1), 139–65.

Szabó, Árpád (1978). *The Beginnings of Greek Mathematics*, Synthese historical library 17, Dordrecht and Boston: Reidel. (Originally published as

Anfänge der griechischen Mathematik. München und Wien: R. Oldenbourg (1969)).

Truesdell, Clifford (1973). The scholar's workshop and tools. *Centaurus* **17**, 1–10.

Unguru, Sabetai (1975). On the need to rewrite the history of Greek mathematics. *Archive for History of Exact Sciences* **15**(1), 67–114.

Unguru, Sabetai (1979). History of ancient mathematics: Some reflections of the state of the art. *Isis* **70** (4), 555–65.

Unguru, Sabetai, and David E. Rowe (1981). Does the quadratic equation have Greek roots? A study of geometrical algebra, application of areas, and related problems. *Libertas Mathematica* **1**, 1–49.

Unguru, Sabetai, and David E. Rowe (1982). Does the quadratic equation have Greek roots? A study of geometrical algebra, application of areas, and related problems (continued). *Libertas Mathematica* **2**, 1–62.

Van der Waerden, Bartel L. (1954). *Science Awakening*, Groningen: Noordhoff.

Van der Waerden, Bartel L. (1975). Defence of a shocking point of view. *Archive for History of Exact Sciences* **15** (3), 199–210.

Venuti, Lawrence (1995). *The Translator's Invisibility: A History of Translation*. London: Routledge.

Vickers, Brian (ed) (1987). *English Science: Bacon to Newton*. Cambridge: Cambridge University Press.

Wardhaugh, Benjamin (2010). *How to Read Historical Mathematics*. Princeton: Princeton University Press.

Weil, André (1978). Who betrayed Euclid? (Extract from a letter to the editor). *Archive for History of Exact Sciences* **19** (2), 91–93.

Weil, André (1980). History of mathematics: why and how. In: *Proceedings of the International Congress of Mathematicians (Helinski, 1978)*. Helsinki: Acad. Sci. Fennica. (Reprinted in A. Weil, *Oeuvres Scientifiques. Collected Papers, Volume 3*. New York, Heidelberg, Berlin: Springer (1980), 434–42.)

Whiteside, Derek T. (1961). Patterns of mathematical thought in the later Seventeenth Century. *Archive for History of Exact Sciences* **1**, 179–388.

Zeuthen, Hieronymus Georg (1886). *Die Lehre von den Kegelschnitten im Altertum*. Copenhagen: A.F. Höst & Sohn. (Originally published in 1885 as: Kegelsnitlaeren in Oltiden. *Kongelig Danske videnskaberens Selskabs Skrifter, 6th ser.*, **1** (3), 1–319.)

Zeuthen, Hieronymus Georg (1896). *Geschichte der Mathematik im Altertum und Mittelalter*. Copenhagen: A.F. Höst & Sohn.

2

From reading rules to reading algorithms: textual anachronisms in the history of mathematics and their effects on interpretation

Karine Chemla
SPHERE (CNRS – University of Paris)

Abstract: This chapter is devoted to what I call textual anachronism. By this expression, I refer to forms of anachronism that lead to interpreting ancient texts on the basis of anachronistic assumptions with respect to how these texts made sense for the ancient actors. This is, for instance, the case when historians take the textual components that they find in ancient documents (like a mathematical problem, an algorithm, a proof, and a diagram), as they would take what they consider to be modern counterparts, and interpret these components on that basis. I argue that this form of anachronism has caused misinterpretations of several kinds, which I identify and analyze, using examples from four historical contexts. I further discuss the historiographical implications of this type of anachronism. I conclude with the thesis that one can limit the effects of this form of anachronism by using a historical approach to forms of text. Moreover, a historical approach of this kind sheds light on the fact that anachronism has a history that might also be interesting to consider.

2.1 Introduction

Conceptual anachronism has probably been the form of anachronism that has most preoccupied the historians of mathematics concerned with methodology and practicing conceptual history. Indeed, this kind of anachronism undermines the very foundation of conceptual history. If it were necessary to demonstrate this assertion, it could be done most efficiently by considering a single example: the sketch of a history of fractions that David Eugene Smith placed at the beginning of the chapter of his *History of Mathematics* entitled "Artificial numbers." The sketch reads as follows: "Gradually, however, the notion of a unit fraction developed; then came the idea of a general fraction (. . .)."[1] The formulation bespeaks a reading of fractions that occur in the oldest writings testifying to quantities of this kind, which is confirmed by Smith's historical treatment: Smith interprets fractions attested to by documents written in Egypt in the early second millennium BCE as "unit fractions," that is, as fractions of the type $1/n$. In other words, he reads them using the modern concept of fraction, thereby providing them with a numerator and a denominator. It is precisely *this* rewriting that transforms these numbers into particular fractions, by comparison with fractions such as m/n, which Smith refers to as "general." Smith then relies on this anachronistic reshaping of the concept to which ancient documents testify, to recount the history of fractions in terms of a development from one stage to the other. However, fractions to which papyri found in Egypt attest had no numerator, and accordingly arithmetic operations were carried out on them in ways that were different from how the same operations were executed on fractions understood as having a numerator and a denominator. Inasmuch as we accept the idea that meaning is use, the uncovering of this conceptual diversity suffices to undermine the scheme of a linear development that Smith proposed. Moreover, it sheds light on historiographical issues that the anachronistic rewriting concealed.[2]

In general, the forms taken by conceptual anachronism are subtler, but they have equally detrimental effects on conceptual history, as is amply demonstrated in other chapters in this book. This chapter will concentrate on a wholly different kind of anachronism, which in my view has also had significant consequences for the practice of conceptual history. However, to my knowledge, it has not yet been discussed. I will refer to the type of anachronism in question as textual. It is, for instance, in play

1 Smith (1925, p. 209). This chapter was first presented during the "Anachronism(s) in the History of Mathematics: The Seventh Biennial Bacon Conference," conference organized by Niccolò Guicciardini on April 13–14, 2018, at Caltech. I benefited from the feedback of all participants, and also from the remarks that Niccolò Guicciardini and Richard Kennedy sent me after the conference. It is my pleasure to extend my thanks to all of them.

2 Benoit et al. (1992) discusses the various concepts of fraction attested in the ancient world. Chemla (1992) addresses the historiographical issues in the global history of fractions that this alternative interpretation raises.

when historians read their sources, and the textual elements composing them (like a mathematical problem, a procedure of computation, a proof, a diagram and so on), as they read modern counterparts. Indeed, a reading such as this implicitly posits that these various textual components constitute ahistorical modes of writing mathematics. In this chapter, I intend to examine various forms taken by textual anachronism and to show why they are detrimental for the interpretation of mathematical sources. As is often the case, it is probably a constant engagement with ancient texts that has made me more sensitive to this problem than I would have become, had I worked only on documents from the nineteenth and twentieth centuries. Probably too, the problem is more significant for ancient history, and its consequences more dramatic. However, I am convinced that the difficulty is general.

In what follows, I will mainly analyze cases in which anachronisms of this type can be identified in how different actors of the past read ancient Chinese mathematical texts. Indeed, these are the sources with which I am the most familiar, and only a great deal of familiarity with the sources allows one to detect anachronisms in the ways they have been read and analyze their consequences on the interpretation of the sources and, more generally, on historiography. Later, however, for reasons that will become clearer, I will have to make an incursion into the historiography of Babylonian mathematics. The writings under consideration in both cases are mainly composed of mathematical problems and procedures. My main focus will thus be on how texts of procedures (and incidentally problems) were read in different historical settings. I have chosen to concentrate on four contexts, and in fact more precisely on four sets of actors. This will constitute a first step towards a *longue-durée* history of the reading of ancient mathematical texts, which I intend to undertake in the future. In what follows, I systematically aim to uncover the assumptions about the texts of problems and procedures that the various readings bespeak and analyze the kinds of anachronisms that these readings represent.

Section 2.2 will concentrate on Edouard Biot (1803–1850), the son of physicist Jean-Baptiste Biot (1774–1862). After scientific studies, in 1822 Edouard passed the competitive entrance exam for the Ecole Polytechnique. However, he did not enter that school, and turned instead to railway engineering. From 1825 he worked with the Seguin brothers in the context of the company that built the Lyon to Saint-Etienne railway.

On being dismissed from this enterprise in 1833, Edouard began studying Chinese language and literature with Stanislas Julien (1797–1873), who since 1832 had held the related chair at the Collège de France in Paris. Edouard soon became the first lay sinologist in Europe to publish on the history of mathematics in China (Martija-Ochoa, 2002; Chemla, 2014). The second actor, whose reading of ancient mathematical sources I analyze in Section 2.3, Mikami Yoshio (三上義夫, 1875–1950), was born in Japan shortly after the Meiji restoration. He engaged in a historical study of Chinese mathematical sources after an initial training that focused mainly on English and mathematics. In his youth, being specifically interested in the philosophy of non-Euclidean geometry, he entered into correspondence with various American mathematicians and philosophers, and in particular with George Bruce Halsted (1853–1922), who encouraged him to turn to the study of mathematics in East Asia (Sasaki, 2002; Mizuno and Chemla, in press). Mikami's *The Development of Mathematics in China and Japan*, published in 1913, reveals an approach to Chinese texts that is drastically different from Biot's. These two men will illustrate for us two quasi-opposite types of anachronistic readings of problems and rules. However, their approaches are similar in that they apparently never addressed the issue of how the text of a rule means what they thought it meant. By contrast, Section 2.4 focuses on two groups of actors who did consider this question, but on the basis of different assumptions. They will illustrate two other types of anachronism in the interpretation of rules. We will first examine aspects of the reading of Chinese mathematical texts around Joseph Needham (1900–1995), a scholar known for having put forward China as a topic in the history of science, notably thanks to his multi-volume *Science and Civilisation in China*, whose publication began in 1954. Needham, who worked as a biochemist before becoming a historian of science, published on mathematics in China exclusively with his collaborator Wang Ling 王鈴 (1917–1994). We will see how anachronism is at play in how Needham and Wang Ling tacitly assumed that Chinese texts of procedure meant the computations they thought these texts meant. Secondly, Needham and Wang Ling's mode of reading will be contrasted with the radically new approach to texts and procedures that began being practiced in the 1970s under the influence of the historical incursions of algorithm specialist Donald Knuth. This analysis will lead me to draw a counter-intuitive conclusion. Indeed, I will suggest that the latest developments in science sometimes allow historians to read ancient sources in less anachronistic ways, in a sense that I will explain in conclusion (Section 2.5).

Ideally, to compare readings carried out by different historians, we should observe how these scholars approached the same ancient texts. I will proceed in this way only partly. Indeed, only some of the ancient mathematical sources available enable us to observe all the facets of anachronism that I intend to address. However, we do not have the interpretations of these specific texts by all the historians under consideration. I will thus rely on a corpus of texts that can best enable us to compare various historians' modes of reading. Texts of procedures contained in the canonical book completed in the first century CE, under the title *The Nine Chapters on Mathematical Procedures* (*Jiuzhang suanshu*, 九章算術, hereafter *The Nine Chapters*), will play a central role in this corpus.[3] However, for reasons relating to both the history of Chinese mathematical sources and Biot's working conditions, this book was not available to him, and we will thus have to analyze his approach to other similar mathematical sources. Let us explain why.

2.2 Edouard Biot's approach to Chinese mathematical texts: a literal interpretation

As a sinologist who never had the opportunity to go to China, Biot depended for his research on the collections to which he had access in Paris. His first publication on mathematics (an appendix to a review published by his father (Biot, 1835, pp. 270–273) bore on the first mathematical book in Chinese he could find in the *Bibliothèque Royale*. This was the *Suanfa tongzong* 算法統宗, whose title Edouard rendered as "Complete treatise on the art of counting" (*Traité complet de l'art de compter*, hereafter *Complete Treatise*), in the more substantial article that he devoted to this book later (Biot, 1839). Biot was aware that this work had been completed at the end of the sixteenth century in China, but that it drew on an older book entitled, precisely, *The Nine Chapters*, whose chapters gave their titles to those into which *Complete Treatise* classified its content (Biot, 1839, pp. 196–98; 201). However, *The Nine Chapters* was not

[3] A note of caution about the transliteration of Chinese is required. Today, pinyin is the standard transliteration, and it is the one I use. However, in the quotations, the reader will encounter different types of transliteration that were used in the past. Reading will hopefully not be hindered by these differences.

available in Paris at the time. This explains why, in this case, we have to describe Biot's approach to other Chinese mathematical sources, that is, sources he could read, but sources that were nevertheless quite close to *The Nine Chapters.*

Edouard Biot read the book several times in search of "original documents;" that is, evidence of innovation (p. 196). More precisely, his reading has the aim of identifying "the most advanced mathematical notions" (p. 194–195). This valuing of originality, especially for knowledge that he felt would be "somewhat significant" (p. 196), constitutes a first form of anachronism, regarding value, for which we will see more examples below. Edouard considered he had already listed all but one of such notions in his 1835 appendix. In the 1839 article, he thus felt there was no point in printing the translation of *Complete Treatise*, although he had "already almost completed" it (p. 196, p. 199). Instead, he made do with the publication of a table of contents. Another related reason made the translation worthless for him: one could not accurately date the "first use of these rules in China" (p. 196). For the purpose of "allowing the reader to assess the state of Chinese knowledge in modern times," the table of contents would suffice "at least for today" (p. 199).

As Edouard put it, a translation would "present only a sequence of *applications* of *rules* that are for the most part *elementary*" (p. 196, my emphasis). We have already seen his lack of interest in the "elementary." What is more, in his view the mathematical problems represented "applications," while "rules" (that is, procedures) offered "ways of operating" ("manières d'opérer," p. 217) to solve them. For him, this reading was consistent with his interpretation of the title. Edouard writes: "As its title indicates, the *Souan-fa-tong-tsong* is a collection of *practical rules of calculation*" (p. 193, my emphasis). Edouard's interpretation of the title as *Complete treatise on the art of counting* can be compared with later renderings, like *A Systematised Treatise on Arithmetic* (Mikami, 1913, p. 110), *Systematic Treatise on Arithmetic* (Needham and Wang, 1959, p. 51), *General Source of Computational Methods* (*Source générale des méthodes de calcul*, in Martzloff (1997, p. 20; 1987, p. 23), respectively). The comparison suggests that Edouard's interpretation, in particular his interpretation of *suanfa* as "art of counting," played a key part in shaping his perception of the rules as "practical."

His reading of the book that the assessment above summarizes is also consistent with the perception of its contents that the table conveys. This is clear from Biot's synopsis of Book IV, which reads as follows:[4]

On grains and coinage, second chapter (of the ancient book *Kieou-tchang* (NoT: *The Nine Chapters*))

P. 1. – Proportional numbers for the various prices of commodities.
P. 1 to 3. – On grains, rice, wheat, gold. – Eight questions.
P. 3 to 4. – On foodstuffs intended for the administration, and on the waste accepted during transportation. – Three questions.
P. 4 to 10. – Calculations related to capacity measures, warehouses and underground pits (silos). – Sixteen questions.
P. 10. – Rule for measuring salt in separate heaps. – One question.
P. 10 to 19 v. – System of scales and weights. – Fourteen questions.
P. 19 v. – Rules for smelting iron and copper in ore. – Three questions.
P. 20 v. – System of length measurements. – Nine questions.
P. 23 v. – On fractions of the main measurement units for miscellaneous sellable goods. – Three questions.

Biot does not mention the fact that all these problems concern issues of proportionality. However, the first sentence of the chapter in *Complete Treatise* makes it clear that all these specific problems have the aim of "mastering transformations 以御變易" (Chapter IV, p. 1). They are, so to speak, paradigms (Chemla, 2009). Instead, Biot considers that the problems have the aim of providing means to fulfill the specific tasks that they present. In other words, Biot reads the problems at face value. One might object that in a sense, he too reads paradigms in problems, since a singular problem refers to the way of fulfilling a task in general. However, the difference between reading a problem as a paradigm for a mathematical issue and as a paradigm for a specific task is clear if we consider the class of problems for which a given problem stands in each case. Biot's table of contents bespeaks this difference. Given the gap between this interpretation and the statement of the *Complete Treatise* just translated, we can consider that Biot's uptake of the problems does not correspond to the intention of their author. This constitutes the first illustration of the issue I address in this chapter: anachronism consists here in taking the text of a problem as what it might have meant for a

[4] Biot (1839, p. 206). Edouard Biot (1839, p. 200) indicates the copy of the book on which he draws for his table and to which the page numbers refer. The book is in the Bibliothèque Nationale de France, under the title 增補算法統宗全書 *Zeng bu suan fa tong zong quan shu. Le Suan fa tong zong augmenté*. It has been digitized, see: `https://gallica.bnf.fr/ark:/12148/btv1b9002942c.r=算法統宗rk=21459;2` (last accessed July 12, 2019), which corresponds to the number 350 in the Fourmont collection, donated to the Bibliothèque Royale, that Edouard mentioned.

reader in Biot's own context. We see how, here too, it is this anachronistic mode of reading that turns procedures into practical rules, and problems into applications (the ancient commentaries on *The Nine Chapters* never adopt an interpretation of this kind).

Edouard applied the same type of reading to the titles of the chapters in *Complete Treatise*, which, as we have seen, derived from *The Nine Chapters*. As a result, for him, some of them "related only to practical operations," while the others were "relatively vague, as usual for the titles of Chinese books" (p. 197). He likewise considered that "the titles of the various questions" were "vague" and "unclear." For this reason, as we will see below, in the table he chose to "add necessary explanations," in italics (p. 199). This anachronistic assessment once again bespeaks values which, like some of his fellow scientists, he had abided by (in this case, those of precision and clarity). One might be tempted to note that Biot apparently did not pause to consider whether these defects could have been caused by his interpretation. However, here Edouard added an interesting nuance: vagueness and obscurity, he asserted, are attributes of "the literal meaning" (p. 199). The point is that his translations systematically put forward the literal meaning, at the expense of understandability.

If we observe how Edouard rendered "the titles of the various questions" in his table, we see that for him, a translation had to be literal. This seems to suggest that, in his view, the Chinese text had no technical terminology, or that if it did, this terminology should not be translated by the related French terminology. Only in explanations that he added to the text and systematically featured in italics was the technical meaning clarified. For instance, he translated the term *shangchu* 商除, which refers to a type of division in which one discusses (*shang*) digit by digit the value of the quotient (also *shang*), by "examined division (division examinée)" (p. 202). He also translated *kai pingfang* 開平方 as "solving an equal square (résoudre un carré égal)" and explained: "*extraction of the square root*" (p. 203, italics in the original). Similarly, *kai lifang* 開立方 became "solving a straight square, that is to say, *a cube* (résoudre un carré droit, c'est-à-dire *un cube*)" and was explained as "*extraction of the cube root*" (p. 203, italics in the original). In a sense, this choice for the translation represents another form of anachronism in the approach to the text. This way of rendering the original wording into French seems to indicate that, for Edouard, the literal reading of a nineteenth-century scholar would give the French reader a more genuine access to how the text meant what he thought it meant for a Chinese user of this text at the time the text was written.

In fact, Biot opted for the same type of literal translation when he translated mathematical texts fully. This is what appears from an analysis of the translation of *The Gnomon of the Zhou* that he published in 1841 (Biot, 1841). Indeed, the reasons that, as we have seen, Edouard adduced for not translating *Complete Treatise* did not hold for this book that was devoted to astronomy and mathematics. Biot had embraced the view that *The Gnomon of the Zhou* dated from the twelfth century BCE (Biot, 1839, p. 196). We think today that it was completed around the beginning of the common era. Moreover, for Edouard, this writing attested to the knowledge "of the fundamental property of the right triangle, (. . .) remarkable by its anteriority of six centuries with respect to Pythagoras's discovery" (Biot, 1841, pp. 593–594). The valuing of "originality" thus played its part in making the book appear to Edouard to be worth translating (Biot, 1841, p. 598). To explain how, in this case too, Biot translated literally, we will examine his rendering of the passage devoted to precisely this "property."

Before we do so, however, a warning is required, to clarify my purpose. The text whose translation we will now analyze is extremely difficult, and even today, specialists do not agree on its interpretation.[5] Although, like Biot, all specialists concur that it states something similar to the "Pythagorean theorem," the way in which the text states this proposition and whether or not it refers to a proof remain issues of contention. My analysis of Biot's translation intends only to go deeper into the translator's workshop. It aims, in particular, to remind us of the historicity of the conditions in the context of which translations were carried out and of the historical dimensions of the practice of translating. In Edouard's translation, the text reads as follows (we keep the entries of the footnotes of the original text to comment on them later):[6]

[5] Chemla (2005) gives an overview of several translations of the same passage published in recent decades. The reader interested in the text will find there a full translation of the passage under consideration and its ancient commentaries.

[6] Biot (1841, p. 600) also for the footnotes mentioned below. The French translation reads as follows (the paragraphs respect the original layout):

> "Divisez le *kuu*:
> Vous ferez le *keou* ou largeur égal à 3;
> Le *kou* ou longueur égal à 4;
> La ligne qui unit les angles (*king-yu*) égale à 5.[2]

Divide the *kuu*:

You will make *keou* or width equal to 3;

the *kou* or length equal to 4;

The line linking the angles (*king-yu*) equal to 5.[2]

Outside the rectangular figure (*fang*), take the half: this will be a *kuu*.[3]

Encompass *or* unite, and together compute,[4] you will obtain perfectly 3, 4, 5.

The two *kuu* together make a length of 25. This is what one calls the sum of the *kuu*.[5]

Obviously, Biot did not translate all the terms, but, for some of them, only offered a transliteration. A few lines before the quotation, at the first occurrence of the first term he left untranslated (*kuu*), Biot explained it as follows: "Bas. 6806. Literal. *The ruler* (NoT, or *The rule*, since the two words are indistinguishable in French), *the instrument to make straight lines and squares*" (italics in the original). For this literal meaning, Biot referred the reader to "Bas.," that is, the Chinese–Latin dictionary that had been prepared by Basilio Brollo (1648–1704), arguably between 1692 and 1701, and that Chrétien-Louis-Joseph de Guignes (1759–1845) had eventually managed to publish in 1813, in the form of a Chinese–French–Latin dictionary.[7] This represented the first multilingual dictionary with Chinese ever printed. Interestingly, to comment on the meaning of the text Biot was translating, he quoted the related entry selectively, and from its Latin section,[8] mentioning only the part of the literal sense that he thought suitable for the context. However, what the "division" of the *kuu*, interpreted along these lines, might mean is unclear. This might explain why in his footnote 2, indicated above, Biot repeated this explanation, adding: "and also 'try square (équerre)'." Noteworthy is the fact that this other meaning does not come from de Guignes's dictionary, and Biot gives no reference for it. In fact, shortly after de Guignes's publication, in 1815, Robert Morrison (1782–1834) began publishing a Chinese–English dictionary (Morrison, 1815). In the second volume of part I, published in 1822, at the entry for the

En dehors de la figure rectangulaire (*fang*), prenez la moitié: ce sera un *kuu*.[3]
Englobez ou réunissez, et ensemble calculez,[4] vous obtenez parfaitement 3, 4, 5.
Les deux *kuu* ensemble font une longueur de 25. C'est ce que l'on appelle la somme des *kuu*.[5]"

7 See de Guignes (1813). The character 6806, *kuu* (in pinyin, *ju*) 矩, is on p. 477. On Basilio Brollo's dictionary, see Bussotti (2015). On its first successful printing by de Guignes, upon Napoleon's order, and after several failed attempts, on the criticisms that the dictionary met, and the accusations of plagiarism subsequent to its publication, see Landry-Deron (2015). This might explain why Biot refers to the dictionary as he does.

8 This incidentally suggests that he might want to use the two meanings of the term "règle;" that is, "rule" and "ruler."

same character, one reads precisely: "A square used by carpenters." In 1822, Jean-Pierre Abel-Rémusat (1788–1832), the holder of the chair for Chinese at the Collège de France before Julien, considering the limits of de Guignes's dictionary, had advised cross-referencing with Morrison's.[9] Biot might have followed this advice when he mentioned this additional meaning of *kuu*, needed in this context.[10] In any event, we see that to make sense of the texts he dealt with, Edouard relied closely on the dictionaries available at the time, and again on the "literal" meanings they provided.

This is confirmed by the other parts of his footnote 2. To begin with, in this footnote Edouard pointed out the values mentioned in the Chinese text (3, 4, 5), to identify the topic of the passage as the right triangle. It was only then, following this insight, that he mentioned the other meaning of *kuu* as "try square," even though he refrained from translating this term. Likewise, in the same footnote, Biot asserted that *keou* and *kou* were "used here as two special terms" (p. 600), and accordingly he did not translate them, making do with a mention of their meaning "in the ordinary style," as found in the dictionary.[11] He went on to explain that, in the context, they "designated," respectively, the base and the height of the right triangle, but following closely what he found in the dictionaries available to him, this was not how he translated them. Similarly, footnote 2 is the place where, after a word-for-word explanation, Edouard interpreted the "line linking the angles" as the hypotenuse, but, in the main text, he gave a literal translation. This way of rendering the original text is thus in keeping with what we have observed for the translation of the table of contents of *Complete Treatise*. The next sentence in the translated text of *The Gnomon of the Zhou* also involves *kuu*. Edouard interpreted this sentence in a way that led him to assert, in footnote 3, that in this context, *kuu* was a "right triangle." Perhaps, he thought this was a meaning he could derive from that of "try square," which Morrison's dictionary gave. However, as far as I can tell, to this day, the meaning of "right triangle" for *kuu* has not been

9 (Landry-Deron, 2015, p. 435).

10 (Biot, 1839, p. 202) refers to Part 3 of Morrison's dictionary (English–Chinese dictionary, 1822, article "Weights and measures", pp. 464–466). Note that Biot knew English (Chemla, 2014).

11 Biot gives the literal meaning of *gou* (股, in pinyin *gu*) from de Guignes's dictionary (character 8414, p. 588). Without accounting for this move, he does not follow the dictionary for *keou* (勾, in pinyin *gou*, character 932, p. 63), and instead glosses it with the character 鉤, whose meaning he gives (characters 11400, p. 794, and 11433, p. 797). (Biot, 1839, p. 212) translates the terminology for the right triangle and explains it in the same way.

established, which illustrates the dangers awaiting those who at the time attempted to deviate even a little from dictionaries.

The following lines of the translation reflect another form of anachronism that presides over Edouard's practice of translation. Footnote 4 makes it clear that to translate the main text, he combined the use of dictionaries with glosses given by the third-century commentary with which *The Gnomon of the Zhou* was handed down. Indeed, it shows that Biot still referred to de Guignes's dictionary for translations of characters with which the commentator explained the ancient text, and in this way produced the translation that features in the main text. In his view, the commentary was a valuable guide for the interpretation of the canon, to the point that one could substitute the former for the latter. Biot's approach thereby illustrates yet another form of textual anachronism. Indeed this was, and is still, a widespread form of anachronism in the study of ancient Chinese writings and beyond.

The way in which Biot interpreted a remark by the commentator sheds light on another respect in which his approach to the text translated was anachronistic. The main text contained the character *pan* 盤, which de Guignes's dictionary translated as "plate, basin," and (Morrison, 1815, Part I, vol. 2, pp. 703–704) as "A tub-like vessel whether made of wood or metal; a bathing tub; a vessel to contain rice (. . .)." In footnote 4, Biot gave the number of this character in de Guignes's dictionary (6570, p. 462), but chose rather to follow what he understood to be the commentator's suggestion, that is, an interpretation of *pan* as "reduce or compute." This explains why in the text, we read: "together compute." The translation derived from Biot's "literal" explanation that was based on the commentator's gloss, and reads: "*treat by the pan* (basin *for computations*)" (p. 600, italics in the text). In other words, here, the commentator's explanation prevailed over those given by dictionaries. However, Biot added to this explanation a layer of meaning deriving from the dictionaries and also from his previous work on mathematical texts. Indeed, in *Complete Treatise*, the term *pan* occurred in the expression *suanpan* (where *suan* means "computation"), to refer to the abacus. Following his usual practice, in this other context Biot had translated the expression literally as "box for computations (caisse à calculs)." Mixing the meaning "basin," from the dictionary, and the interpretation "box for computations," from *Complete Treatise*, Biot considered that the computation meant by *The Gnomon of the Zhou* made use of "the box for computations that one uses to count with balls." In other words, due to the occurrence of the word *pan*, Biot assumed that the calculating tool to which the sixteenth-century book referred had been used in the twelfth century BCE (the time when for Biot, *The Gnomon of the Zhou* had been compiled). This is the first anachronism that this interpretation reveals. But there is more. To draw this conclusion, Biot had to assume that the meaning of Chinese words had not changed over this time span. In fact, this hypothesis permeated Edouard's reading through and through. Reading an ancient text with a dictionary of the Chinese language that did not distinguish between modern spoken Chinese and written classical Chinese, as Biot did, implied the tacit assumption that over so many centuries in China, the meaning of terms had undergone no major change.[12] This remark illustrates yet another form of textual anachronism, which echoes a belief Biot formulated in 1839: "In this eminently stationary country, exact sciences have not made a single step forward" (Biot, 1839, p. 200). This belief informed the working method, and as a result, the belief found its way into the conclusions, where it seemed to have been proved.

To return to the translation of *The Gnomon of Zhou* quoted above, I think we will all agree on the fact that the literal translation Biot offered does not make much sense. This illustrates cogently his own remark that the sense of vagueness and obscurity that the reader might get, when perusing the French translation, was correlated with the choice to render the Chinese in this way (as I have highlighted above, Biot noticed that vagueness and obscurity were attributes of "the literal meaning").[13] Still, there were enough hints in the text interpreted in this way to allow Edouard to state that it referred to the "fundamental property of the right triangle." For this, he relied on the commentary, to assert, for instance, that "the text indicates the extraction of a square root." The numerical values also played a key part in favor of this interpretation. For instance, reading in the passage of *The Gnomon of the Zhou* under examination that a sum yielded 25, he commented on the sentence in footnote 5 as follows:

[12] In the entries of Morrison's dictionary, modern meanings and ancient writings are given side by side.

[13] Even today, some specialists privilege literal translation in a similar fashion. See, for instance, Martzloff (1997, p. 118 of the 2006 revision; 1987, p. 103), for a translation and (only in the English translation) the theory behind it.

This obviously designates the sum of the two squares of 3 and 4, 9 and 16, which is equal to 25, square of the hypotenuse.

Before the entry of footnote 5, the translation faithfully mirrors two occurrences of *kuu* in the Chinese text, although they could not be explained with any of the literal meanings and further interpretations that Biot had already adduced. Biot did not comment on them: the mention of the numbers seemed sufficient to Edouard to convince himself that the meaning of the passage was clear. However, the way in which he reached his understanding had a consequence on his assessment of the text, since he went on as follows:

> Let us note that here and in what follows, there is no proof of the theorem of the square of the hypotenuse for all right triangles. There is only the indication of simple numbers that verify this theorem for a particular case.

Ironically, Biot seems to have been unaware that as a historian, he was reproducing the same operation that he claimed for his actors, when, from a particular case (or at least, what, in the original text, appeared to him to be so), he drew the following general conclusion:

> This is always the case of the mathematical notions that the Chinese books contain. One finds rules expressed with numbers; never does one find any proof.[14]

As we have seen above with respect to problems, this assertion highlights the historiographical consequences deriving from textual anachronisms. Biot's conclusions on rules rested on an interpretation of rules taken as if their texts were modern texts. His expectation for proofs similarly relied on modern ideas on texts for proofs, as did the understanding of theory that led him to assert: "the titles of the various questions (. . .) evoke no theoretical idea" (p. 197). The search for theory dictated how he would publish the table of contents, as he explained: "for problems a bit significant, I will make sure to indicate whether the process followed in the solution is by trial and error, or whether it rests on a rigorous method" (Biot, 1839, p. 199).

This emphasis is especially manifest in his treatment of the solution

[14] (Biot 1841, 600–601). In 1852, when Protestant missionary Alexander Wylie (1815–1887) was based in Shanghai (for the context, see Han, 1998; Chen, 2017), he published a major work on the history of mathematics in China (Wylie, 1897). In it, he translated the same passage, also with difficulty, but with a completely different practice of translation (p. 163). He interpreted that it stated Pythagorean theorem with full generality (p. 164). Chemla (2009) discusses the expression of generality in Chinese texts of this kind.

of algebraic equations, for which *Complete Treatise* provides evidence. In the table of contents, in line with what was described above, Biot translated the related technical terms literally and provided the explanations in italics (Biot, 1839, pp. 208–210). These explanations assert that some problems can be "represented" by algebraic equations. However, for Biot, *Complete Treatise* offered no solution for them. As he stated, "*One extracts the value of x by trial and error, decomposing it into tens and units, and without solving the equation*" (italics in the original). In other words, in Edouard's expectation, a numerical solution was not a "rigorous solution." Probably, only a solution by radicals would be. This was nevertheless the example that led Biot to conclude: "In fact, to believe that the Chinese *ever possessed* any real *theory* of the exact sciences, one would have to contradict the repeated assertions of Parennin, Gaubil, Verbiest, and of so many distinguished men who succeeded each other in the eighteenth century in the China missions. (. . .) The Chinese people are *completely practical* and focused on what is *material*" (Biot, 1839, p. 200, my emphasis). As above, anachronistic expectations, this time with respect to "theory," inspired general conclusions, that at the same time rested on and supported the thesis of the unchanging character of the Chinese [le peuple chinois].

This concludes our examination of how Biot read rules, as well as problems. We have seen that several forms of anachronism were involved in his practice: anachronism in values (clarity, precision, rigor), anachronism in what counts as a proof or a theory, and finally, textual anachronisms, on which from now on, we will concentrate. Among the latter, we have seen anachronism in the approach to the Chinese language, in the use of a commentary to interpret the text commented upon, in what counts as a general statement, and in the interpretation of mathematical problems and rules. Biot approached problems and rules as he would approach their modern counterparts, and assessed them on the basis of a literal interpretation. Moreover, except for *The Gnomon of Zhou*, he said nothing about how the rules referred to computations or about the computations the rules indicated, summarizing the text into mathematical facts. To these conclusions, we must add the mention of an exception, which shows Biot's ambivalence, and which will also bring us to the main topic on which below, we will compare various historians' readings of rules, that is, square (and also cube) root extraction.

This exception occurs at the end of the 1839 article on *Complete*

Treatise (Biot, 1839, pp. 215–216), and shows Edouard's awareness that some characters do have technical meanings to be found outside dictionaries. To begin with, Biot advises the reader to pay great attention to the opening pages of *Complete Treatise*, because there, he stresses, the author explains the technical terminology used. The alternative route, he admits, would be to derive these meanings from an understanding of the solutions. The illustrations Edouard then gives show that he addresses key issues of translation. His discussion centers on the two characters *shi* 實 and *fa* 法. In addition to their literal meanings, found in de Guignes's dictionary (for instance, for *shi* "*sum, mass*" and for *fa* "*method, rule or system*" (italics in the original)), Biot explains that the former corresponds to both multiplicand and dividend, and the latter to both multiplier and divisor.[15] Translating differently, in relation to the modern meaning that the term took in each of its occurrences, is not a solution that Biot considers appropriate. Somehow, using anachronistic terms, we might say that he sometimes shows an awareness of the fact that technical terms might reflect categories used by the actors. In this case, Biot thus creates locutions to the effect that they could be used to "translate" these terms in all contexts, that is, "*mass number*" for *shi*, and "*operator number or factor*" for *fa*. One such context will be the extraction of square and cube roots, and Biot does not apply his suggestion in his table. However, in addition to the remarks on these two terms, he adds his regrets for not having inserted into his publications layouts of computation that might have helped the reader read the original text. Indeed, these rules are difficult to read, and these difficulties reveal properties of the text of rules that various historians would address differently. Let us now turn to them.

2.3 Mikami on rules and problems: a mathematical and contextual interpretation

In contrast with Biot, who had very restricted access to ancient mathematical writings from China, the other historians whose readings of rules and problems we will evoke below could consult most of the books that we have at our disposal today.[16] In particular, they had access to the

15 This is true for *Complete Treatise*, but not for the earlier writings that we will consider below.

16 A major exception to this assertion must be mentioned: in recent decades, mathematical manuscripts older than the oldest received books so far known have been found. I will not deal with them here.

oldest known canonical text devoted to mathematics: *The Nine Chapters*, which we mentioned above and whose completion I date from the first century CE.[17] Since the procedures contained in this canon will enable us to observe a wide spectrum of anachronisms with respect to the reading of rules, we will concentrate on the way in which the historians examined have approached *The Nine Chapters*. We have seen that *Complete Treatise* had a structure deriving from that of *The Nine Chapters*. In a sense, this will also enable us to compare Biot's reading to the others'.

Mikami was a prolific historian of mathematics, who mainly worked in Japan, but published in a broad range of international venues on the mathematics of China and Japan, as well as on comparative history. The issues he addressed and his working practices changed over the decades in which he wrote on the history of mathematics (Mizuno and Chemla, in press). Here, we will limit ourselves to an observation of his first major publication on the topic: his 1913 book entitled *The Development of Mathematics in China and Japan*. Mikami was conversant enough in English to write directly in this language. However, nobody seems to have polished his writing in depth, which explains the oddities that the reader will notice when we quote him below.

In contrast with Biot, for Mikami the problems were transparent. The difference is striking if we quote his comment on the chapter of *The Nine Chapters* corresponding to Book IV in *Complete Treatise*. As we have seen, for Biot, the main point of the chapter was the different types of problems solved. Mikami (1913, p. 11) simply wrote:

> The second section (. . .) treats of questions in simple percentage and proportion.

Similarly, in Chapter 8 of *The Nine Chapters*, Mikami (1913, p. 18) read only the mathematical issues. He synthesized as follows:

> The eighth section, on the *fang-ch'eng*, is in actuality a consideration of systems of linear simultaneous equations in three or more unknown quantities.

One might argue that, despite their differences, from a textual viewpoint Mikami's reading is as anachronistic as Biot's. After all, the ancient authors did use problems, and, even though they performed opposite

[17] In what follows, quotations from *The Nine Chapters* are drawn from the critical edition and French translation in Chemla and Guo (2004).

readings of these problems, neither of these historians attempted to restore the meaning of this practice from the actors' viewpoint.[18] They judged problems from their perspective as observers. For Mikami, only the mathematical "rules" given to solve the problems mattered,[19] and he considered that some problems in Chapter 8 represented types of cases for the rules presented (Mikami, 1913, p. 20).

The main point for us is that Mikami also read "rules" in ways wholly different from Biot. Let us characterize his practice as a historian in this respect. A first facet of Mikami's reading is that in some cases, he accounted for the list of operations that constituted the rule, by transforming it into a sentence. For the area of the triangular field, for instance, the related problems in *The Nine Chapters* give a base (called "width," that is, East–West dimension) and the corresponding height ("straight length," that is, North–South dimension), and the "procedure" solving them reads:

> One halves the width and one multiplies by it the straight length.

Mikami (1913, p. 10) referred to this procedure as follows:

> The area of a triangle is given as half its base multiplied by its altitude.

We see that Mikami respected the structure of the text. The fact of taking half of one of the dimensions before multiplying by the other is meaningful in the sense that the third commentator on *The Nine Chapters*, Liu Hui, proved the correctness of the procedure through an interpretation of this structure of the formulation. However, Mikami did not seem to consider it necessary to translate the exact terminology of the rule.

In other cases, Mikami read "rules" as being equivalent to formulas, such as in the case of the area of the circle, in which he writes:

> Four different rules are recorded for the area of a circle, which are equivalent to $\frac{1}{2}c \times \frac{1}{2}d,\ \frac{1}{4}cd,\ \frac{3}{4}d^2, and \frac{1}{12}c^2$, where c and d are written for the circumference and the diameter. In these rules π is obviously taken as equal to 3 (Mikami, 1913, pp. 10–11).

Figure 2.1 shows one example among several others, in which the rule is bluntly reformulated as a formula. However, elsewhere Mikami

[18] Chemla (2009) addresses precisely this issue.

[19] This is manifest in his summary of Chapter 1 (Mikami, 1913, p. 11).

"There is a string hanging from the top of a tree, its end measuring 3 feet (a) on the ground. When however it is tightly stretched, it reaches a point 8 feet (b) from the base of the tree. How long is the string (x)?"

Solution. $x = \frac{1}{2}\left(\frac{b^2}{a} + a\right).$

Figure 2.1 Mikami transforms the "rule" into a formula. Source: Mikami (1913, p. 22).

(1913, p. 23) nuances his practice, warning his reader that he has thereby reshaped the original text:

In the text of the "Nine Sections" this rule is of course expressed like all other rules in full words, not in algebraical expressions.

This second facet of his practice of reading has a noteworthy historiographic correlate: this anachronistic rendering meets with Mikami's thesis that the "birth of algebra" should be "ascribe[d]" to "Chinese mathematicians in a comparatively early part of their history" (Mikami, 1913, p. 36). Further, when discussing the rules of false double position, Mikami (1913, p. 17) makes explicit his belief that, although "no explanation is tried in the 'Arithmetic in Nine Sections' for the construction of this rule," as well as for other rules, "we can well see that the process is quite algebraical."

This highlights a third facet of his practice of reading: in contrast with Biot, for whom texts should be taken at face value and who thus underscores the universal absence of proof in Chinese texts, lack of evidence does not undermine Mikami's conviction that reasoning lay at the basis of these rules. Further, in this case, for him the reasoning is also algebraic. In fact, *The Nine Chapters* was handed down with ancient commentaries, the oldest of which being that completed by Liu Hui in 263, which we mentioned above. Mikami's assertion, also in relation to the rules of false position, that "we have no means of knowing by what kind of reasoning the ancient Chinese had been led to the above way of solution" has a revealing implication: he does not consider the reasoning that these commentaries contain as valid evidence for approaching ancient reasoning. More generally, for Mikami, the commentators are considered *qua* later authors, whose rules are sometimes mentioned, but not *qua* commentators. This might derive from an anachronistic perception of genres of text and the nature of the author.

Interestingly, the rewriting of a rule from *The Nine Chapters* under the form of a formula does not prevent Mikami from seeing the rule as equivalent to a "geometrical proposition" (Mikami, 1913, p. 22). In such cases, the property of the rule that it gives a list of computations whose use allows the practitioner to yield a numerical result remains in the shadow, and is even lost in translation. However, as Mikami cannot but notice, a rewriting of a rule as a formula or a proposition is not always possible. This remark has two interesting consequences for us. First, it means that Mikami's reading compels him to account in different ways for texts that in *The Nine Chapters* are all referred to using the term of "procedure 術 *shu*." In a sense, this reveals the textual anachronism of his way of approaching the text *qua* text. Second, this divide between procedures that the interpretation elicits occurs precisely with respect to the specificity of the procedure that we have just mentioned, namely, that the list of operations it gives yields a numerical value. Let us thus focus on those rules that do not lend themselves to reformulation.

Some of these are key rules for us, since they reveal important specificities of this way of writing mathematics to which *The Nine Chapters* attests. Mikami manifested his awareness of this issue, when, encountering a problem that is solved by a rule in which one might read a quadratic equation, he wrote:

> The solution of this problem is very remarkable, because it is *not given in a rule as usual, that is equivalent to a formula*. It is only indicated that the answer should be obtained by evolving[20] the root of an expression which expresses nothing but the equation, $x^2 + (20 + 14)x - 2 \times 20 \times 1775 = 0$ (Mikami, 1913, p. 24, my emphasis).

The context makes it clear that here, for Mikami, the "rule" does not correspond only to the statement of the operation equivalent to the equation, but also includes the solution of this equation.[21] To his eyes,

[20] NoT: "Evolving," instead of "extracting," is terminology that we find in English algebra books of the nineteenth century (Chen, 2017). The use of this term, and the discussion that follows, show that Mikami understands how, in this context, the solution of the equation is an algorithm that derives from square root extraction. We return to square root extraction below. However, we cannot discuss this specificity of quadratic equations in *The Nine Chapters*. For more explanation, see (Chemla and Guo, 2004, pp. 689–693).

[21] In this, his reading is not coherent with how he reads rules that prescribe division. We cannot develop this point here. In brief, the prescription of a quadratic equation is similar to that of a division. Mikami reads the former as including the execution of the "operation," whereas the

the procedure in *The Nine Chapters* gives an "expression" that is the equation, and additionally refers to a process yielding "its" root. This root will be obtained digit by digit (this is precisely the type of procedure that, as we have seen above, Biot thought was carried out by trial and error), and as a result the algorithm cannot be reformulated into a formula. This circumstance accounts for why Mikami placed this remark in relation to this procedure. Several "rules" in *The Nine Chapters* share the feature of not being "equivalent to a formula," notably the algorithms for solving systems of linear equations and for extracting square roots. How did Mikami read "rules" of this kind?

In these two key cases, in contrast with Biot, Mikami gave translations as well as explanations of the original text. They provide us with evidence on his practice of interpretation for these "rules." In *The Nine Chapters*, the general "rule" for solving systems of linear equations is formulated in the context of the first problem in Chapter 8. This problem puts three qualities of grain (high-, medium-, and low-quality) into play and, as usual, associates the data with numerical values. The "rule" in *The Nine Chapters* mentions the three qualities of grain of the problem explicitly, and also describes how to arrange the numerical values given by the problem on a calculating surface and, then, how to compute by reference to this layout. However, the rule does not mention any explicit numerical values. Moreover, *The Nine Chapters* contains illustration of neither the initial layout, nor of the successive tables of numbers on the calculating surface throughout the execution of the operations. The general "rule" is formulated in a paradigmatic way, in the sense that it consists of the specific sequence of computations solving the first problem in the chapter.

Mikami's translation of the problem (1913, pp. 18–20) pushes it towards a more abstract formulation than it actually has. The different qualities of grain become "three classes," and Mikami renders the rule into English, referring to the "first," the "second," and the "third class," and to the different columns that the numerical values shape on the calculating surface. Interestingly, it becomes clear later that Mikami uses the term "class" in relation to positions on the calculating surface. His explanations gloss the operations prescribed in the text and restore all the successive tables through which the resolution goes. In other words,

prescription of the division is transformed into a formula, which does not include the execution of the operation.

we can answer the question of how he reads this rule in two ways. First, his interpretation includes following the solution step by step, and numerically, with the paradigm constituted by the first problem. Second, and conversely, the explanations show how essential the reference to the solution of a specific problem is in interpreting the terms referring to operations in a way that differs from their common meaning. It is in this sense that the text of the rule is taken precisely as a list of operations, and not as a formula or as a proposition. Mikami emphasizes this feature as follows:

> The above *process*, as will be seen on a first glance, does not deviate seriously from our *procedure* in solving the simultaneous system ... (NoT: and here Mikami transforms the outline of the problem into three linear equations) (Mikami 1913, p. 19, my emphasis).

The comparison Mikami outlines is thus now between the process to which the "rule" referred and the modern "procedure." Neither Biot nor Mikami use the term algorithm. We will return to this fact below. One detail of the translation is revealing of how Mikami understands the text. To explain it, we need to get into the detail of the computation. The general procedure, which in *The Nine Chapters* is placed immediately after the formulation of the first problem, proceeds by placing on the calculating surface the coefficients of the first linear equation in a column to the right, those of the second equation in the same order in a column to its left, and those of the third similarly in the left column. We might assume that the first problem represents the general case in which no cell in this initial table is empty and the execution of the procedure does not require the introduction of negative coefficients. The procedure first eliminates the top position of the middle and the left column using the right column (exactly as we would eliminate between equations). In the computations on the calculating surface, the column that is the object of the elimination is replaced by the column in which the elimination was carried out. Then, the purpose is to eliminate the middle coefficient of the new left column using the new middle column. This is where the key clue appears in the text. Indeed, in my view, the first clause of the sentence introducing this third elimination can be interpreted in two ways. Either it reads: "If the (coefficient associated to the) medium-(quality) grain in the middle column was not exhausted, ... 以中行中禾不盡者,"[22] or it reads

[22] This is the translation adopted in Chemla and Guo (2004, p. 619).

"Then with what remains of the second class in the middle column. . . ." The latter is Mikami's translation (1913, p. 19). Choosing between these two interpretations amounts to understanding the nature of the text in two radically different ways. Indeed, the crucial point is this: if the procedure were written down by reference to all the problems, in which the initial table was full and eliminations between columns led to no negative coefficients, it might happen that, for some problems, and only for some, the cell of the calculating surface designated by this clause is empty (the coefficient equal to 0) at that point of the computation. The first interpretation of the clause makes the assumption that the procedure is written down in order to cover all possible cases (for which a solution exists) within this framework. By contrast, Mikami's translation assumes that the text of the procedure solves the problem of *The Nine Chapters* specifically (or those that are like this one in a more narrow sense). Indeed, the cell in question is *not* empty at this point of the computations for the first problem in Chapter 8. The assumption about how the procedure expresses a general algorithm differs depending on how one interprets this clause. The former interpretation considers that the text is more prescriptive, whereas Mikami's interpretation is both descriptive and prescriptive. We will return to this contrast, since it is essential for a key textual anachronism that will be discussed below.

The procedure that *The Nine Chapters* gives for square (and in fact also cube) root extraction shares with the previous one the property that it cannot be reformulated into a formula. However, it is expressed in a way that is quite different from what we have described above, since its formulation is abstract and makes no reference to any specific problem. In correlation with these features, Mikami recorded his initial difficulties in making sense of it and he explained how he overcame these problems:

> The rules given in the section before us for the evaluation of the square and cube roots are enunciated in obscure languages hardly intelligible to an unprepared mind. But when we compare these descriptions with those in the works of Sun-Tsu and Ch'en Luan, and when we reflect on the ways in which the calculating pieces have been manipulated in later ages, all mysteries are cleared up without leaving any trace of doubt (Mikami, 1913, pp. 12–13).

From these lines, we draw three key pieces of information. Firstly, we see here how cautious Mikami was, refraining from assimilating right away the computing tool used at the time of *The Nine Chapters* and that described in *Mathematical Classic by Master Sun*. This holds true more generally in his book, but let us rather focus on the main issue at stake here. Secondly, Mikami's declaration shows that, when he encountered this rule, he was unable to interpret its text as a stand-alone document. One might put forward the assumption that this difficulty derived, among other things, from an anachronistic approach to the text. Below, we will understand this point better and thereby find evidence showing precisely a textual anachronism of the type on which this chapter concentrates in how Mikami approached the text. Thirdly, it was only when Mikami was able to relate this text to later mathematical texts that also dealt with root extraction that he was able to interpret the "rule" from *The Nine Chapters*. In the pages where he treats the old canonical text, Mikami only translates the rules for square and cube root extraction, waiting for the pages that he devotes to *Mathematical Classic by Master Sun* (*Sunzi suanjing* 孫子算經) to which the quotation above refers, to explain the procedures that were meant (Mikami, 1913, pp. 29–31). In a first and simple sense, there is anachronism since Mikami relies on a later text to interpret an older text. However, Mikami is clear about the difference between the two procedures, and points to slight differences between them (although he does not dwell on them). Why, then, does the later text help him interpret the older one?

The key point is that in *Mathematical Classic by Master Sun* the "rules" for square root extraction in question are formulated like the procedure solving systems of linear equations in *The Nine Chapters*, that is, each by reference to a specific problem. Mikami translates and explains these rules exactly in the same way as he had proceeded with this other procedure.[23] In this later context, the rule accordingly consists of a sequence of operations, mentioning (in *Mathematical Classic by Master Sun*, in contrast with *The Nine Chapters*) the specific auxiliary values computed during the procedure and making the positions where numerical values are placed on the calculating surface explicit. Mikami is able to rely on this additional information to interpret the rule (that is, to interpret both its text and the process on the calculating surface to which the text refers). But there is more. In relation to the fact that the

[23] A difference should be emphasized. The general procedure solving systems of linear equations is described in *The Nine Chapters* with reference to a specific problem. However, its statement does not mention the specific numerical values of the problem or those computed during the execution of the rule. In *Mathematical Canon by Master Sun*, these numerical values are systematically made explicit. In both cases, Mikami's explanations include the numerical values, computed on the basis of a paradigm.

rule is formulated by reference to a paradigm, the handling of its text becomes simpler. Mikami needs only to follow the operations of the rule linearly, step by step, from beginning to end, to understand the process. On this basis, he gets the information needed to interpret the text in *The Nine Chapters*.

The key point is that the handling required to use the text of the "rule" in *The Nine Chapters* is completely different, and this relates precisely to the fact that the text is formulated with no reference to a specific example. If we now observe how Mikami translated this older text, we know that he understood the computations that were meant, but that he failed to understand *how* the text referred to these computations, that is, what we have called the uptake that the text required. This is where the textual anachronism lies: Mikami also read this text linearly, exactly as he had read the related procedure from *Mathematical Classic by Master Sun*, when the text of *The Nine Chapters* required another form of approach. Let us quote his translation to explain this point. It reads as follows (we number the paragraphs into which Mikami cuts the text to make later reference easier, however these numbers are neither in the original text nor in Mikami's translation):[24]

1. Arrange the number to be evolved in the dividend class.
2. Borrowing a unit calculator, advance it every two columns (NoT: here Mikami inserts a footnote explaining that this refers to the dissection of the number whose root is sought into groups of two digits), and consider what number should be obtained (for the first digit in the root).
3. With the obtained number multiply the borrowed unit, and arrange the result in the divisor class.
4. Then carry out the division (with the dividend and the divisor).
5. The division being over, double the divisor and take it as the determinate divisor.
6. Draw back the new divisor one column.
7. Again borrow a unit calculator and advance it as before.
8. Again consider what is to be the next figure in the root.
9. Multiply it by the borrowed unit, and subjoin the product to the next lower column of the determinate divisor. Then carry out the division.
10. Make the subjoining for a second time and proceed in the same manner as before.

[24] (Mikami, 1913, p. 13). We will not try to interpret the procedure here. For this, I refer the reader to Chapter 4 in Chemla and Guo (2004). We only examine textual features of the translation. In *The Nine Chapters* the text for cube root extraction has exactly the same properties as those on which we will dwell below, and Mikami's translation (1913, p. 14) shows features identical to those discussed for square root extraction.

It is not our purpose in this chapter to dwell on the detail of the interpretation. Our interest lies only in the approach to the text of the rule that this translation reveals. A simple browsing through the text shows that Mikami does not translate technical terminology as does Biot. He uses technical terms like "number to be evolved" (積 *ji*), "dividend" (*shi*), and "divisor" (*fa*). We remember in particular that Biot had translated the expression for square root extraction as "solving an equal square." Moreover, he had shaped specific terms that could translate *shi* and *fa* in all the rules in which they occurred and in which, for us, they would take different meanings. Mikami proceeds differently. His translation shows the relationship between root extraction and division that the ancient text expresses using terminology in a specific way. Indeed, Mikami's translation mirrors the fact that the terms linked to division are used to describe the procedures for root extraction. However, in their first occurrences, Mikami translates *shi* and *fa* as "dividend class," and "divisor class," respectively. What we have seen above suggests that he seems to interpret the use of these terms, in the context of root extraction, as referring to positions on the calculating surface. The translation of cube root extraction supports this conclusion.

Several points where Mikami's translation deviated from the original text will highlight anachronism in the way he assumed the text was prescribing operations. We will thus concentrate on these points.

In steps 6 and 10, the same two characters occur in the corresponding sentences of Chinese text: 折 *zhe* and 下 *xia*. However, the related translations do not manifest the recurrence of the same terminology, since they read, respectively: "Draw back the new divisor one column," and "proceed in the same manner as before." This has a crucial consequence: In the Chinese text, the expression "as before" clearly refers the reader back precisely to the point placed above in the text where *zhe* and *xia* occur, and from which the list of operations should be repeated again. The English text does not. The translation thus suggests that Mikami was not aware of how precise the text was in the way it pointed to an earlier sentence.

The translation of step 6 mentions a "new divisor," when in the Chinese text there is only "divisor." Here too, the original text uses terms in a way that Mikami did not follow faithfully in his translation, suggesting that his approach to the use of the ancient technical terminology was somewhat anachronistic. We will return to this issue shortly, since the same problem is even more manifest in how Wang and Needham dealt with the same text.

Moreover, immediately before step 6, and also in step 10 before the sentence "proceed in the same manner as before," the original text twice has a clause "if one divides again. . . *fu chu* 復除" that Mikami does not translate. This is not by chance: the same remark holds true for his translation of the rule for cube root extraction (p. 14). Why does the clause appear dispensable to him?

In fact, all the features of the translation just listed could be explained by a single assumption: although the rule in *The Nine Chapters* was *not* formulated using a specific example, Mikami still interpreted the text *qua* text, as similar to the one in *Mathematical Classic by Master Sun*. In that other context, the text of the procedure was prescriptive and descriptive at the same time: it prescribed the operations required to extract the root of a given number and described their effect on the paradigm in the context of which it was presented (that is, the results of the successive operations for the specific numerical example). For such a way of writing a procedure, conditionals are useless: one knows how many digits the root of the given number has, and one goes on until the last digit has been dealt with. Accordingly, the text in *Mathematical Classic by Master Sun* has no conditional of this kind. When it goes from one digit of the root to the next, without asking whether one should search for the next digit, it simply says, in Mikami's translation: "Again, arrange 80 for the root next to the previous digit."[25] Interestingly, *Mathematical Classic by Master Sun* also contains a description of the same procedure to solve systems of linear equations as was mentioned above. There, the text of the procedure presents the same characteristics as we just pointed out for the root extraction in the same book. In other words, in contrast with *The Nine Chapters*, it mentions the numerical values yielded by the successive operations for a given paradigm. Accordingly, it does not have any statement comparable to the one in *The Nine Chapters*, whose interpretation (as a conditional or not) we discussed above.[26] We have emphasized that for systems of linear equations, Mikami translated the related key sentence of *The Nine Chapters* in a way showing that he interpreted the rule as written by reference to the paradigm (and hence the sentence as not expressing a conditional). He does exactly the same

[25] Mikami (1913, p. 30). For a critical edition, see Qian (1963, vol. 2, p. 301).
[26] See Qian (1963, vol. 2, p. 319).

for the "rules" for root extraction, translating them as if, like those in *Mathematical Classic by Master Sun*, for a specific paradigm, they were at the same time prescribing and describing a list of operations to be followed from beginning to end, step by step. At the end of the text, Mikami noticed a mention that one could repeat the procedure, and he translated it as if it did not make explicit how exactly and in which circumstances. In line with what we have seen, he probably interpreted it as referring to the third digit. In my view, this reading derives from an anachronistic uptake of the text. Accordingly, the translation does not show the intentions the actors had in producing, for this rule, a text that was different from those for the other rules. Indeed, the formulation of the text without reference to a paradigm suggests that the text was different. Nor does the translation highlight the work the actors carried out for this purpose, and their achievement in writing this text.

Finally, Mikami left another feature of the text untranslated. In his step 3, he writes "multiply the borrowed unit," whereas the original text has "multiply *once* the borrowed unit 以一乘所借一算." Here, too, this additional character, which one might perceive as useless, reveals a textual operation that the actors carried out. With his anachronistic assumption that the text referred to a specific computation, Mikami did not grasp, let alone translate, this addition. Which uptake does the text require? This is the question that the two subsequent sets of actors address and to which we now turn.

2.4 How a text refers to the computation meant: Joseph Needham, Wang Ling, and Donald Knuth

In 1955, Wang and Needham published a new translation of the texts of the rules (which they referred to as "procedures") for square and cube root extraction contained in *The Nine Chapters*. We will thus concentrate on these procedures to compare, not so much their interpretation, as their mode of approach to the text with Mikami's. For the sake of brevity, we will limit ourselves to some salient features.

There is one feature of Mikami's treatment of the root extractions on which we have not dwelt. In the case of root extractions, like elsewhere, Mikami treated the problems in the context of which *The Nine Chapters* presented the "rule" for square extraction as transparent. Only the text of the "rule" mattered. This stands in contrast with the key feature of his interpretation of the solution of systems of linear equations that we have described above. By contrast, Wang and Needham shape a textual *dispositif* to translate and explain the "procedure," bringing into play the numerical value of the first problem (of square root and cube root extraction, respectively) in *The Nine Chapters*.

First, Wang and Needham cut the procedures into phases, which are in turn cut into steps. Second, they translate the text step by step (their steps being different from Mikami's). They add explanations to the translation in footnotes, in the main text, and also between the words of the translation itself, with a complex system of parentheses (Wang and Needham, 1955, p. 350). The explanations in the main text include an interpretation of the text of the procedure – with respect to the numerical value given in the specific problem being tackled – with the related computations. They also include representations of the successive states of the calculating surface during the process of root extraction. Finally, in the end, Wang and Needham also discuss the procedures by reference to geometrical diagrams. The numerical value of the first problem thus plays a key part in the explanations and representations of the related procedure. In other words, although Wang and Needham are also aware of the similarity between the procedures of *The Nine Chapters* and those of *Mathematical Classic by Master Sun* (p. 364–365), they restore the layout and computations, drawing on a direct interpretation of the text of *The Nine Chapters*. This practice compelled them to address the question of how the text of the procedure referred to the computations in a more systematic fashion.

Interestingly, unlike Mikami, they did translate the clause marking the transition of the procedure from one digit to the next one (which I interpret as "if one divides again. . . "). However, there is a correlation between the *dispositif* they shaped to carry out their reading (in particular, the fact that they read the procedure in relation to a paradigm), and how they interpret this clause. Let us consider the translation of its first occurrence (in what corresponds to Mikami's step 5):

Step 7

(a) After the division has been made, the divisor, *Fa*, is doubled, to form the *Ting-fa*,
(b) The *Ting-fa*[1] is cut short (i.e. moved back by one digit) [and this is the (first) fixed divisor, *Ting-fa*$_1$] in preparation for the next division operation.

The interpretation of the clause as "in preparation for the next division operation" shows that Wang and Needham considered the text was formulated in relation to a process of computation in which the root has at least two digits (and not only one). They repeated the same interpretation in step 12, which they translated as referring to the beginning of the computation for the third digit of the root. For them, the subsequent step (13) specifically prescribed that one should deal with the third digit (and not with any following digit) in the same way as one had dealt with the second one (Wang and Needham, 1955, p. 355). They thus interpreted this last sentence as a global reference to the steps used to deal with the second digit, beginning, however, precisely with step 8, in which the first occurrence of the character that they translated as "proceed" occurred. Accordingly, in line with this prescription, they added steps 14 to 16 to the text. In other words, like Mikami's, but in a more explicit fashion, Wang and Needham's translation showed an interpretation of the procedure as referring specifically to the extraction of the root of a given number (for them, that given in the first problem), and not to all possible extractions. Exactly the same phenomenon accounts for their treatment of the cube root extraction (Wang and Needham, 1955, pp. 356–364). The explanation in the context of a paradigm rubs off on the translation, which is accordingly shaped to refer to a case similar to that of the illustration chosen. However, nothing in the original text warrants this assumption. In fact, as I will argue, the type of generality with which the original text was formulated was surreptitiously lost. This illustrates how an anachronistic expectation with respect to how the text expresses a procedure entails a historiographical loss. It also affects the interpretation of the text. Moreover, Wang and Needham assumed in practice that the text had to be read linearly, from beginning to end. For them, as for Mikami, the procedure was both prescriptive and descriptive. Noteworthy is the fact that in contrast to Mikami's translation of the verbs mainly in the imperative mood, Wang and Needham alternate the imperative and the indicative moods (as can be seen in the quotation above). I will also argue that other assumptions about the way in which the text referred to the procedure and accordingly how it should be handled help us eliminate difficulties that most exegetes faced in interpreting it.

Here is an example of a difficulty that Wang and Needham had, and that highlights yet another facet of their anachronistic expectation about how the text made sense. It is manifest when we examine another interesting feature of the passage from the translation quoted above. In step (7a), a term is introduced to designate a quantity: *Ting-fa*, which Wang and Needham translated, only between parentheses, as "fixed divisor," and which corresponds to Mikami's "determinate divisor." In step (7b), an operation having been applied to the *Ting-fa*, between square brackets, Wang and Needham refer to the result as *Ting-fa*$_1$. In step 10, the translation mentions a computation being made on *Ting-fa*$_1$, which produces what, between parentheses, is referred to as *Ting-fa*$_2$. The term *Ting-fa*$_2$ will occur in the translation of step 12. However, in *The Nine Chapters*, it is the same term that recurs in these different steps. The same phenomenon recurs for other terms used in the original text of the procedure: whether it be in the translation, in the explanations, or in the tables representing what happens on the calculating surface, Wang and Needham translate different occurrences of the same term in different ways. In fact, we had also mentioned a phenomenon of this kind in Mikami's translation: it corresponds to the only occurrence of the term for "divisor" that, against the Chinese text, Mikami translates as "new divisor." Probably because Wang and Needham systematically confronted the translation and the actual computation on a given number, they were led to address this issue explicitly. Clearly, the problem that they think they solve in this way is the following: the same terms (*Ting-fa* is an example) were used in different sentences of the procedure, and, depending on the sentence in which they occurred, the same term referred to different numerical values. However, Wang and Needham's understanding of how the text made sense implied that the same term should refer to a single reality. Here again, we see the impact of reading the text as referring to an actual process of computation. Accordingly, in Wang and Needham's eyes, different values should correspond to different terms. They thus translate the same term differently, to match the fact that the term changes reference. This comes out quite clearly when they comment on this phenomenon. Indeed, they were so convinced that terms must have been used in this way, that they discussed material devices that might have been used to differentiate between *Ting-fa*$_1$ and *Ting-fa*$_2$ (p. 365):

> (. . .) judging by the constant changes in the names of headings, we may be fairly sure that the ancient mathematicians did not make much use of them. Distinctions such as that between *Ting-fa*$_1$ and *Ting-fa*$_2$ etc. could easily have been made by placing a separate counting-rod in an oblique position, or using counting-rods of different colours, according to methods afterwards used for distinguishing positive and negative numbers.

However, by doing so Wang and Needham introduce in this respect a gap between the original Chinese text and its translation, which, I argue, derives from a textual anachronism in their way of reading the procedure. Indeed, whether one examines the square root or the cube root extraction, the procedures in *The Nine Chapters* use the same terms throughout, and one can claim to have a plausible interpretation of how the text makes sense only if one accounts for it as it stands. Interestingly in this case, Mikami, on the one hand, and Wang and Needham, on the other, all understood the process of execution. However, Mikami did not address the issue of how the text referred to the procedure explicitly. Moreover, although Wang and Needham did, they failed to describe how actors derived the computations required from the text as given in *The Nine Chapters*.

Finally, Wang and Needham, like Mikami, spared no effort to make sense of the "once" that occurs in their step 5 and that we have seen Mikami left untranslated (Wang and Needham, 1955, pp. 352–353, 358, 361). I consider their interpretation in this respect as problematic, since it requires that they suggest editorial changes to the text (see an example on p. 358). One point is worth noting, however: again, their interpretation depended significantly on the assumption that the text referred to the procedure for a specific number. This assumption, as we will now see, was undermined by the introduction of a completely new understanding of these texts *qua* texts, and hence of interpreting how they referred to computations. We will also see the historiographical consequences that this new approach brought with it.

The context in which new readings of ancient mathematical sources of this kind developed is that of the rise of algorithm theory as a subdiscipline of mathematics, in relation to the emergence, multiplication, and wide dissemination of computers. Donald Knuth's multi-volume classic, entitled *The Art of Computer Programming*, the first volume of which appeared in 1968, is emblematic of the establishment of this new subdiscipline. At the same time, the wide access to personal computers familiarized a large community of users with programming, and languages to do so. These events went along with the emergence of new types of mathematical texts, with which users wrote down algorithms in different kinds of computer languages, and whose use became widespread. Algorithms thus grew into a new research topic. What is essential for us is that the fact of writing them down required the introduction of new ways of writing mathematics. In the first section of the first chapter of the first volume of his classic book, entitled "Algorithms," Knuth (1968, pp. 1–9) developed an analysis of the key properties of texts of algorithms and of the key operations that these texts could bring into play. In particular, he described the assignment of variables, and conditions that ordain how the list of operations should be followed, both resources being absent from other ways of writing mathematics in other subdisciplines at the time.

The important fact for us is this: in this context, Donald Knuth simultaneously developed a new way of looking at mathematical texts from the past from this perspective. I am indebted to Wu Wenjun for having introduced me to Donald Knuth's work in 1981. Like Knuth, Wu had embraced two directions in his research. He was developing automated proof, and further, he had begun looking at ancient Chinese mathematical texts from the perspective of algorithms (Hudecek, 2014). Indeed, the word "algorithm" was now used in the history of mathematics to refer to the way in which ancient mathematical texts were written, and the viewpoint from which one might look at the work to which they testified.

Interestingly, Knuth's explorations in the history of mathematics took two paths. He was interested in the ways of proceeding to fulfill mathematical tasks that authors of the past had devised. This is a historical viewpoint that is most represented in *The Art of Computer Programming*. However, in an article devoted to "Ancient Babylonian Algorithms," Knuth (1972) followed another line of inquiry, which allows us to return to the theme of this chapter. In particular, Knuth began to closely examine textual operations that the ancient scribes had carried out to write down texts for algorithms. Interestingly, Knuth referred to this as "Babylonian programming." At this point, the new ways of writing down texts for algorithms that had emerged opened new possibilities for imagining how ancient actors might have written down texts for procedures. Accordingly, Knuth (1972) established the use of an "assignment statement" in ancient texts of procedure. Moreover, he looked for "conditionals and iterations," and reported finding only "very little evidence of this in the Babylonian texts" (p. 674). Knuth also raised the issue of the type of generality texts of procedures had, and he further formulated hypotheses accounting for why cuneiform texts of procedures often mentioned specific numerical values. Noteworthy is the fact that, for him, this practice of using numbers might have been intended "as an aid to exposition, in order to clarify the general method" (p. 672). The contrast with Biot's interpretation of numbers occurring in texts of procedures is telling.

Knuth opened the way to entirely new possibilities of reading ancient texts of procedures. Following suit, Wu Wenjun embraced the same project for ancient Chinese mathematical texts. Wu (1987) summarizes the main issues that his own historical research into algorithms from ancient China had addressed. In my understanding, they fall under the first of the two directions followed by Knuth that I have identified above. In other words, Wu Wenjun was mainly interested in the mathematical ideas brought to play in algorithms and in the properties that, as a result, the procedures presented. He did not dwell on how ancient actors had shaped texts to write down algorithms. Nor did he examine the textual operations carried out to write down the procedures to which ancient Chinese texts of procedure attest.

I argue that following in the footsteps of Knuth (1972), however, was essential for the introduction of new assumptions with respect to how texts of procedures found in ancient Chinese mathematical books had to be read.[27] Let us examine in turn how this new way of regarding texts brought to light the textual operations the actors used to compose the texts of the algorithm for root extraction to which *The Nine Chapters* attests.

To begin with, in line with Knuth's identification of an assignment of variables in cuneiform texts, it became clear that the reason why terms like "divisor," and "fixed divisor," were used without change throughout the texts for root extraction was that they referred to the content of positions on the calculating surface, at the time when the content of these positions was fetched in the computation. In the computation, the same term could thus refer to different values unequivocally. There was no need to add "new" to "divisor," or numbers to *Ting-Fa*. Authors simply used an assignment of variables. This example illustrates clearly how assumptions about the way in which texts make sense might vary. The fact that some of these assumptions create problems for the interpretation indicates that they are most probably anachronistic. We might look at the diversity of historians' approaches to texts of procedures from another angle: it shows that in different contexts, actors used different types of

27 Chemla (1987) brings these ideas into play to offer a new interpretation of the texts of procedures for root extraction in *The Nine Chapters*.

operations to write down procedures. These operations should also more generally be part of what history of mathematics dwells on, not least for the tools it would give to historians to tackle texts produced in different contexts. The texts for root extractions in *The Nine Chapters* give more evidence supporting these claims.

Above, I have frequently highlighted the fact that neither Mikami nor Wang and Needham knew exactly what to do with the clause "if one divides again. . .*fu chu* 復除." Mikami simply did not translate it, while Wang and Needham rendered it as "in preparation for the next division operation." In fact, if we understand this clause as a conditional ruling over the flow of computation (a possibility introduced by modern ways of writing algorithms), we get an interesting interpretation not only of the sentence, but also of the structure of the whole text of procedure. At the first occurrence of this clause, after the first digit of the root has been dealt with, if this answer is no, the computation stops. In other words, the text of the procedure allows users to address the case of a root with a single digit as well. If the answer is yes, however, one goes on with the operations placed immediately after the expression "if one divides again;" that is, *zhe* and *xia*. Moreover, the second occurrence of the clause, at the end of the text (Mikami's step 10), is again followed by the two characters *zhe* and *xia*, and in cases where the answer is yes, it refers the user back exactly to the first occurrence of the two characters *zhe* and *xia*, the idea being that the practitioner takes up using the text from that point onwards. If the answer is no, the user can interrupt her computations at this point. In other words, the conditional can be interpreted as articulated with the use of an iteration. In fact, seen from this perspective, the points of the list of computations where the two conditionals are placed in the text are quite striking (Chemla, 1987).

If we follow this interpretation, the conclusion is that the text covers all possible cases for root extractions.[28] It is general, not simply because, with a paradigm, it illustrates all possible cases, but because it prescribes what to do in all possible cases. The text is not written like other texts that we have evoked, that is, with reference to a paradigm (we can think of the procedure solving systems of linear equations). It should thus come as no surprise that it must be interpreted differently. However, the conclusion about the mode of generality of the text goes hand in hand

28 A final sentence placed after what Mikami translated also covered the case when computations do not end with the digit of the units. This goes beyond the scope of this chapter.

with another conclusion regarding the handling of the text. In contrast with previous interpretations that read the text mainly from beginning to end, as prescribing and describing the operations for a generic case, the new interpretation rests on the understanding that for any root to be extracted, the reader had to navigate through the text, relying on the conditionals, and repeat a sequence of operations as many times as the answer to the second conditional was yes. The text thus appears to have been written in such a way that through this specific handling, one could derive the computations required for any case. The conclusion sheds light on the fact that the actors did not only invent a mathematical procedure, but they also invented a specific type of text for writing down procedures.

What are the historiographical consequences of this new interpretation of texts for root extractions? We have seen that it opens the way to an understanding of the text as it was written, without requiring the addition of specifications to it, let alone making changes to it. We have also seen that it suggests that actors had worked on and developed the types of text with which they wrote procedures. This alone suffices to explain that reading the texts they composed requires that we contextualize these texts. Applying any anachronistic reading to them does not allow us to interpret and analyze them. Finally, we have seen that this interpretation drew our attention to a new way in which the actors wrote general procedures. However, there is more.

To write the text for root extractions, as described by the last interpretation, implies that actors had carried out specific work on lists of operations *qua* lists of operations. For instance, they elaborated a list of operations that could deal with all succeeding digits after the first one, and they formulated this list of operations once and for all, whichever digit was dealt with. This brings to light the algebraic facet of this work with operations. To achieve the latter goal, one might imagine that the treatments of successive digits were compared, reworked, and unified into a single list of operations.

If we take a step back, we discover that this last conclusion about the work carried out on lists of operations accounts for other phenomena that we have so far left unattended.

First, we have seen that the texts of procedures for square (and cube) root extractions were written using the terminology for division. This, too, implies that the lists of operations carrying out these different arithmetic operations were compared and that the way in which root extraction procedures were written expressed a way of understanding the relationship between division and root extraction.

Second, this is the point where we can return to the sentence that Mikami, on the one hand, and Wang and Needham, on the other, had problems translating: "multiply *once* the borrowed unit." In fact, this "once" in the text for extracting square roots corresponds to the "twice" that occurs in exactly the same position in the text for cube root extraction. This further confirms the actors' interest in comparing lists of operations and shows another way in which they expressed the result of their reflections about the relationship between them, using the texts of procedures.

We thus see that the writing of the iteration fits within a wider landscape. Various specificities of the texts written by the actors reflect from several perspectives research questions they had about the relationship between operations. Ways of writing (in this case, a single procedure with iterations or different procedures in relation to one another) are in continuity with ways of working and ways of thinking. This is a key reason why anachronism in the approach to texts of procedures has dramatic historiographic consequences.

From this perspective, we understand another tacit assumption that was underlying the interpretations that we have analyzed above: they all assumed that a text of procedure had the aim of executing a task and should be interpreted with respect to this task. We now see that this assumption does not allow us to make sense of some features of the text that require looking beyond a given procedure. Moreover, this assumption hides mathematical work that the actors were engaged in and that is meaningful for a history of algebra. It also conceals important features of the actors' conceptions about the organization of mathematical knowledge.

2.5 Conclusion

In this chapter, I have analyzed the type of interpretation of ancient texts of procedures (and, to a lesser extent, of problems) that four groups of scholars have offered. In particular, I have examined the resources that were needed to produce these interpretations, including the access to documents, dictionaries, and mathematical knowledge – and not forgetting the epistemological values these individuals put into play. The comparison between the four types of interpretation showed how in each context, the scholars under consideration read the texts as if they had been written in modern times. This anachronism of a textual type led them to bring various assumptions into play. We have seen how these assumptions had the effect that exegetes left elements of the original text untranslated and changed the way in which the terms were used in ancient writings. This approach to texts of the past had an impact not only on the interpretation of these texts, but also more widely on the historiography.

Against all expectations, the most recent type of approach in line with Knuth's reading of texts of algorithms might be the one that is the least anachronistic. At least, we have seen that it could account for features of the texts that previous attempts failed to accommodate. More generally, it provided an interpretation that did not require any change to the ancient text. We have also seen how this interpretation highlighted features in the original sources that are coherent with other clues about mathematical activity, in the context in which these texts of procedures were composed.

This brings out an important fact: the advancement of science is accompanied by the introduction of new types of text (at least, the actors perceive them as new), which offer new resources for historians to imagine how the ancient texts might have been written. This conclusion has a bearing on our topic. If actors shape types of texts for the sake of (and in relation to) their mathematical practice, this explains why historians may encounter texts that make sense in ways different from those they are used to. An approach to ancient sources that does not address the issue of the textual practices that were used in the context in which the documents were written is doomed to fail. This remark urges historians to work on ancient documents not only to describe the results and practices to which they attest, but also to find in them evidence about how the actors wrote and read the textual components with which they were working. In brief, it urges us to consider historicizing the ways in which actors dealt with their texts and to incorporate their shaping of new textual resources as an integral part of their mathematical activity.

References

Benoit, Paul, Karine Chemla, and Jim Ritter, editors, (1992). *Histoire de Fractions, Fractions d'Histoire*. Basel: Birkhäuser Verlag.

Biot, Edouard (1839). Table générale d'un ouvrage chinois intitulé 算法統宗

Souan-fa-tong-tsong, ou TRAITE COMPLET DE L'ART DE COMPTER (Fourmont, no. 350), traduite et analysée par M. Ed. Biot. *Journal Asiatique 3e série* **7** (mars), 193–217.

Biot, Edouard (1841). Traduction et examen d'un ancien ouvrage chinois intitulé TCHEOU-PEI, littéralement "Style ou signal dans une circonférence;" par M. Edouard Biot. *Journal Asiatique 3e série* **11** (juin), 593–639.

Biot, Jean-Baptiste (1835) Memoirs of John Napier of Merchiston, etc; Mémoires sur Jean Napier de Merchiston (Premier Article, Deuxième Article). *Journal des Savants* Mars 151–162; Mai 257–273.

Bussotti, Michela (2015). Du dictionnaire chinois–latin de Basilio Brollo aux lexiques pour le marché: deux siècles d'édition du chinois en Italie et en France. *T'oung Pao* **101**, 363–406.

Chemla, Karine (1987). Should they read FORTRAN as if it were English? *Bulletin of Chinese Studies* **1**, 301–316.

Chemla, Karine (1992). De la synthèse comme moment dans l'histoire des mathématiques. *Diogène* **160**, 97–114.

Chemla, Karine (2005). Geometrical figures and generality in ancient China and beyond: Liu Hui and Zhao Shuang, Plato and Thabit ibn Qurra. *Science in Context* **18**, 123–166.

Chemla, Karine (2009). On mathematical problems as historically determined artifacts: Reflections inspired by sources from ancient China. *Historia Mathematica* **36**, 213–246.

Chemla, Karine (2014). L'histoire des sciences dans la sinologie des débuts du XIXe siècle: Les Biot père et fils. (Preprint handed out for the conference). Link: `https://halshs.archives-ouvertes.fr/halshs-01509318/document`. Revised version forthcoming in: *Jean-Pierre Abel-Rémusat et ses Successeurs: Deux Cents Ans de Sinologie Française en France et en Chine* 雷慕沙及其繼承者: 紀念法國漢學兩百週年學術研討會, Pierre-Etienne Will (ed). Paris: Académie des Inscriptions et Belles-Lettres.

Chemla, Karine and Guo Shuchun (2004). *Les Neuf Chapitres: le Classique Mathématique de la Chine Ancienne et ses Commentaires*. Paris: Dunod.

Chen Zhihui 陳志輝 (2017). Weilieyali de zhongxi daishuxue bijiao ji qi shixue yuanyuan 偉烈亞力的中西代數學比較及其史學淵源 (Alexander Wylie's Comparisons of Algebra between China and Europe, and its Historiographical Origins). *Ziran kexueshi yanjiu* 自然科學史研究 *Studies in the History of Natural Sciences* **36**, 502–518.

de Guignes, Chrétien-Louis-Joseph (1813). *Dictionnaire Chinois, Français et Latin*. Paris: Imprimerie Impériale.

Han Qi 韓琦 (1998). Chuanjiaoshi weilieyali zaihua de kexue huodong 傳教士偉烈亞力在華的科學活動 (The Missionary Alexander Wylie and his scientific activities in China). *Ziran bianzhengfa tongxun* 自然辯證法通訊 *Journal of Dialectics of Nature* **20**, 57–70.

Hudecek, Jiri (2014). *Reviving Ancient Chinese Mathematics: Mathematics, History and Politics in the Work of Wu Wen-Tsun*. Abingdon: Routledge.

Knuth, Donald (1968). *The Art of Computer Programming*. Reading, MA: Addison-Wesley.

Knuth, Donald (1972). Ancient Babylonian algorithms. *Communications of the ACM* **15**, 671–677.

Knuth, Donald (1976). Ancient Babylonian algorithms. *Communications of the ACM* **19**, 108.

Landry-Deron, Isabelle (2015). Le dictionnaire chinois, français et latin de 1813. *T'oung Pao* **101**, 407–440.

Martija-Ochoa, Isabelle (2002). *Edouard et Jean-Baptiste Biot. L'astronomie chinoise en France au XIXe siècle*. Paris: University Paris Diderot.

Martzloff, Jean-Claude (1987). *Histoire des Mathématiques Chinoises*. Paris: Masson.

Martzloff, Jean-Claude (1997) (rev. ed. 2006). *A History of Chinese Mathematics*: with Forewords by Jacques Gernet and Jean Dhombres. [Translator: Stephen S. Wilson]. Heidelberg: Springer-Verlag.

Mikami, Yoshio (1913). *The Development of Mathematics in China and Japan, vol. XXX*. Abhandlungen zur Geschichte der Mathematischen Wissenschaften mit Einschluss ihrer Anwendungen. Leipzig: B.G. Teubner; New York: G.E. Stechert & Co.

Mizuno, Hiromi and Karine Chemla (in press). Mikami Yoshio (1875–1950): From the philosophy of mathematics to cultural history of mathematics. In: *Writing Histories of Ancient Mathematics: Reflecting on Past Practices and Opening the Future, 18th to 21st Centuries*, Karine Chemla, Agathe Keller, and Christine Proust (eds). Dordrecht: Springer. (A preprint version was handed down at the conference held in Paris, October 24–28, 2016; see `https://sawerc.hypotheses.org/conferences/conference-octobre-2016`.)

Morrison, Robert (1815–1822). *Dictionary of the Chinese Language, in Three Parts. Part the first, Containing Chinese and English, Arranged According to the Radicals; Part the Second, Chinese and English Arranged Alphabetically; and Part the Third, English and Chinese*, Vol. I & II-Part I. Macao: East India Company's Press.

Needham, Joseph, and Wang Ling (1959). Section 19: Mathematics. In: *Science and Civilisation in China*, Joseph Needham (ed). Cambridge: Cambridge University Press, 1–168.

Qian Baocong 錢寶琮 (1963). Suanjing shishu 算經十書 (Qian Baocong jiaodian 錢寶琮校點) (*Critical punctuated edition of The Ten Classics of Mathematics*). Beijing 北京: Zhonghua shuju 中華書局.

Sasaki Chikara (2002). Mikami Yoshio. In: *Writing the History of Mathematics: Its Historical Development*, Joseph Dauben and Christoph J. Scriba (eds). Basel: Birkhäuser, pp. 484–486.

Smith, David Eugene (1925). *History of Mathematics. Volume II: Special Topics of Elementary Mathematics*. Boston: Ginn and Company.

Wang Ling and Joseph Needham (1955). Horner's method in Chinese mathematics: its origin in the root-extraction procedures of the Han dynasty. *T'oung-pao* **43**, 345–401.

Wu Wen-Tsun (1987). Recent studies in the history of Chinese mathematics. In: *Proceedings of the International Congress of Mathematicians, Berkeley, California, USA, August 3–11, 1986*, Andrew M. Gleason (ed). Providence, RI: American Mathematical Society, pp. 1657–1667.

Wylie, Alexander (1897). Jottings on the science of the Chinese: Arithmetic. (Papers originally published in the *North China Herald*, August–November 1852: 108, 111, 112, 113, 116, 117, 119, 120, 121). Reprinted in: *Chinese Researches*, 159–194. Shanghai.

3

Anachronism and anachorism in the study of mathematics in India

Kim Plofker
Union College

Abstract: The foremost historiographic challenge in interpreting pre-modern Indian mathematics is arguably not anachronism so much as anachorism, the blurring of geographical or cultural rather than chronological distinctions. For example, historians struggle constantly with ways to avoid or explain calling Indian analyses of right-triangle relations "Pythagorean," or using the term "Diophantine equations" for the type of problems designated in Sanskrit as *kuṭṭaka* or *varga-prakṛti*. Nonetheless, the combination of anachronism and anachorism provides the study of Indian mathematics with a powerful lens, that clarifies even as it distorts. This chapter will address such trade-offs between popular misconceptions and deeper insights, especially in the application of concepts from the historiography of early modern European calculus to infinitesimal methods used in Sanskrit mathematics of the early to mid-second millennium.

3.1 Introduction

Historians of mathematics nowadays routinely warn their readers of the temptations and dangers of anachronistic thinking in studying their sub-

[a] From *Anachronisms in the History of Mathematics: Essays on the Historical Interpretation of Mathematical Texts*, edited by Niccolò Guicciardini © 2021 Cambridge University Press.

ject.[1] Sabetai Unguru stressed the particular vulnerability of mathematically trained historians to these hazards:

The history of mathematics typically has been written as if to illustrate the adage "anachronism is no vice." Most contemporary historians of mathematics, being mathematicians by training, assume tacitly or explicitly that mathematical entities reside in the world of Platonic ideas where they wait patiently to be discovered by the genius of the working mathematician. Mathematical concepts, constructive as well as computational, are seen as eternal, unchanging, unaffected by the idiosyncratic features of the culture in which they appear. [...] Various forms of the same mathematical concept or operation are not considered merely mathematically equivalent but also historically equivalent. Indeed mathematical equivalence is taken to represent historical equivalence. [...] We cannot know what went through Euclid's mind when he wrote the *Elements*. But we can determine what Euclid could not have thought when he compiled his great work. He, most likely, did not employ concepts or operations for which there is no genuine evidence either in his time or in the works of his predecessors [...] he could not have used mathematical devices and procedures which were invented many hundreds of years after his death (Unguru, 1979, pp. 555–556).

Nick Jardine (2000, p. 252), on the other hand, maintains that

[...] unfaithfulness to the categories of past agents does not always constitute vicious anachronism, that is, historically incoherent interpretations of past deeds and works.

These disputes apply not only to anachronism *per se* but to the related concept sometimes called "anatopism" or, in Jardine's case, "anatropism," i.e., the assumption of the "historical equivalence" of similar concepts from different mathematical traditions (Jardine, 2000, p. 253):

Interpretative anachronism applies categories from one period to deeds and works from a period from which those categories were absent. It is a species of a more general kind of displacement, namely the imposition of categories originating in one culture or society onto deeds or works of a culture to which those categories are alien. We may call this mode of interpretation 'anatropism',

[1] The nature of the threat perceived in anachronism has evolved with the changing meaning of the term. In *Histoire des mathématiques* (Montucla *et al.*, 1798–1802, Part I, Book IV, p. 217), as in *De Thalès à Empédocle* (Tannery, 1877, p. 42), "l'anachronisme" means simply an error or inconsistency in chronological fact (such as identifying Euclid of Alexandria; with the philosopher Euclid of Megara (Neugebauer, 1929, p. 6)). Gino Loria's reviews of history of mathematics works in 1900 used "anacronismo" both in this sense of a straightforward timeline mistake (Loria, 1900, p. 51) and in its subsequent meaning of a historically absurd incongruity (Loria, 1900, p. 117). Changing historiographic perspectives on the role of anachronism in history of science and elsewhere are discussed in, e.g., Jardine (2000) and Pollmann (2017, pp. 47–72).

exploiting the fact that the ancient Greek *tropos* could be applied to social and cultural orientation as well as to physical direction.

The present discussion employs instead the synonym “anachorism” for its traditional connotation of incongruity with “the character” or “the spirit” of the “country” on which the category is imposed. The object of this choice is not to revive obsolete cultural-essentialist notions such as “national character,” but to emphasize that different scientific traditions are distinguished by differences in what might be termed their “institutional culture” as well as in their language and geographical location.

Mathematics as a discipline may be said to be especially vulnerable to anachorism as well as anachronism, as it is so largely devoted to identifying ways in which two concepts or statements can be thought of as equivalent. Moreover, the history of mathematics attests to many direct transmissions of important concepts and methods, between different time periods as well as between different cultures. So mathematical ideas occurring in different contexts where they were independently developed, and/or differently conceptualized, can seem deceptively similar. Some of these similarities are in fact demonstrably influenced by historical transmission, such as the definition-theorem-proof structure inherited by modern mathematics ultimately from Hellenistic Greek ancestors, or the medieval Arabic studies of conic sections that explicitly built on that same Hellenistic tradition. Others are the result of parallel evolution in separate traditions, such as the focus on “Pythagorean” characteristics of right triangles in both Old-Babylonian and ancient Chinese mathematics (neither of which had anything to do with the Greek Pythagoreans), or the solutions to so-called “Diophantine” problems derived independently in the Greek mathematics of late antiquity and in the Sanskrit mathematics of medieval India.

A naive form of anachronism/anachorism is almost inevitable when these distinctions are ignored, particularly in popular journalism about research developments in the history of mathematics. This approach appeals to popular attention by emphasizing the “prescient” brilliance of the discoverers, which in turn magnifies the perceived significance of the findings.[2] But should a distinction be drawn between “vicious” (or

[2] A recent example discusses research on Late-Babylonian cuneiform tablets that employed trapezoidal figures as a kind of velocity/time diagram, under the title “Math whizzes of ancient Babylon figured out forerunner of calculus” (Cowen, 2016). The same author, writing on the

at least counterproductively naive) and legitimate anachorism, similar to Jardine's analysis of anachronism?

The counterproductive sort of anachorism, in its most blatant form, is tediously familiar from innumerable popular articles, blog posts, etc., categorically declaring ancient Indian scientists to be the original discoverers of binary numbers, differential and integral calculus, transfinite cardinals, wavelengths and speed of light, aeronautics, and so on. Any such assertion is easily (if not always effectively) contradicted by pointing out the superficiality or flimsiness of the direct resemblance between the ancient and modern concepts thus equated. But that does not necessarily make the resemblance altogether meaningless or uninteresting, particularly in the context of mathematical ideas. Although Indian scientists of antiquity may not have been thinking explicitly of what we now call "binary numbers" any more than they were explicitly thinking of what we now call "wavelengths of light," their mathematical thinking cannot be quarantined in a box labeled "pre-modern" or "non-western" to preserve it from historical "impurities."

The remainder of this chapter discusses various ways in which anachronistic/anachoristic "cross-contamination" between different mathematical traditions in the historiography of Indian mathematics may lead to improved understanding of it. We use as our reference point the "strong form" of this approach advocated recently by P.P. Divakaran (2018, p. vi) in rejecting

> [. . .] the long-held notion that mathematicians in India arrived at their insights in mysterious ways that are different from how "everyone else" did it. Aryabhata, as anyone who has read him should have known, had no doubts: the "best of gems" that is true knowledge can be brought up from "the ocean of true and false knowledge" only by the exercise of one's intellect, by means of "the boat of my own intelligence." Others have said the same in less poetic language. The first vindications of this ringing endorsement of the rational mind, through the reconstruction of the actual proofs and demonstrations, are another gain from the current analytical studies. [. . .] The hope is that we can begin to chip away, little by little, at the encrusted layers of the mythology – and, occasionally, ill-informed bias – that still has some currency: that Indian mathematics was, in some strange and exotic way, not quite of the mainstream. If this book has an overriding message, it is that, conceptually and technically, the indigenous mathematics of India is in essence the same as of other mathematically advanced

well-known Old-Babylonian tablet Plimpton 322, relating to Pythagorean triples, announced "This ancient Babylonian tablet may contain the first evidence of trigonometry" (Cowen, 2017).

cultures – how can it be otherwise? – and that its quality is to be adjudged by the same criteria.

This message reminds us that the line between eschewing anachronism/anachorism on the one hand, and embracing essentialist clichés of "cultural difference" on the other, can be perilously fine. It calls on us to re-examine, for instance, Jardine's previously quoted criticism of "the [anatropic] imposition of categories originating in one culture or society onto deeds or works of a culture to which those categories are alien." In what way and to what extent can any mathematical category be confidently characterized as "alien" to any culture explicitly engaged with another mathematical category clearly related to it? We will thresh out some of these questions in the following examples of historical analysis of developments in Indian mathematics.

3.2 The "error" of division by zero in medieval Sanskrit algebra

By the early seventh century CE at the latest, Sanskrit mathematics incorporated a sophisticated set of rules for determining unknown quantities, including the arithmetic manipulation of positive, negative, and zero terms in equations, equivalent to much of what we now call elementary algebra. This subject was called in Sanskrit texts *bīja-gaṇita* "seed-computation" or *avyakta-gaṇita* "computation [with] unknown [quantities]." Its invention is often popularly attributed to the astronomer-mathematician Brahmagupta of Bhillamāla in Rajasthan (who may have been a court astronomer under the Gurjara monarch Vyāgramukha (see Pingree, 1970–1994, vol. 4, p. 254), because Brahmagupta's 628 CE treatise *Brāhmasphuṭasiddhānta* contains the earliest surviving systematic description of its rules. However, the history of this topic in Indian *gaṇita* ("mathematics," "computation") appears to extend considerably further back than Brahmagupta's time.[3]

What has generally been regarded as the most "strange and exotic" (to borrow Divakaran's phrasing) feature of Brahmagupta's algebra rules is his apparent acceptance of division by zero as a legitimate operation,

[3] Popular surveys such as Wallin (2002) assert Brahmagupta's priority in statements like "Brahmagupta, around 650 AD, was the first to formalize arithmetic operations using zero [. . .] Brahmagupta wrote standard rules for reaching zero through addition and subtraction as well as the results of operations with zero." The tradition of earlier Indian work on which Brahmagupta's treatise drew is outlined in, e.g., Hayashi (2013).

included in his elegantly concise summary of operations with negative, positive, and zero quantities:[4]

dhanayor dhanam ṛṇam ṛṇayor dhanarṇayor antaraṃ samaikyaṃ kham |
ṛṇam aikyaṃ ca dhanam ṛṇadhanaśūnyayoḥ śūnyayoḥ śūnyam ||

[The sum] of two positives [is] positive; [the sum] of two negatives [is] negative; [the sum] of a positive and a negative [is] the difference [of their magnitudes]; the sum of equal[-magnitude positive and negative quantities is] zero. The sum of a negative or a positive and zero [is] negative or positive [respectively]; [the sum] of two zeroes [is] zero.

ūnam adhikād viśodhyaṃ dhanaṃ dhanād ṛṇam ṛṇād adhikam ūnāt |
vyastaṃ tadantaraṃ syād ṛṇaṃ dhanaṃ dhana ṛṇaṃ bhavati ||

A lesser[-magnitude positive quantity] subtracted from a greater [is] positive, [as is] a negative [subtracted] from a positive [or] a greater[-magnitude negative] from a lesser negative. [When] their difference [is] reversed, positive becomes negative and negative positive. [That is, changing the signs or order of the operands in the subtraction changes the sign of the result.]

śūnyavihīnam ṛṇam ṛṇaṃ dhanaṃ dhanaṃ bhavati śūnyam ākāśam |
śodhyaṃ yadā dhanam ṛṇād ṛṇaṃ dhanād vā tadā kṣepyam ||

A negative [or] positive [quantity] minus zero becomes negative [or] positive [respectively]; zero [minus zero is] zero. When a positive is subtracted from a negative or a negative from a positive, [their magnitudes are] added.

ṛṇam ṛṇadhanayor ghāto dhanam ṛṇayor dhanavadho dhanaṃ bhavati |
śūnyarṇayoḥ khadhanayoḥ khaśūnyayor vā vadhaḥ śūnyam ||

The product of a negative and a positive [is] negative; the product of two negatives [is] positive; the product of positives is positive. The product of zero and a negative, or of zero and a positive, [or] of zero and zero [is] zero.

dhanabhaktaṃ dhanam ṛṇahṛtam ṛṇaṃ dhanaṃ bhavati khaṃ khabhaktaṃ kham |
bhaktam ṛṇena dhanam ṛṇaṃ dhanena hṛtam ṛṇam ṛṇaṃ bhavati ||

A positive divided by a positive [or] a negative divided by a negative is positive; zero divided by zero [is] zero. A positive divided by a negative [is] negative; a negative divided by a positive is negative.

khoddhṛtam ṛṇaṃ dhanaṃ vā tacchedaṃ kham ṛṇadhanavibhaktaṃ vā |

A negative or positive divided by zero, or zero divided by a negative or positive, [has] that [as its] divisor.

[4] *Brāhmasphuṭasiddhānta* 18.30–35. The Sanskrit original quoted here follows Dvivedi (1902, pp. 309–10). Here and henceforth, except where otherwise noted, the accompanying translations of Sanskrit quotes are the present author's.

These statements about quantities divided by zero are routinely criticized as a mathematical mistake, part of a flawed attempt to "extend arithmetic" of nonzero numbers to include operations with zero. See, for example, the remark in Wallin (2002): "The only error in his rules was division by zero, which would have to wait for Isaac Newton and Gottfried Wilhelm Leibniz to tackle." Similarly, in a well-known website devoted to the history of mathematics (O'Connor and Robertson, 2000) one finds a summary of Brahmagupta's algebra:[5]

> He also gives arithmetical rules in terms of fortunes (positive numbers) and debts (negative numbers):
>
> *A debt minus zero is a debt.*
> *A fortune minus zero is a fortune.* [...]
>
> Brahmagupta then tried to extend arithmetic to include division by zero:
>
> *Positive or negative numbers when divided by zero is a fraction the zero* [sic] *as denominator.*
> *Zero divided by negative or positive numbers is either zero or is expressed as a fraction with zero as numerator and the finite quantity as denominator.*
> *Zero divided by zero is zero.*
>
> Really Brahmagupta is saying very little when he suggests that n divided by zero is $n/0$. He is certainly wrong when he then claims that zero divided by zero is zero. However it is a brilliant attempt to extend arithmetic to negative numbers and zero.

All of these interpretations, including the present author's own translation, lean very heavily on the intended meaning of the Sanskrit term *taccheda*, literally "[having] that [as] divisor." A more detailed description of related concepts is found in the arithmetic treatise *Līlāvatī* of Bhāskara a half-millennium or so after Brahmagupta (*Līlāvatī* 45–46; Āpaṭe, 1937, I.39):

> *yoge khaṃ kṣepasamaṃ vargādau khaṃ khabhājito rāśiḥ* |
> *khaharaḥ syāt khagunaḥ khaṃ khaguṇaś cintyaś ca śeṣavidhau* ||

Zero in a sum [produces a result] equal to the addend, [but] in a square etc.

[5] This article combines an anachronistic assessment of Brahmagupta's division by zero with an equally common form of anachronism in the opposite direction: namely, its translation of the Sanskrit terms *dhana* "positive" and *ṛṇa* "negative" as "fortune" and "debt," respectively. While these mathematical technical terms do appear to have been originally borrowed from the context of financial credits and debits in ancient India, they had been naturalized in a more abstract mathematical context long before Brahmagupta used them. To translate them over-literally as "fortune" and "debt" in the context of medieval Sanskrit mathematics is as anachronistic as it would be to translate their modern English equivalents "positive" and "negative" as "agreed" and "denied" respectively, based on their etymological origin in the late Latin *positivus* and *negativus* of medieval philosophy and law.

[i.e., in a square, square-root, or other powers], zero. A quantity divided by zero should be [considered] "zero-divided," [but a quantity] multiplied by zero [is] zero; [it is] to be considered "zero-multiplied" in remaining calculations.

śunye guṇake jāte khaṃ hāraś cet punas tadā rāśiḥ |
avikṛta eva jñeyas tathaiva khenonitaś ca yutaḥ ||

When zero is the multiplier, if zero is the divisor then again the quantity is understood [as] unaltered, just as [when] diminished or increased by zero.

A. Padmanabha Rao (2017, pp. 26–27) points out the sophistication of this concept as interpreted in the sixteenth-century commentary of Gaṇeśa on these verses of the *Līlāvatī* (Āpaṭe, 1937, I.39) which he translates as follows:

Further operations pending multiple of zero be given a second thought. That is, if a number is multiplied by zero and there are further operations remaining then the multiple of zero is not to be construed as zero, but 0 is to be kept by the side of the number (as a mere symbol).

If, having done all the operations, there is further operation of division by zero then the denominator and the numerator being equal, they be cancelled (0 being treated as a mere symbol). If, however, there is no division by zero then the multiple of zero is to be treated as zero.

This interpretation inspires the following exposition (Padmanabha Rao, 2017, p. 20):

Normally a multiple of 0, $x0$, is 0.

Rule 1 If, however, further (mathematical) calculations are there, multiple of zero, $x0$, should be regarded as not zero. (0 is therefore not zero.) [. . .]

Rule 2 If a multiple of zero be followed by further operation of division by zero, i.e., $\frac{x0}{0}$, it is to be understood that the multiplicand x remains unaltered i.e., $\frac{x0}{0} = x$. (Thus $x0$ is treated as not zero, and the indeterminate form is avoided.) [. . .]

Rule 3 If, however, division by zero is not there, $x0$ is zero [. . .]

The meaning of these rules is illustrated in Bhāskara's subsequent verse, a worked sample problem (*Līlāvatī* 47cd; Āpaṭe, 1937, I.40):

khenoddhṛtā daśa ca kaḥ khaguṇo nijārdha-
yuktas tribhiś ca guṇitaḥ svahatis triṣaṣṭiḥ ||

What [quantity] multiplied by zero, added to its own half and multiplied by three, and divided by zero, [is] sixty-three?

As Padmanabha Rao explains (2017, pp. 20–21), this example

[...] illustrates the *infinitesimal* nature of multiple of zero [...]
In mathematical terms it amounts to solving the equation:

$$\frac{\left(x0 + \frac{x0}{2}\right) \cdot 3}{0} = 63$$

It is obvious that Bhāskarācārya could not have meant 0 = zero in this case, for, if it were so, x, which is to be determined, would itself be eliminated and the problem becomes purposeless. Since there are further calculations, by **Rule 1**, $x0$ is not 0. On simplification we get $\frac{x0}{0} = 14$. Therefore by **Rule 2** of the above algorithm, $x = 14$.

Note that this deeper investigation of division by zero in medieval Indian algebra does not eliminate, and may even increase, the potential for historical anachronism/anachorism. In the first place, of course we cannot take it for granted that Brahmagupta in the seventh century understood the rules of algebra in exactly the same way as Bhāskara in the twelfth, much less Gaṇeśa in the sixteenth.[6]

Second, instead of a superficial critique of some of these rules as mere "error" and "certainly wrong," we now have a more nuanced understanding of them built upon post-medieval concepts such as "infinitesimal" and "indeterminate form," which emerged in the context of European mathematicians' creation of calculus in the seventeenth through nineteenth centuries (Miller, 2019). Nothing in the Sanskrit text of the rules explicitly invokes the idea of an infinitesimal quantity (although Padmanabha Rao (2017) argues convincingly that this concept is unmistakably manifested elsewhere in the Sanskrit mathematical corpus, and we shall encounter it again in the following section). Nor does the procedure of multiplying and dividing the same quantity by zero correspond exactly to the notion of an "indeterminate form" of a limit. In refining our assessment of Indian division by zero to consider it an "infinitesimal" technique instead of merely a "wrong" one, have we simply replaced one instance of Jardine's "imposition of categories" by another?

I would argue that while it is necessary to be cautious in the application of mathematical concepts across eras or cultures, we gain more than we lose by this reinterpretation. It is not possible, nor should we try,

6 We must bear in mind, however, that the linguistic and textual continuity of the Sanskrit mathematical tradition makes such a hypothesis quite reasonable. To suppose that Gaṇeśa understood Brahmagupta's ideas of algebra is, at the very least, no more far-fetched than to suppose that Fermat understood Diophantus's ideas of arithmetic.

to "fix" Brahmagupta's or Bhāskara's uses of zero so that they become perfectly consistent by the standards of modern analysis. But the hypothesis that they were seeking to formulate algebra rules in a way that took into account arithmetic with very small quantities is more historically plausible than the idea that they simply did not notice any problem with assertions like $\frac{0}{0} = 0$. That, I think, is the easy reliance on the label of "strange and exotic" that we must guard against: our willingness, born partly of our desire to avoid mathematical anachron/chorism, to accept that mathematicians of a different time or place believed for some unknown reason things that seem to us obviously wrong or meaningless.

3.3 Mādhava's infinitesimal "calculus" in fourteenth-century Kerala

One of the classic topics for exploring anachronism in the history of mathematics is the development of infinitesimal analysis or calculus. Pre-modern sources are repeatedly analyzed and debated to determine whether their versions of infinitesimal quantities or instantaneous changes qualify as "true" calculus. See, for example, Carl Boyer's *The Concepts of the Calculus* (republished under the title *The History of the Calculus and Its Conceptual Development*), which uses the modern concept of limit as its criterion for distinguishing calculus *per se* from its various "forerunners" (Boyer, 1949, p. 11):

> It is the purpose of this essay to trace the development of these two concepts [i.e., derivative and integral] from their incipiency in sense experience to their final elaboration as mathematical abstractions, defined in terms of formal logic by means of the idea of the limit of an infinite sequence.

Later explorations such as Katz (1995) and Bressoud (2002) apply the same question of what "counts" as calculus to Sanskrit and Arabic sources, particularly the remarkable fourteenth- through sixteenth-century work on infinite series in the Kerala region of south India.

In the Niḷā River basin in central Kerala in the late fourteenth century, a mathematician-astronomer named Mādhava derived many remarkable results involving what we would call power series for trigonometric functions. The most famous of these is the so-called "Mādhava–Leibniz" series for π, a special case of the power series for inverse tangent. Its

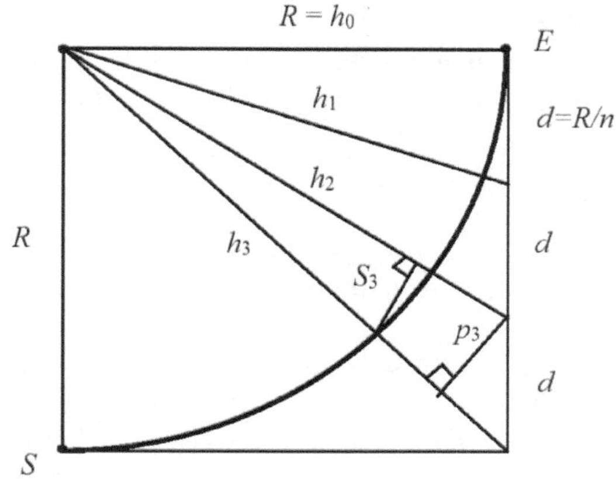

Figure 3.1 The circle quadrant of radius R inscribed in the southeast corner of a square of side $2R$, on which the derivation of Mādhava's result for the circumference is based.

modern version is commonly expressed as

$$\frac{\pi}{4} = 1 - \frac{1}{3} + \frac{1}{5} - \frac{1}{7} + \frac{1}{9} - \cdots$$

But its original Sanskrit version took the form of an expression for the circumference C of a circle in terms of its known radius R, which we may write in modern notation as

$$\frac{C}{8} = R - \frac{R}{3} + \frac{R}{5} - \frac{R}{7} + \frac{R}{9} - \cdots$$

Mādhava's work on trigonometry and series is now known mostly from the writings of later followers of his school, especially in their commentaries on the above-mentioned *Līlāvatī* of Bhāskara. In the early to mid-sixteenth century one of these commentators, Śaṅkara by name, a student of a student of the son of Mādhava's own student, described an elaborate geometric and algebraic rationale for Mādhava's series. We will briefly survey its main features.[7]

Śaṅkara's, and presumably Mādhava's, demonstration of this result uses equal segments of a square side to build approximations to unequal arcs of the circle octant, as shown in Figure 3.1. A circle quadrant with radius R, inscribed within a square of side R, extends between

[7] The explanation and translations shown here are adapted from Plofker (2009, pp. 223–9), and the sources cited therein.

the South and East points S and E. The square's eastern side is divided into some number n of equal segments of length d, whose lower endpoints are connected to the center of the circle by the "hypotenuse-lines" $h_1, h_2, h_3, \ldots, h_n$. In Śaṅkara's words, considering the entire circle inscribed in a square of side $2R$:

> For demonstrating this, having placed a square whose four sides are equal to a desired diameter, draw a circle in it by means of a compass [whose opening] is equal to half of the diameter and draw the east-west and north-south line(s). Then the circumference touches the middle of each of the four sides of the square in all four directions. Having placed as many dots as you wish with equal intervals at that end of the eastern side which is south from the end of the east-west line, draw lines from the center of the circle to the dots.

Perpendicular line segments $p_1, p_2, \ldots, p_n$ are then dropped from the end of each corresponding hypotenuse-line onto the hypotenuse-line below it (only p_3 is shown in the figure). The lengths of the successive h_k and p_k must then be determined:

> Then add the square of the distance between the end of that line and the end of the east-west line to the square of the half-diameter, and take the square root of each. The measure of the length of the hypotenuse-line is produced.... That [distance] passing from the end of the left line to the hypotenuse [which is] the southern line, is placed with its [perpendicular] direction opposite to its (the hypotenuse's). This is the edge of the segment.... This figure is similar in shape to the figure whose hypotenuse is the southern line and edge is the half-diameter.

To paraphrase in modern notation, for $1 \leq k \leq n$, $h_k = \sqrt{R^2 + (kd)^2}$. And each h_k is the hypotenuse of a right triangle with horizontal leg R and vertical leg $k \cdot d$. Consequently, by similar triangles,

$$\frac{p_k}{d} = \frac{R}{h_k} \quad \Longrightarrow \quad p_k = \frac{dR}{h_k}.$$

Each hypotenuse h_k is also the hypotenuse of a smaller right triangle with shorter leg p_{k+1}, which is traversed by a small arc of the circumference. The Rsine of each of these small arcs, i.e., its (modern) sine times the assumed radial length R, is represented by another perpendicular line segment s_k, as in the s_3 shown in the figure. Śaṅkara explains the determination of the length of s_k as follows:

> Having multiplied the distance derived in this way by the half-diameter, divide

by [the length of] the left one of two adjacent lines. The result is the R sine of the part of the circumference between the lines.

That is, each s_k segment cuts off another right triangle with hypotenuse R, smaller than but similar to the right triangle containing the corresponding p_k:

$$\frac{s_k}{R} = \frac{p_k}{h_{k-1}} \qquad \Longrightarrow \qquad s_k = \frac{p_k R}{h_{k-1}} = \frac{dR^2}{h_k\, h_{k-1}}.$$

Here Śaṅkara departs from ordinary right-triangle geometry to argue for the "infinitesimal" approximations that will ultimately produce a value for $C/8$:

[That] is also a part of the circumference when the segment is small. Therefore, multiply each segment by the square of the half-diameter.... There, though it should be divided by the product of the two adjacent lines, even if it is divided by the square of one of those two, the error will not be large in the calculation of a part of the circumference.... Therefore [this] is said assuming the division of the parts multiplied by the square of half of the diameter by the squares of the southern lines.

In other words, if n is large enough, we can assume $s_k \approx dR^2/{h_k}^2$, and consequently

$$\begin{aligned}\frac{C}{8} &\approx s_1 + s_2 + \cdots + s_n \\ &\approx \frac{dR^2}{{h_1}^2} + \frac{dR^2}{{h_2}^2} + \cdots + \frac{dR^2}{{h_n}^2}.\end{aligned}$$

The rest of Śaṅkara's rationale, which we will summarize briefly in modern notation, consists of an ingenious rewriting of the s_k expressions to make them both accurate and computable. The first step is a sort of decomposition technique to replace the unknown denominators ${h_k}^2$ with constants. In general, for a quantity x with a multiplier m and a divisor q (all positive), we can write

$$\begin{aligned}x \cdot \frac{m}{q} &= x + \frac{x(m-q)}{q} \\ &= x + \frac{x(m-q)}{m} \cdot \frac{m}{q}\end{aligned}$$

and repeat this decomposition process indefinitely, always replacing the factor m/q by an added term containing the factor $(m-q)/q$, which we then rearrange in the form $\frac{m-q}{m} \cdot \frac{m}{q}$, as follows:

$$\begin{aligned} x \cdot \frac{m}{q} &= x + \frac{x(m-q)}{q} = x + \frac{x(m-q)}{m} \cdot \frac{m}{q} \\ &= x + \frac{x(m-q)}{m} + \frac{x(m-q)}{m} \cdot \frac{(m-q)}{q} \\ &= x + \frac{x(m-q)}{m} + \frac{x(m-q)^2}{m^2} \cdot \frac{m}{q} \\ &= x + \frac{x(m-q)}{m} + \frac{x(m-q)^2}{m^2} + \frac{x(m-q)^2}{m^2} \cdot \frac{(m-q)}{q} \\ &= x + \frac{x(m-q)}{m} + \frac{x(m-q)^2}{m^2} + \frac{x(m-q)^3}{m^3} \cdot \frac{m}{q} \\ &\vdots \end{aligned}$$

Ultimately, we will have replaced our original simple multiplication and division with an infinite series of terms in successive powers of $(m-q)/m$:

$$x \cdot \frac{m}{q} = x + \frac{x(m-q)}{m} + \frac{x(m-q)^2}{m^2} + \frac{x(m-q)^3}{m^3} + \cdots$$

Thus if we regard our s_k approximation expression as the minuscule segment length d times the multiplier R^2 over the divisor ${h_k}^2$, we can write:

$$s_k \approx d \cdot \frac{R^2}{{h_k}^2} = d + \frac{d(R^2 - {h_k}^2)}{R^2} + \frac{d(R^2 - {h_k}^2)^2}{(R^2)^2} + \frac{d(R^2 - {h_k}^2)^3}{(R^2)^3} + \cdots$$

By the original Pythagorean relationship between R and h_k,

$$R^2 - {h_k}^2 = -({h_k}^2 - R^2) = -(kd)^2$$

so

$$s_k \approx d - \frac{d(kd)^2}{R^2} + \frac{d(kd)^4}{R^4} - \frac{d(kd)^6}{R^6} + \cdots$$

Adding up the corresponding terms of all the s_k expressions, Śaṅkara derives a new, infinite expression for the octant arc $C/8$, which we can represent as

$$\begin{aligned}
s_1 &\approx d - \frac{d(1d)^2}{R^2} + \frac{d(1d)^4}{R^4} - \frac{d(1d)^6}{R^6} + \cdots \\
+\, s_2 &\approx d - \frac{d(2d)^2}{R^2} + \frac{d(2d)^4}{R^4} - \frac{d(2d)^6}{R^6} + \cdots \\
+\, s_3 &\approx d - \frac{d(3d)^2}{R^2} + \frac{d(3d)^4}{R^4} - \frac{d(3d)^6}{R^6} + \cdots \\
&\ddots \\
+\, s_n &\approx d - \frac{d(nd)^2}{R^2} + \frac{d(nd)^4}{R^4} - \frac{d(nd)^6}{R^6} + \cdots \\
&\approx \frac{C}{8} \\
&\approx s_1 + s_2 + \cdots + s_n \\
&\approx nd - \frac{d}{R^2}\sum_{k=1}^{n}(kd)^2 + \frac{d}{R^4}\sum_{k=1}^{n}(kd)^4 - \frac{d}{R^6}\sum_{k=1}^{n}(kd)^6 + \cdots \\
&= R - \frac{d}{R^2}\sum_{k=1}^{n}(kd)^2 + \frac{d}{R^4}\sum_{k=1}^{n}(kd)^4 - \frac{d}{R^6}\sum_{k=1}^{n}(kd)^6 + \cdots
\end{aligned}$$

Śaṅkara then turns to an induction-like argument to express the sums of powers of the successive multiples $\{1d, 2d, 3d, \ldots, nd\}$ when n is very large, beginning with the formulas

$$\sum_{k=1}^{n} k = \frac{n(n+1)}{2} \approx \frac{n^2}{2} \quad \text{and} \quad \sum_{k=1}^{n} k^2 = \frac{n(n+1)(2n+1)}{6} \approx \frac{2n^3}{6} = \frac{n^3}{3}$$

to infer the general closed form

$$\sum_{k=1}^{n} k^j \approx \frac{n^{j+1}}{j+1}, \qquad \text{whence} \qquad d\sum_{k=1}^{n} d^j k^j \approx d^{j+1}\frac{n^{j+1}}{j+1} = \frac{R^{j+1}}{j+1}.$$

Replacing each sum in the preceding expression for $\frac{C}{/}8$ by one of these simple quotients gives

$$\begin{aligned}
\frac{C}{8} &\approx R - \frac{d}{R^2}\sum_{k=1}^{n}(kd)^2 + \frac{d}{R^4}\sum_{k=1}^{n}(kd)^4 - \frac{d}{R^6}\sum_{k=1}^{n}(kd)^6 + \cdots \\
&\approx R - \frac{1}{R^2}\frac{R^3}{3} + \frac{1}{R^4}\frac{R^5}{5} - \frac{1}{R^6}\frac{R^7}{7} + \cdots,
\end{aligned}$$

which is clearly trivially equivalent to the "Mādhava–Leibniz series" introduced at the beginning of this section.

This material, far more technically involved than the division-by-zero rules discussed in the previous section, brings to the fore again the issue of anachronism and anachorism. While it would be possible – given sufficient space and time – to follow Śaṅkara's/Mādhava's rationale in the verbal expository format of its original Sanskrit (and Malayalam) verse and prose rather than in modern symbolic notation, it is hard to see how we would recognize its truth without connecting it to the notions of infinite series and infinitesimally small quantities so familiar from modern calculus. In particular, these notions allow us to recognize, in Śaṅkara's "when the segment is small" and similar caveats, a mathematical equivalence deeper than finite geometric approximation of arcs and line segments by quantities nearly but not quite equal to them.

It should be emphasized here that as far as we know, there is no direct historical connection between the discoveries of these results in the Indian and European contexts. The work of Mādhava and his Niḷā school of course pre-dates the late-seventeenth-century activity of Newton and Leibniz, and there is at present no persuasive evidence for inferring transmission of Mādhava's ideas to Europe before nineteenth-century colonial scholars encountered them. So the question of how to relate Mādhava's power series to the category of "calculus" depends not on any hereditary relationship (except insofar as distant ancestors in Greek, Indian, and/or Islamic methods of trigonometry and algebra contributed to both of them), but on the parallel evolution of similar ideas.

Numerous scholars have approached this question with a cautious emphasis on the differences in the context and fate of these ideas within their different traditions. Victor Katz (1995, pp. 173–174) sums up:

> How close did Islamic and Indian scholars come to inventing the calculus? [...] some of the basic ideas of calculus were known in Egypt and India many centuries before Newton. It does not appear, however, that either Islamic or Indian mathematicians saw the necessity of connecting some of the disparate ideas that we include under the name calculus. There were apparently only specific cases in which these ideas were needed.
>
> There is no danger, therefore, that we will have to rewrite the history texts to remove the statement that Newton and Leibniz invented the calculus. They were certainly the ones who were able to combine many differing ideas under the two unifying themes of the derivative and the integral, show the connection between them, and turn the calculus into the great problem-solving tool we have today.

Likewise David Bressoud (2002, pp. 2, 3):

> No. Calculus was not invented in India. But two hundred years before Newton or Leibniz, Indian astronomers came very close to creating what we would call calculus. Sometime before 1500, they had advanced to the point where they could apply ideas from both integral and differential calculus to derive the infinite series expansions of the sine, cosine and arctangent functions [. . .]
>
> The expansions of the sine, cosine, and arctangent had been passed down through several generations of disciples, but they remained sterile observations for which no one could find much use.

I argue against the hypothesis of transmission (Plofker, 2009, pp. 251–252):

> [H]ow would Sanskrit texts on infinitesimal methods for trigonometry and the circle, transmitted no earlier than the late sixteenth century, have inspired the quite different European infinitesimal techniques used at the start of the seventeenth century? Mathematicians like Kepler and Cavalieri focused instead on mechanical questions such as centers of gravity and areas and volumes of revolution, as well as general problems of quadrature explicitly associated with the work of Hellenistic forerunners and of Renaissance geometers such as Nicholas of Cusa. The infinite series that Mādhava discovered did not appear in Latin mathematics until late in the seventeenth century, as suggested by the names – Newton, Leibniz, Gregory – attached to them by earlier historians.

These arguments bring out a rather paradoxical distinction between calculus and the "ideas of calculus": the Niḷā-school scholars somehow "know" and "apply" the latter without ever having succeeded in "inventing" the former. This suggests that "calculus" as a mathematical phenomenon is being defined as much by its historical context as by its technical content. It is not any specific breakthrough, concept or technique that constitutes calculus so much as the eventual cohesion of these breakthroughs, concepts and techniques into a particular strand of early modern European mathematics, closely tied to particular ancient and new problems involving mechanics and curves. Under these rules, is there any possible non-anachronistic way to associate the term "calculus" with any mathematics outside early modern Europe? And if not, what shall we use instead to help understand and describe the trigonometric infinite series of the Niḷā school?

Divakaran (2018, pp. 2–3) strongly advocates a different approach, based on the model of a different cohesion within the Indian mathematical tradition:

The central concern of the Niḷā school was the conceptualisation and execution of a method of dealing exactly with geometrical relationships in nonlinear problems, specifically problems of rectification and quadrature involving circles and spheres, by resorting to a process of 'infinitesimalisation' through division by unboundedly large numbers. In other words, the fundamental achievement of Mādhava and the Niḷā school was the invention of calculus for trigonometric functions (as well as, along the way, for polynomials, rational functions and power series). [. . .] In the process, it sharpened and deepened algebraic methods traceable in a formal sense to Brahmagupta, defined and made respectable entirely new types of mathematical objects (infinite series) and introduced methods of proof (mathematical induction) unknown till then.

He continues (pp. 261–262):

The fact that these series were also among the spectacular early successes of European calculus served to give their discovery by Mādhava, almost three centuries earlier, a special cachet among historians (when they finally got round to them). But this benchmarking against a very particular set of series common to both traditions had also the effect that it took the spotlight away from the fact that the breakthrough that led to them was the same as in 17th century Europe, the invention of calculus. And there were other notable advances, modest only in comparison with the precise articulation of the infinitesimal philosophy – the metaphysics of calculus in the words D'Alembert used later in Europe – in which the new mathematics was embedded, that did not get their due. [. . .] the conceptual advance that underpinned the Nila discovery of the arctangent and the π series is none other than the fundamental principle of infinitesimal calculus, in the particular form of a reduction to simple quadrature of a rectification problem.

and (p. 348):

The fact however is that historians are divided on what exactly Madhava's programme represents – calculus or not? – in the grand scheme of the evolution of mathematics and it is the purpose of this section to try and understand the basis, mathematical and historiographic, of the continuing ambivalence in certain circles about the true import of the Mādhava revolution. [. . .]

It goes without saying that the benchmark against which to set the calculus of the Nila school has to be the vastly more general form in which European calculus was visualised and formulated by its founders; and that has its own pitfalls. Nowhere is the dilemma of the objective historian of science more acute than here: how can one deliberately unlearn all that one has learned of a discipline that has grown so vigorously and in so many unanticipated directions to where it is today, beyond anything that Newton and Leibniz, leave alone Madhava, might have imagined?

This statement of the historiographic "dilemma" re-frames the historian's task in a significant way. Instead of treating the historical evolution of European infinitesimal analysis as the "benchmark" against which its Indian counterpart is measured to see if it qualifies as "calculus," we must begin by inquiring more critically what calculus is. And this inquiry needs to be informed by relevant developments in both European and Indian mathematics (as well as, ideally, in any other mathematical tradition where related ideas emerged). In short, the task before us is to renounce both naive anachron/chorism (immediately declaring something to be "calculus" because it bears some resemblance to early modern European calculus) and naive avoidance of anachron/chorism (declaring something not to be "calculus" because it doesn't bear a strong enough resemblance to early modern European calculus) in favor of clarifying our criteria for the definition of calculus.

3.4 Conclusion

The foregoing examples illustrate the difficulty of distinguishing between the benefits of anachron/choristic thinking and its disadvantages. On the one hand, it's clear that the much-lauded universality of mathematical truth does not obviate the need to understand the particularity of different mathematical traditions. As historians, we cannot simply invoke a modern mathematical concept to reduce an earlier related concept, in the words of the old joke, to a problem already solved. On the other hand, excluding earlier sources from our modern conceptual categories cannot be taken for granted as a neutral or self-explanatory act. It implies historically specific criteria for how we define those categories, criteria that sometimes take shape only in the process of exclusion itself rather than in the initial construction of the categories.

It is likely that the more superficial problems of anachron/chorism in understanding the history of Indian mathematics will continue to recede as more Sanskrit (and other Indic-language) sources are translated, more historical details traced, and more sophisticated interpretations proposed for the ways in which Indian mathematicians themselves perceived the aims and meaning of their work. In the process, though, deeper problems involving them will emerge: how to re-define concepts originally shaped solely by our experience of modern and/or western mathematics, how to justify the use of the same category label for works from different languages and eras, and many more. Given that the history of mathematics has yet to fully resolve such problems even for traditions as closely related as modern English and ancient Greek mathematics, or even modern English and early modern Latin mathematics, the prospect is somewhat daunting. But if it ever becomes possible to achieve something that could justly be called a "grand scheme of the evolution of mathematics" as a human endeavor, it will be found along this path.

References

Āpaṭe, Vasant Anant (1937). *Līlāvatī*, 2 vols. Anandashrama Sanskrit Series 107, Puṇe.

Boyer, Carl (1949). *The Concepts of the Calculus: a Critical and Historical Discussion of the Derivative and the Integral.* New York: Hafner. Republication: Boyer C. (1959). *The History of the Calculus and its Conceptual Development.* New York: Dover.

Bressoud, David (2002). Was calculus invented in India? *The College Mathematics Journal* **33** (1), 2–13.

Chatterjee, Bina (1981). *Śiṣyadhīvṛddhida Tantra* of Lalla with the commentary of Mallikārjuna Sūri, 2 vols. New Delhi: Indian National Science Academy.

Cowen, Ron (2016). Math whizzes of ancient Babylon figured out forerunner of calculus. *Science.* `sciencemag.org/news/2016/01/math-whizzes-ancient-babylon-figured-out-forerunner-calculus`. Accessed 2 March 2021.

Cowen, Ron (2017). This ancient Babylonian tablet may contain the first evidence of trigonometry. *Science.* `sciencemag.org/news/2017/08/ancient-babylonian-tablet-may-contain-first-evidence-trigonometry`. Accessed 2 March 2021.

Divakaran, P.P. (2018). *The Mathematics of India: Concepts, Methods, Connections.* Singapore: Springer.

Dvivedi, Sudhakar (1902). *Brāhmasphuṭasiddhānta.* Benares: The Pandit.

Geslani, Marco *et al.* (2017). Garga and early astral science in India. *History of Science in South Asia* **5** (1), 151–191.

Hayashi, Takao (2013). Algebra in India: *Bījagaṇita.* In: *Encyclopaedia of the History of Science, Technology and Medicine in Non-Western Cultures,* H. Selin (ed). Berlin: Springer, 47–48.

Jardine, Nicholas (2000). Uses and abuses of anachronism in the history of the sciences. *Hist. Sci.* **38** (3), 251–270.

Katz, Victor J. (1995). Ideas of calculus in Islam and India. *Mathematics Magazine* **68** (3), 163–174.

Kuppanna Sastry, T.S. (1993). *Pañcasiddhāntikā of Varāhamihira.* Madras: PPST Foundation.

Loria, Gino (ed) (1900). Recensioni ed annunzi. *Bollettino di Bibliografia e Storia delle Scienze Matematiche* Feb.–Mar. 1900, 41–92.

Miller, Jeff (2019). Earliest known uses of some of the words of mathematics (I). `jeff560.tripod.com/i.html`. Accessed 2 March 2021.

Montelle, Clemency (2011). *Chasing Shadows*. Baltimore, MD: Johns Hopkins University Press.

Montelle, Clemency and Plofker, Kim (2018). *Sanskrit Astronomical Tables*. Cham: Springer.

Montucla, Jean Éttiene *et al.* (1798–1802). *Histoire des Mathématiques*. Paris: H. Agasse.

Neugebauer, Otto (1929). *Über vorgriechische Mathematik*. Leipzig: Teubner.

Neugebauer. Otto and Pingree, David Edwin (1970–1971). The *Pañcasiddhāntikā of Varāhamihira*, in two volumes. Copenhagen: Royal Institute.

O'Connor, John and Robertson, Edmund F. (2000). Brahmagupta. `www-groups.dcs.st-and.ac.uk/history/Biographies/Brahmagupta.html`. Accessed 2 March 2021.

Padmanabha Rao, A.B. (2017). Ideas of infinitesimals of Bhāskarācārya in *Līlāvatī* and *Siddhāntaśiromaṇi*. *Indian Journal of History of Science* **52** (1), 17–27.

Pingree, David Edwin (1970–1994). *Census of the Exact Sciences in Sanskrit*, series A, vols. 1–5. Philadelphia: American Philosophical Society.

Plofker, Kim (2005). Relations between approximations to the Sine in Kerala mathematics. In: *Contributions to the History of Indian Mathematics*, Gerard G. Emch *et al.* (eds). Delhi: Hindustan Book Agency, 135–152.

Plofker, Kim (2009). *Mathematics in India*. Princeton, NJ: Princeton University Press.

Pollman, Judith (2017). *Memory in Early Modern Europe, 1500–1800*. Oxford: Oxford University Press.

Rao, S. Balacandra and Venugopal, Padjama (2008). *Eclipses in Indian Astronomy*. Bangalore: Bharatiya Vidya Bhavan.

Raynaud, Dominique (2016). *A Critical Edition of Ibn al-Haytham's* On the Shape of the Eclipse. Cham: Springer.

Śāstrī, Bapu (ed) (1989). *The Siddhāntaśiromaṇi* (rev. Gaṇapati Deva Śāstrī, Kashi Sanskrit Series 72). Varanasi: Chaukhambha Sanskrit Sansthan.

Shukla, Kripi Shankar (1986). *Vaṭeśvarasiddhānta and Gola of Vaṭeśvara*, in two volumes. New Delhi: Indian National Science Academy.

Sidoli, Nathan (2007). What we can learn from a diagram: The case of Aristarchus's "On The Sizes and Distances of the Sun and Moon". *Annals of Science* **64** (4), 525–547.

Stephenson, F. Richard (2006). Babylonian timings of eclipse contacts and the study of the Earth's past rotation. *Journal of Astronomical History and Heritage* **9** (2), 145–150.

Stephenson, F. Richard and Steele, John M. (2006). Astronomical dating of Babylonian texts describing the total solar eclipse of s.e. 175. *Journal for the History of Astronomy* **37** (1), 55–69.

Tannery, Paul (1877). *Pour l'histoire de la science Hellène: de Thalès á Empédocle*. Paris: Félix Alcan.

Unguru, Sabetai (1979). History of ancient mathematics – some reflections on the state of the art. *Isis*, **70** (4), 555–565.

Van Brummelen, Glen (2009). *The Mathematics of the Heavens and the Earth: The Early History of Trigonometry*. Princeton, NJ: Princeton University Press.

Wallin, Nils-Bertil. (2002). The History of Zero. Yale Global Online. `yaleglobal.yale.edu/history-zero`. Accessed 2 March 2021.

Whish, Charles M. (1834). Hindu quadrature of the circle and the infinite series of the proportion of the circumference to the diameter exhibited in the four Sastras, the Tantra Sangraham. *Philosophical Transactions of the Royal Society of Great Britain and Ireland* **3** (3), 509–523.

4

On the need to re-examine the relationship between the mathematical sciences and philosophy in Greek antiquity

Jacqueline Feke
University of Waterloo

Abstract: Scholars tend to assume that the mathematical sciences and philosophy were distinct disciplines in antiquity, as they are today. From the fourth century BCE onward, mathematicians and philosophers did distinguish themselves. They criticized each other's work and, in some areas of the Greek world, strong rivalries developed between philosophers and mathematicians. I argue, however, that the distinction between philosophers and mathematicians did not entail that their fields of inquiry were distinct. This chapter examines the relationship between the mathematical sciences and philosophy from the perspective of the practitioners of the mathematical sciences, in particular Archytas of Tarentum, Hero of Alexandria, and Claudius Ptolemy. I argue that these practitioners viewed the relationship between the mathematical sciences and philosophy as more complex, where the mathematical sciences are not only in a relationship with philosophy but, even stronger, they are forms of philosophy. Moreover, the mathematical sciences answer some of the most fundamental questions of philosophy, e.g., how to obtain knowledge, how to form a just society, and how to attain the good life.

4.1 Introduction

What was the relationship between philosophy and the mathematical sciences in the ancient Greek world? Generally speaking, scholars treat these two areas of study as distinct. This separation is reflected especially in the division of academic labor, where historians of the ancient exact

[a] From *Anachronisms in the History of Mathematics: Essays on the Historical Interpretation of Mathematical Texts*, edited by Niccolò Guicciardini

sciences tend to work on problems and texts that are distinct from the ones that scholars of ancient philosophy address, and these two sets of specialists tend to inhabit different university departments. This division of labor and subject matter is no doubt due to historically contingent factors, and yet it is widely assumed that ancient Greek philosophy is similar to modern philosophy: that it asked similar questions and used similar methods. This assumption that ancient philosophy is like today's philosophy yields a very narrow portrait of what ancient philosophy consisted in and, consequently, what problems and texts are studied within the academic field of ancient philosophy. Moreover, this assumption restricts our comprehension of the relation ancient philosophy held to other types of discourses and intellectual practices.

The assumption that philosophy and the mathematical sciences were distinct fields of inquiry more often is implied rather than directly expressed in academic literature. For examples, one may look to the reputable histories of ancient Greek philosophy written in the mid- to late-twentieth century.[1] W.K.C. Guthrie, for instance, wrote the highly respected six-volume *A History of Greek Philosophy*. Guthrie pays particular attention to issues of anachronism, and he even makes the bold statement, "We cannot remind ourselves too often of the difference between philosophy as it was understood in the period here treated, and as it is most often understood today, at least in our own country" (Guthrie, 1962–1981, vol. 2, p. xiii). He explains that in selecting the various subject matters of his treatise, his criterion was not whether some idea was philosophical but instead whether a historical actor was a philosopher (Guthrie, 1962–1981, vol. 2, pp. xii–xiii). Because some ancient philosophers also studied mathematics, Guthrie includes some discussion of the mathematical sciences, especially as it appears in the texts of Plato and his predecessors, notably the Pythagoreans. Yet, occasionally Guthrie reveals a tacit assumption that philosophy as a field does not include the mathematical sciences. For instance, he states the following (Guthrie, 1962–1981, vol. 1, p. 2):

> Without belittling the magnificent achievements of the Greeks in natural philosophy, metaphysics, psychology, epistemology, ethics and politics, we shall find that because they were pioneers, and therefore much nearer than ourselves

[1] I would like to thank Matthew Watton for his research assistance on these twentieth-century histories of ancient Greek philosophy, funded by my Social Sciences and Humanities Research Council Insight Development Grant.

to the mythical, magical or proverbial origins of some of the principles which they accepted without question, we can see these origins clearly; and this in turn throws light on the dubious credentials of some of the principles which gain a similarly unquestioned acceptance among many today.

Guthrie enumerates the subdisciplines of ancient philosophy, but he omits the mathematical sciences. Correspondingly, he remarks that most people would agree that the middle of the fifth to the end of the fourth century BCE constitutes the zenith of Greek philosophy (Guthrie, 1962–1981, vol. 1, p. 7). This period can be considered the height of ancient Greek philosophy only if the definition of philosophy remains narrow and excludes the mathematical sciences.

Like Guthrie before him, Pierre Hadot pays special attention to problems of anachronism in his survey called *What is Ancient Philosophy?* He proclaims, "In this book, I intend to show that a profound difference exists between the representations which the ancients made of *philosophia* and the representation which is usually made of philosophy today – at least in the case of the image of it which is presented to students, because of the exigencies of university teaching" (Hadot, 2002, p. 2).[2] Nevertheless, several lines later he lists the areas at which ancient philosophy excelled: "Obviously, there can be no question of denying the extraordinary ability of the ancient philosophers to develop theoretical reflection on the most subtle problems of the theory of knowledge, logic, or physics" (Hadot, 2002, pp. 2–3).[3] Although Hadot omits the mathematical sciences in his methodological statement, he does mention them in passing, especially as an area of study within the Academy.[4] Nonetheless, in Hadot's treatment, the mathematical sciences remain something different from philosophy, even if some philosophers studied them to some degree. When describing the study of the sciences, including astronomy, at the great Alexandrian library, Hadot juxtaposes science and philosophy: "The library of Alexandria gathered together the entire corpus of philosophical and scientific literature, and provided a

[2] In the original edition, this passage reads: "J'ai l'intention de montrer dans mon livre la différence profonde qui existe entre la représentation que les Anciens se faisaient de la *philosophia* et la représentation que l'on se fait de nos jours, habituellement, de la philosophie, tout au moins dans l'image qui en est donnée aux étudiants à cause des nécessités de l'enseignement universitaire" (Hadot, 1995, p. 17).

[3] "Évidemment, il ne s'agit pas de nier l'extraordinaire capacité des philosophes antiques à développer une réflexion théorique sur les problèmes les plus subtils de la théorie de la connaissance ou de la logique ou de la physique" (Hadot, 1995, p. 17).

[4] See especially Hadot (2002, pp. 61–62; 1995, pp. 100–101).

fertile environment for great scholars – the physician Herophilus and the astronomer Aristarchus of Samos, for example" (Hadot, 2002, p. 93).[5] The sciences and philosophy require juxtaposition, because it is assumed that they are distinct, even if they were studied side by side. In response to this prevailing assumption, I call for a re-examination of the distinction between philosophy and the mathematical sciences in the ancient Greek world.

First, what were "mathematics" and the "mathematical sciences" in Greek antiquity? Two trends have emerged for how to define ancient mathematics. The more traditional historiography treats mathematics as high level, and studies in this genre analyze the most advanced contributions to the mathematical sciences in the Greek world, such as by Euclid, Archimedes, and Ptolemy. For an example, one may look to the scholarship of Reviel Netz, a prominent historian of Greek mathematics. Netz defines mathematics narrowly. He has stated of Greek mathematics, "It is a genre predicated on the primacy of proof" (Netz, 2016, p. 87). According to Netz, the two determinants of what makes an ancient Greek text mathematical are the diagram and the formulaic language, used in conjunction with the diagram in a proof. Correspondingly, according to Netz, a mathematician is "whoever has written (or perhaps merely produced orally) an argument showing the validity of some claim, using the techniques we identify with Greek mathematics (a lettered diagram, a specific mode of language use)" (Netz, 1997, p. 4). The principal feature of Greek mathematics, then, is the demonstrative proof.

The other historiographical trend reacts against the definition of Greek mathematics as only high level. Serafina Cuomo, for example, characterizes mathematics more broadly to encompass fields that one might not typically classify as mathematics. In the introduction to her survey *Ancient Mathematics* (Cuomo, 2001, p. 1), she articulates her approach:

> So, I will cast the net wider than usual, and give an account not just of the advanced, high-brow practices, but also of 'lower' and more basic levels of mathematics, such as counting or measuring. I think such a choice will pay off in at least two senses: we will achieve a better-balanced picture, with a full spectrum of activities rather than an isolated upper end; and, since counting and measuring affect a greater section of the population than squaring the parabola

[5] "... et, dans cette même ville, la Bibliothèque rassemblait toute la littérature philosophique et scientifique. De grands savants exerçaient là leur activité: le médecin Hérophile, l'astronome Aristarque de Samos, par exemple" (Hadot, 1995, p. 148).

or trisecting the angle, looking at them will give us more insight into everyday, everyperson, 'popular' views of mathematics.

In concert with this approach, Cuomo presents a wide array of subject matters, including architecture, land distribution, and the use of numbers in trade and commerce. According to Cuomo, the criterion for whether a text or practice is mathematical is not whether the historical actors actually made use of mathematics, but rather that they claimed that their work had to do in some way with mathematics.[6]

Although this second historiographical approach is especially useful in combatting the shortcomings of anachronism, in this chapter I will treat the scope of mathematics more narrowly and focus only on high-level mathematics. It is necessary to focus the field of inquiry in this way, for the problem that is addressed here is a problem of demarcation. Similar to how Karl Popper tackled the question of what science is by contrasting it with pseudo-science, in order to assess the distinction between philosophy and the mathematical sciences, one must emphasize the type of mathematics that was close to, similar to, or even identical with philosophy (Popper, 1963). In this way, the greater context is still relevant, but the type of mathematics addressed is that which was in the closest relation to the relevant context, in this case philosophy. Hence, this chapter will examine the advanced mathematical sciences, including astronomy, number theory, and harmonics – as well as areas of so-called practical mathematics, such as mechanics – and their relation to philosophical discourses.

In solving this problem of demarcation, it is important to distinguish the relationship between practitioners – philosophers and mathematicians – from the relationship between the areas of inquiry: philosophy and mathematics. Unlike the relationship between the areas of inquiry, the relationship between the practitioners was rather clear cut. As Zhmud and Kouprianov (2018, p. 469) have observed, mathematicians were distinguished from philosophers from the fourth century BCE onward. In other words, by the end of the classical period of Greek history, a group of practitioners studying high-level mathematics distinguished themselves and, for the most part, a sharp distinction held between individuals who identified as philosophers and those individuals who identified as mathematicians.

[6] Cuomo (2001, p. 2; 2000, p. 10).

Of course, some philosophers – notably Platonic and Neopythagorean philosophers – studied mathematics to some degree, but these philosophers did so as part of their philosophical practice. In the *Republic*, Plato includes mathematics as a key component of the education of the philosopher king. Socrates argues that there is a right way and there is a wrong way to study mathematics. The right way turns the soul towards the intelligible realm and prepares it to engage in dialectic, the study of the Forms; the wrong way remains focused on the sensible realm, the objects of perception. Enforcing this distinction, Socrates criticizes those individuals who study mathematics in the wrong way. For example, he mocks the practitioners of geometry:

Socrates: Now, no one with even a little experience of geometry will dispute with us that this science is itself entirely the opposite of what is said about it in the accounts of its practitioners.

Glaucon: How so?

Socrates: Well, they say completely ridiculous things about it because they are so hard up. I mean, they talk as if they were practical people who make all their arguments for the sake of action. They talk of squaring, applying, adding, and the like; whereas, in fact, the entire subject is practiced for the sake of acquiring knowledge.

Glaucon: Absolutely.

Socrates: Mustn't we also agree on a further point?

Glaucon: What?

Socrates: That it is knowledge of what always is, not of something that comes to be and passes away.

Glaucon: That's easy to agree to, since geometry is knowledge of what always is.

Socrates: In that case, my noble fellow, it can draw the soul toward truth and produce philosophical thought by directing upward what we now wrongly direct downward.[7]

[7] Plato, *Republic* VII, 527a–b, trans. Reeve (2004): Οὐ τοίνυν τοῦτό γε, ἦν δ' ἐγώ, ἀμφισβητήσουσιν ἡμῖν ὅσοι καὶ σμικρὰ γεωμετρίας ἔμπειροι, ὅτι αὕτη ἡ ἐπιστήμη πᾶν τοὐναντίον ἔχει τοῖς ἐν αὐτῇ λόγοις λεγομένοις ὑπὸ τῶν μεταχειριζομένων. Πῶς; ἔφη.
Λέγουσι μέν που μάλα γελοίως τε καὶ ἀναγκαίως· ὡς γὰρ πράττοντές τε καὶ πράξεως ἕνεκα πάντας τοὺς λόγους ποιούμενοι λέγουσιν τετραγωνίζειν τε καὶ παρατείνειν καὶ προστιθέναι καὶ πάντα οὕτω φθεγγόμενοι· τὸ δ' ἐστί που πᾶν τὸ μάθημα γνώσεως ἕνεκα ἐπιτηδευόμενον
Παντάπασι μὲν οὖν, ἔφη.
Οὔκουν τοῦτο ἔτι διομολογητέον;
Τὸ ποῖον;
Ὡς τοῦ ἀεὶ ὄντος γνώσεως, ἀλλ' οὐ τοῦ ποτέ τι γιγνομένου καὶ ἀπολλυμένου.
Εὐομολόγητον, ἔφη· τοῦ γὰρ ἀεὶ ὄντος ἡ γεωμετρικὴ γνῶσίς ἐστιν.
Ὁλκὸν ἄρα, ὦ γενναῖε, ψυχῆς πρὸς ἀλήθειαν εἴη ἂν καὶ ἀπεργαστικὸν φιλοσόφου διανοίας πρὸς τὸ ἄνω σχεῖν ἃ νῦν κάτω οὐ δέον ἔχομεν.

Socrates distinguishes the practitioners of geometry, the mathematicians, from the philosophers, and he criticizes the mathematicians for studying mathematics in the wrong way. According to Socrates, one should study mathematics only inasmuch as it illuminates the nature of metaphysical phenomena.

This contention that mathematics has a limited and merely propaedeutic value took hold in the Greek philosophical tradition. Centuries of philosophers studied mathematics either not at all or only to a limited degree, inasmuch as it contributed to their studies in philosophy. Two Platonic philosophers from the early second century CE, for example, are remembered principally for their contributions in mathematics – Theon of Smyrna and Nicomachus of Gerasa – and yet their texts are limited in their aim and mathematical content. The sole surviving text of Theon of Smyrna is *On Mathematics Useful for the Understanding of Plato*. It is a didactic text, a handbook presenting students with an introductory account of the mathematical sciences that Plato alludes to in his corpus, especially the *Republic*, *Timaeus*, and *Epinomis*, taken to be authentic. Theon seems to have designed his text to survey all five of the mathematical sciences mentioned in the *Republic* – arithmetic, geometry, stereometry, astronomy, and harmonics – but only the sections on arithmetic, harmonics, and astronomy survive. Despite its dedication to mathematics, Theon's text is not a technical treatise. He demarcates a limited scope for the text, articulated from the very first line: "Everyone would agree that it is not possible to understand the mathematical material in Plato without being oneself practised in that discipline. . . ."[8] The aim of the text is simply to illuminate what Plato had written on mathematics.

Theon does, however, go beyond Plato. For instance, in the section on astronomy he discusses the epicyclic and eccentric models of planetary motion. These hypotheses did not emerge until approximately 200 BCE, but Theon describes them and attributes them to Plato.[9] Even so, Theon does not address contemporary advances in astronomy. As Alexander Jones has noted, Theon examines only single-anomaly theories, or hypotheses that account only for the synodic anomaly – which produces the

8 Theon, *On Mathematics Useful for the Understanding of Plato* 1.1–3, trans. Dillon (1977): Ὅτι μὲν οὐχ οἷόν τε συνεῖναι τῶν μαθηματικῶς λεγομένων παρὰ Πλάτωνι μὴ καὶ αὐτὸν ἠσκημένον ἐν τῇ θεωρίᾳ ταύτῃ, πᾶς ἄν που ὁμολογήσειεν

9 On the origin of the epicyclic and eccentric models, see Evans and Carman (2014). For Theon's exegetical strategy, see Petrucci (2018).

five planets' retrograde motion – and not the zodiacal anomaly, by which a planet's motion varies as it moves through the zodiac. It is unlikely that a mathematician living in Theon's time with knowledge of the latest astronomical literature would have been unaware of the planets' zodiacal anomaly and the need for a more complex theory than the simple epicycles and eccenters described by Theon (Jones, 2015, p. 77). We have evidence that hypotheses accounting for two anomalies circulated as early as the mid-first century CE, and yet Theon does not mention them. Not only does Theon not acknowledge the latest advances in astronomy, but he also criticizes mathematicians for their disagreements over whether the epicyclic or the eccentric model saves the phenomena since, in Theon's preferred hypothesis, the planets trace these circles only accidentally when revolving on their larger spherical shells.[10] Criticizing mathematicians, Theon indicates that he is not a mathematician but a philosopher.

Although Theon declares that studying the mathematical sciences is necessary for understanding Plato's texts, Nicomachus of Gerasa, a contemporary of Theon and a Neopythagorean philosopher, makes a stronger claim.[11] He states that mathematics is necessary for philosophy. Like Theon, Nicomachus composed handbooks on mathematics, of which two survive: *Introduction to Arithmetic* and *Manual of Harmonics*. In the former, he argues that the mathematical sciences contribute to, and are even necessary for, philosophy. He defines philosophy as the love and desire for wisdom, and after defining four mathematical sciences – arithmetic, music, geometry, and astronomy – he proclaims, "Without the aid of these, then, it is not possible to deal accurately with the forms of being nor to discover the truth in things, knowledge of which is wisdom, and evidently not even to philosophize properly . . . ."[12] In other words, without the contribution of the mathematical sciences, it is not possible to study philosophy. Nicomachus's commitment to mathematics is evidenced by his surviving handbooks, but his mathematics is elementary. His texts are not technical treatises but didactic introductions that in late

10 Theon, *On Mathematics Useful for the Understanding of Plato*, 178.24–179.6. See Jones (2016, p. 477).

11 Theon, *On Mathematics Useful for the Understanding of Plato*, 1.10-2.2.

12 Nicomachus, *Introduction to Arithmetic* 1.3.3, trans. D'Ooge (1926): οὐχ ἄρα τούτων ἄνευ δυνατὸν τὰ τοῦ ὄντος εἴδη ἀκριβῶσαι οὐδ' ἄρα τὴν ἐν τοῖς οὖσιν ἀλήθειαν εὑρεῖν, ἧς ἐπιστήμη σοφία, φαίνεται δέ, ὅτι οὐδ' ὀρθῶς φιλοσοφεῖν

antiquity became basic school texts, on which Neoplatonic philosophers, including Iamblichus and John Philoponus, wrote commentaries.

It is important to note that at least one Neoplatonic philosopher did make more technical contributions to mathematics. Proclus, in the fifth century CE, wrote a commentary on the first book of Euclid's *Elements* as well as a short text on astronomy. The mathematical sciences remained, however, propaedeutic to dialectic for Proclus,[13] and his choice to write a commentary on Euclid resulted, as Dominic O'Meara has argued, from his characterization of Euclid as a faithful exponent of Plato. In his commentary, Proclus characterizes the goal of the *Elements* as the construction of the geometrical objects of Plato's *Timaeus*, and thus he makes Plato the principal authority in his commentary on the *Elements* (O'Meara, 1989, pp. 156–76). These Platonist philosophers did not have an independent interest in mathematics. They may have derived inspiration or found influence in the contemporary practice of the mathematical sciences for their philosophical systems, but Plato's characterization of mathematics as merely propaedeutic to dialectic limited, with few exceptions, their study of the mathematical sciences.

For entirely different reasons from Plato and his successors, Epicurus, too, sought to limit the study of the mathematical sciences, and astronomy in particular. This agenda may be surprising, since two of Epicurus's three surviving letters are dedicated to conveying the principles of his physics and meteorology, which for Epicurus included the study of things high up in the cosmos, such as the stars. Epicurus presents multiple explanations for the appearances of the sun, moon, and stars in their revolutions about the earth, and he advocates for their study in his attempt to alleviate the fear that human beings had that the phenomena were caused by the intervention of gods. In this way, Epicurus divorces religious beliefs from the study of the natural world. Despite this advocacy for the study of natural phenomena, Epicurus warns against excessive study of the heavens. In his *Letter to Herodotus*, he states the following:[14]

13 See Proclus, *In primum Euclidis elementorum librum commentarii* 42.9-43.21. For an analysis of this passage, see Bonelli (2016).

14 Diogenes Laertius, *Lives* 10.79, trans. Inwood and Gerson (1994): Τὸ δ' ἐν τῇ ἱστορίᾳ πεπτωκός, τῆς δύσεως καὶ ἀνατολῆς καὶ τροπῆς καὶ ἐκλείψεως καὶ ὅσα συγγενῆ τούτοις μηθὲν ἔτι πρὸς τὸ μακάριον τῆς γνώσεως συντείνειν, ἀλλ' ὁμοίως τοὺς φόβους ἔχειν τοὺς ταῦτα κατειδότας, τίνες δ' αἱ φύσεις ἀγνοοῦντας καὶ τίνες αἱ κυριώταται αἰτίαι, καὶ εἰ μὴ προσῄδεισαν ταῦτα· τάχα δὲ καὶ πλείους, ὅταν τὸ θάμβος ἐκ τῆς τούτων προσκατανοήσεως μὴ δύνηται τὴν λύσιν λαμβάνειν καὶ τὴν περὶ τῶν κυριωτάτων οἰκονομίαν.

And what falls within the ambit of investigation into settings and risings and turnings, and eclipses and matters related to these, makes no further contribution to the blessedness which comes from knowledge; but people who know about these things, if they are ignorant of what the natures [in question] are and what the most important causes are, have fears just the same as if they did not have this special knowledge – and perhaps even more fears, since the wonderment which comes from the prior consideration of these phenomena cannot discover a resolution or the orderly management of the most important factors.

In this passage, Epicurus warns against the mathematical study of the heavens, or astronomy, and claims that only the physical causes of phenomena should be studied. In the *Letter to Pythocles* he condemns a certain type of astronomy. Maintaining that multiple explanations of the phenomena should be given, he rejects claims for only a single explanation:[15] "But to supply one cause for these facts, when the phenomena suggest that there are several different explanations, is the lunatic and inappropriate behavior of those who are obsessed with a pointless [brand of] astronomy. . . ." Epicurus not only denounces the type of astronomy that seeks a single explanation of the phenomena, but he also criticizes astronomers. After providing physical explanations of the movements of the sun and the moon, he makes reference to "the slavish technicalities of the astronomers."[16] Similar to Socrates in the *Republic*, Epicurus articulates a right and a wrong way to pursue a mathematical study, and he ridicules the set of individuals – in this case, astronomers – that studies that mathematical science in the wrong way.

Epicureans sought to limit the study of not only astronomy but also geometry. Epicurus's followers notoriously rejected geometry. During his time teaching at Lampsacus, Epicurus converted Polyaenus of Lampsacus, formerly a mathematician, to his philosophical system. In the process, according to Cicero, Epicurus made Polyaenus "unlearn" (*dedocere*) geometry, and ultimately Polyaenus came to believe that "all geometry is false."[17] In general, Epicureans were known for their absolute rejection of geometry, including its most basic principles.[18] David Sedley has argued that Epicurus's criticism of astronomy and geome-

15 Diogenes Laertius, *Lives* 10.113, trans. Inwood and Gerson (1994): τὸ δὲ μίαν αἰτίαν τούτων ἀποδιδόναι, πλεοναχῶς τῶν φαινομένων ἐκκαλουμένων, μανικὸν καὶ οὐ καθηκόντως πραττόμενον ὑπὸ τῶν τὴν ματαίαν ἀστρολογίαν ἐζηλωκότων

16 Diogenes Laertius, *Lives* 10.93, trans. Inwood and Gerson (1994): τὰς ἀνδραποδώδεις ἀστρολόγων τεχνιτείας.

17 Cicero, *De finibus bonorum et malorum* 1.20; ibid., *Academica* 2.106: totam geometriam falsam

18 See Proclus, *In primum Euclidis elementorum librum commentarii* 199.3–200.3.

try developed in the context of the rivalry between his school and the Eudoxan school at Cyzicus, a neighboring city on the Hellespont.[19] Eudoxus was a contemporary of Plato known for his contributions to astronomy and geometry and, during Epicurus's time, a school run by Eudoxus's successors flourished in Cyzicus. The Cyzicenes, like Eudoxus, sought to develop astronomical models that saved the phenomena and, perhaps due to their interest in Babylonian astronomy and astrology, Epicurus deemed them "enemies of Greece."[20]

The distinction – and sometimes rivalry – between mathematicians and philosophers does not necessitate the distinction between the mathematical sciences and philosophy in Greek antiquity. In other words, even if the two sets of practitioners remained distinct, their fields of inquiry may not have been. Not only could some philosophers study mathematics, and some mathematicians study philosophy, but some of these historical actors may have equated high-level mathematics with philosophy, or practiced high-level mathematics in such a way that it constituted philosophy.

For Epicureans and strong adherents to the Platonic tradition, it does seem to be the case that mathematics was distinct from philosophy. Although Plato's philosopher-kings-in-training do need to study the mathematical sciences, they do so only in preparation for studying dialectic and the Forms. Aristotle, however, paved the way for a more complicated history by dividing all theoretical knowledge into three kinds: physics, mathematics, and theology. After Aristotle defines each of these areas of knowledge in the *Metaphysics*, he goes on to call them types of theoretical "philosophy":[21] "There must, then, be three theoretical philosophies, mathematical, physical, and theological. . . ." This trichotomy of theoretical philosophy – into physics, mathematics, and theology – became authoritative in the subsequent ancient Greek philosophical tradition. Ptolemy, for example, discusses this trichotomy of theoretical philosophy in the *Almagest*. In one of his few citations of a philosopher, he states,[22] "For Aristotle divides the theoretical [part of philosophy], too, very fittingly, into three primary genera, the physical, mathemati-

19 See Sedley (1976b,a).

20 Diogenes Laertius, *Lives* 10.8: ἐχθροὺς τῆς Ἑλλάδος. See Sedley (1976b, p. 42).

21 Aristotle, *Metaphysics* VI, 1026a18–19: ὥστε τρεῖς ἂν εἶεν φιλοσοφίαι θεωρητικαί, μαθηματική, φυσική, θεολογική. This translation and others are my own unless otherwise indicated.

22 Ptolemy, *Almagest* 1.1, H5: ὥστε τρεῖς ἂν εἶεν φιλοσοφίαι θεωρητικαί, μαθηματική, φυσική, θεολογική.

cal, and theological." Because in this trichotomy mathematics is a part of philosophy, one may conclude that, for the many philosophers and mathematicians who appropriated this division of theoretical philosophy, the distinction between mathematics and philosophy did not hold, for mathematics is, on this definition, philosophy.

One may still ask, however, whether, despite this classification of mathematics as philosophy, there still remained an implied distinction, or whether these various philosophers and mathematicians still treated mathematics as if it were separate from philosophy. I argue that, even though Aristotle classifies mathematics as a type of philosophy in the *Metaphysics*, he still treats the mathematical sciences as distinct from philosophy. He portrays them as separate, for instance, in one of the more mathematical chapters of the *Metaphysics*: Lambda 8, in which he describes Eudoxus's and Callippus's astronomical models. In this chapter, Aristotle describes astronomy as "that one of the mathematical sciences that is most akin to philosophy."[23] Although it is a mathematical science, astronomy here is not a part of philosophy; it is akin to philosophy and, therefore, distinct from it. In other words, being like philosophy entails that mathematical sciences are not philosophy. Similarly, in *Metaphysics* Gamma, Aristotle compares philosophy to mathematics, and he notes that just as there are many parts of mathematics so, too, there are many parts of philosophy:[24]

> And there are as many parts of philosophy as there are kinds of substances, so that necessarily there must be among them a first [philosophy] and one which follows this. For being falls immediately into genera; and therefore the sciences too will correspond to these [genera]. For 'philosopher' is like 'mathematician'; for [mathematics] also has parts, and there is a first and a second science and other successive ones within the mathematical sciences.

Even though Aristotle defines mathematics as a type of philosophy, in other chapters he treats them as distinct. Consequently, when examining the relationship between the mathematical sciences and philosophy in Greek antiquity, one must look not just at the terminology used but also at whether apparent identifications and distinctions are meaningful ones.

[23] Aristotle, *Metaphysics* XII, 1073b4–5, translation modified from W.D. Ross in Barnes (1984): τῆς οἰκειοτάτης φιλοσοφίᾳ τῶν μαθηματικῶν ἐπιστημῶν

[24] Aristotle, *Metaphysics* IV, 1004a2-9: καὶ τοσαῦτα μέρη φιλοσοφίας ἔστιν ὅσαι περ αἱ οὐσίαι· ὥστε ἀναγκαῖον εἶναί τινα πρώτην καὶ ἐχομένην αὐτῶν· ὑπάρχει γὰρ εὐθὺς γένη ἔχον τὸ ὂν [καὶ τὸ ἕν]· διὸ καὶ αἱ ἐπιστῆμαι ἀκολουθήσουσι τούτοις. ἔστι γὰρ ὁ φιλόσοφος ὥσπερ ὁ μαθηματικὸς λεγόμενος· καὶ γὰρ αὕτη ἔχει μέρη, καὶ πρώτη τις καὶ δευτέρα ἔστιν ἐπιστήμη καὶ ἄλλαι ἐφεξῆς ἐν τοῖς μαθήμασιν.

4.2 Claudius Ptolemy

I argue that for some ancient Greek mathematicians the distinction between philosophy and the mathematical sciences did completely unravel. For example, Ptolemy, as mentioned above, appropriated Aristotle's division of theoretical philosophy into three genera: physics, mathematics, and theology. In the first chapter of the *Almagest*, he defines them, and then he proceeds to evaluate their epistemic success. For Aristotle, because the three theoretical sciences are all types of knowledge, they by definition produce knowledge. Ptolemy disagrees. Having classified these fields as only "parts of philosophy" and not "parts of knowledge," he is able to argue that one of the three sciences produces knowledge but the other two are merely conjectural. He presents this argument as follows:[25]

> From all this we concluded that the first two genera of the theoretical [part of philosophy] should rather be called conjecture than knowledge, the theological because of its complete invisibility and ungraspability, and the physical because of the instability and unclearness of matter, so that, on account of this, [we concluded] never to hope that philosophers will be agreed about them, and that only the mathematical can provide sure and incontrovertible knowledge to its practitioners, if one approaches it rigorously, for its demonstration proceeds by indisputable methods, both arithmetic and geometry…

What Ptolemy frames as his conclusion regarding the epistemic status of the three sciences is in fact a polemical argument. Ptolemy is very critical of philosophers; he claims that they will never come to an agreement. This emphasis on agreement is significant, because for Ptolemy agreement is a feature of knowledge. He defines knowledge in his short epistemological text *On the Kritêrion and Hêgemonikon*, where he states that knowledge corresponds to a judgment that is "most clear and agreed upon."[26] Similarly, in his *Geography* Ptolemy associates that which is "nearest the truth"[27] with what is "more or less agreed upon."[28] Therefore, for Ptolemy, the degree of agreement entails how

[25] Ptolemy, *Almagest* 1.1, H6: ἐξ ὧν διανοηθέντες, ὅτι τὰ μὲν ἄλλα δύο γένη τοῦ θεωρητικοῦ μᾶλλον ἄν τις εἰκασίαν ἢ κατάληψιν ἐπιστημονικὴν εἴποι, τὸ μὲν θεολογικὸν διὰ τὸ παντελῶς ἀφανὲς αὐτοῦ καὶ ἀνεπίληπτον, τὸ δὲ φυσικὸν διὰ τὸ τῆς ὕλης ἄστατον καὶ ἄδηλον, ὡς διὰ τοῦτο μηδέποτε ἂν ἐλπίσαι περὶ αὐτῶν ὁμονοῆσαι τοὺς φιλοσοφοῦντας, μόνον δὲ τὸ μαθηματικόν, εἴ τις ἐξεταστικῶς αὐτῷ προσέρχοιτο, βεβαίαν καὶ ἀμετάπιστον τοῖς μεταχειριζομένοις τὴν εἴδησιν παράσχοι ὡς ἂν τῆς ἀποδείξεως δι' ἀναμφισβητήτων ὁδῶν γιγνομένης, ἀριθμητικῆς τε καὶ γεωμετρίας

[26] Ptolemy, *On the Kritêrion*, La7: τρανωτάτη καὶ ὁμολογουμένη

[27] Ptolemy, *Geography* 2.1.2: ἐγγυτάτω τῆς ἀληθείας

[28] Ptolemy, *Geography* 2.1.2: ὡς ἐπίπαν ὁμολογούμενον

close a proposition is to the truth. In failing to come to any agreement, philosophers fail to attain knowledge. Moreover, in this same passage of the *Almagest*, Ptolemy critiques the two types of theoretical philosophy traditionally studied by philosophers: physics and theology. He argues that physics and theology are conjectural, and that mathematics alone generates knowledge. Therefore, mathematics is the only part of theoretical philosophy to achieve the epistemological objective of philosophy – it is the only part of philosophy to yield knowledge – and, in this way, mathematics is superior to every other part of philosophy.

Ptolemy simply could have left aside physics and theology after establishing the epistemological superiority of mathematics, but instead he goes on to co-opt physics and theology for the mathematician:[29]

> . . . [mathematics] can work in the domains of the other [two parts of the theoretical part of philosophy] no less than they. For this is the best [part of philosophy] to pave the way for theology, as it is the only one able to guess well at [the nature] of the immovable and separated activity . . . [mathematics] contributes to physics not accidentally, for nearly every peculiar attribute of material substance is made apparent from the peculiar qualities of its motion from place to place. . .

Just because theology and physics are fundamentally conjectural does not entail that they should not be studied. According to Ptolemy, mathematics can lend some of its epistemic capability to physics and theology. The contribution of mathematics leads to the best guesses possible about the nature of physical bodies and theological entities. It is through the contribution of mathematics to physics and theology that the distinction between the mathematical sciences and traditional areas of philosophical inquiry especially wears thin. According to Ptolemy, by studying objects mathematically, one can make progress in the other areas of theoretical philosophy.

In *Almagest* 1.1, Ptolemy also argues for the ethical superiority of mathematics. He makes reference to his ethical theory in the first lines of *Almagest* 1.1 and then again near the end of the chapter. Near the beginning, he makes it clear that the good life involves a certain state of the soul:[30]

[29] Ptolemy, *Almagest* 1.1, H.7: πρὸς δὲ τὰς ἄλλας οὐχ ἧττον αὐτῶν ἐκείνων συνεργεῖν. τό τε γὰρ θεολογικὸν εἶδος αὕτη μάλιστ' ἂν προοδοποιήσειε μόνη γε δυναμένη καλῶς καταστοχάζεσθαι τῆς ἀκινήτου καὶ χωριστῆς ἐνεργείας [. . .] πρός τε τὸ φυσικὸν οὐ τὸ τυχὸν ἂν συμβάλλοιτο· σχεδὸν γὰρ τὸ καθόλου τῆς ὑλικῆς οὐσίας ἴδιον ἀπὸ τῆς κατὰ τὴν μεταβατικὴν κίνησιν ἰδιοτροπίας καταφαίνεται

[30] Ptolemy, *Almagest* 1.1, H4-5: ἔνθεν ἡγησάμεθα προσήκειν ἑαυτοῖς τὰς μὲν πράξεις ἐν

Hence, we thought it fitting for them [i.e., the legitimate philosophers], on the one hand, to order their actions according to the applications of the *phantasiai* of these in such a way as never to forget, even in ordinary affairs, to strive for a fine and well-ordered state, and, on the other hand, with leisure to devote the most time to the instruction of theories, being many and fine, but especially those particularly called mathematical.

Ptolemy indicates that the best life is one in which our souls are in a fine and well-ordered state. Near the end of the chapter, he explains what he means by, and how one attains, this fine and well-ordered state:[31]

With regard to virtuous conduct in actions and character, [mathematics], above all, could make clear-sighted men; from the constancy, good order, commensurability, and calm that are contemplated in the case of the divine, it, on the one hand, makes its followers lovers of this divine beauty, and, on the other hand, accustoms and, as it were, reforms their natures to a similar state of the soul.

Ptolemy appropriates a traditional goal of Platonic philosophy: the *telos*, or objective, of becoming godlike: *homoiôsis theôi*. This objective of philosophy appears in Plato's *Theaetetus* and *Timaeus*, and I argue that the account in the *Timaeus* provides the foundation for Ptolemy's version.[32] Timaeus explains how an individual can organize the revolutions of his soul according to the order of the universe:[33]

... the god invented sight and gave it to us so that we might observe the orbits of intelligence in the heavens and apply them to the revolutions of our own understanding, for there is a kinship between them, [even though our revolutions are] disturbed, [whereas the universal orbits are] undisturbed, and so once we have come to know them and to share in the ability to make correct calculations according to nature, we should stabilize the straying revolutions within ourselves by imitating the completely unstraying revolutions of the god.

ταῖς αὐτῶν τῶν φαντασιῶν ἐπιβολαῖς ῥυθμίζειν, ὅπως μηδ' ἐν τοῖς τυχοῦσιν ἐπιλανθανώμεθα τῆς πρὸς τὴν καλὴν καὶ εὔτακτον κατάστασιν ἐπισκέψεως, τῇ δὲ σχολῇ χαρίζεσθαι τὸ πλεῖστον εἰς τὴν τῶν θεωρημάτων πολλῶν καὶ καλῶν ὄντων διδασκαλίαν, ἐξαιρέτως δὲ εἰς τὴν τῶν ἰδίως καλουμένων μαθηματικῶν.

31 Ptolemy, *Almagest* 1.1, H7: πρός γε μὴν τὴν κατὰ τὰς πράξεις καὶ τὸ ἦθος καλοκαγαθίαν πάντων ἂν αὕτη μάλιστα διορατικοὺς κατασκευάσειεν ἀπὸ τῆς περὶ τὰ θεῖα θεωρουμένης ὁμοιότητος καὶ εὐταξίας καὶ συμμετρίας καὶ ἀτυφίας ἐραστὰς μὲν ποιοῦσα τοὺς παρακολουθοῦντας τοῦ θείου τούτου κάλλους, ἐνεθίζουσα δὲ καὶ ὥσπερ φυσιοῦσα πρὸς τὴν ὁμοίαν τῆς ψυχῆς κατάστασιν.

32 For an extended examination of Ptolemy's ethics, see Feke (2018). See also Feke (2012).

33 Plato, *Timaeus* 47b-c, translation modified from Donald J. Zeyl in Plato (1997): θεὸν ἡμῖν ἀνευρεῖν δωρήσασθαί τε ὄψιν, ἵνα τὰς ἐν οὐρανῷ τοῦ νοῦ κατιδόντες περιόδους χρησαίμεθα ἐπὶ τὰς περιφορὰς τὰς τῆς παρ' ἡμῖν διανοήσεως, συγγενεῖς ἐκείναις οὔσας, ἀταράκτοις τεταραγμένας, ἐκμαθόντες δὲ καὶ λογισμῶν κατὰ φύσιν ὀρθότητος μετασχόντες, μιμούμενοι τὰς τοῦ θεοῦ πάντως ἀπλανεῖς οὔσας, τὰς ἐν ἡμῖν πεπλανημένας καταστησαίμεθα.

Timaeus presents both a descriptive and a normative claim. He explains that there exists a kinship between the heavens and the human soul, and he maintains that human beings should imitate the heavens in order to stabilize the orbits of their souls.

Ptolemy appropriates this Platonic doctrine and further mathematizes it. I argue that Ptolemy portrays an ethical process that is more mathematical than what tends to appear in Platonic texts, because the ethical exemplars Ptolemy advances are not simply aethereal bodies; they are not the stars and planets; they are the mathematical objects in the heavens: the movements and configurations of the heavenly bodies, rather than the heavenly bodies themselves. That Ptolemy means these mathematical objects we can infer from the definition of mathematics he presents in *Almagest* 1.1: "the species [of the theoretical part of philosophy] indicative of the quality concerning forms and movements from place to place, and which serves to investigate shape, number, size, and place, time, and suchlike, one may define as 'mathematical'."[34] According to Ptolemy, astronomy is one of the mathematical sciences and, as a type of mathematics, it concerns movements and shapes; in particular, it studies the movements of the heavenly bodies and the configurations of the stars. I argue that when Ptolemy says "from the constancy, good order, commensurability, and calm that are contemplated in the case of the divine," the "divine" may refer to the aethereal bodies, but the qualities contemplated – constancy, good order, and commensurability – are mathematical. Moreover, the objects that reveal these qualities are mathematical; they are the movements and configurations of aethereal bodies, rather than the aethereal bodies themselves. In other words, these qualities of constancy, good order, commensurability, and calm become apparent to the mathematician when he studies astronomical objects, specifically the stars' movements and configurations.

In Ptolemy's account, no other divine objects can serve as ethical exemplars. In order to serve as an exemplar for virtuous conduct, an object must be not only divine – not any god will do – but rather the exemplar must be a divine object or set of objects about which human beings can have knowledge. Therefore, theological objects cannot serve as ethical exemplars in Ptolemy's system, because, as discussed above,

[34] Ptolemy, *Almagest* 1.1, H5-6: τὸ δὲ τῆς κατὰ τὰ εἴδη καὶ τὰς μεταβατικὰς κινήσεις ποιότητος ἐμφανιστικὸν εἶδος σχήματός τε καὶ ποσότητος καὶ πηλικότητος ἔτι τε τόπου καὶ χρόνου καὶ τῶν ὁμοίων ζητητικὸν ὑπάρχον ὡς μαθηματικὸν ἂν ἀφορίσειε

theology is conjectural. Ptolemy defines theology very narrowly in the *Almagest* as concerning only one entity: "the first cause of the first motion of the universe," which recalls Aristotle's Prime Mover.[35] Therefore, because theology – the study of Ptolemy's Prime Mover – is conjectural, and theological objects are fundamentally unknowable, they cannot serve as ethical exemplars. In order to attain a godlike condition of the soul, one must study mathematics, especially astronomy, the mathematical science of divine entities, the movements and configurations of the stars.

Thus, through a combination of his epistemology and his appropriation of Platonic ethics, Ptolemy co-opts ancient virtue ethics for the mathematician. According to Ptolemy, it is only through the study of mathematics – and, according to the *Almagest*, astronomy – that a human being can attain a virtuous condition of the soul and live the good life.[36] Ptolemy not only classifies mathematics as a part of philosophy, but he also ascribes the traditional goals of philosophy to mathematics, and he argues that only mathematics is capable of achieving these goals: generating knowledge and engendering the good life.

4.3 Hero of Alexandria

For another case study, one may look to the corpus of Hero of Alexandria, the first-century CE mechanician. Hero characterizes the relationship between philosophy and the more practical types of mathematics in a similar way as Ptolemy portrays the relationship between the traditional areas of philosophy – physics and theology – and the mathematical sciences.[37] Hero defines mechanics as a type of philosophy, similar to how Ptolemy defines mathematics as a part of philosophy, at the very beginning of the *Belopoeica*, his text on artillery construction:[38]

The greatest and most necessary part of study in philosophy is that which

35 Ptolemy, *Almagest* 1.1, H5: τὸ μὲν τῆς τῶν ὅλων πρώτης κινήσεως πρῶτον αἴτιον

36 Ptolemy indicates in the *Harmonics* that the study of harmonics also can transform the human soul into a virtuous state. See Feke (2018).

37 For an examination of the relationship between Ptolemy's and Hero's arguments concerning philosophy, see Feke (2014).

38 Hero, *Belopoeica* 71.1-72.9: Τῆς ἐν φιλοσοφίᾳ διατριβῆς τὸ μέγιστον καὶ ἀναγκαιότατον μέρος ὑπάρχει τὸ περὶ ἀταραξίας, περὶ ἧς πλεῖσταί τε ὑπῆρξαν ζητήσεις παρὰ τοῖς μεταχειριζομένοις τὴν σοφίαν καὶ μέχρι νῦν ὑπάρχουσιν· καὶ νομίζω μηδὲ τέλος ποτὲ ἕξειν διὰ τῶν λόγων τὴν περὶ αὐτῆς ζήτησιν. μηχανικὴ δὲ ὑπερβᾶσα τὴν διὰ λόγων περὶ ταύτης διδασκαλίαν ἐδίδαξεν πάντας ἀνθρώπους ἀταράχως ζῆν ἐπίστασθαι δι' ἑνὸς καὶ ἐλαχίστου μέρους αὐτῆς, λέγω δὴ τοῦ κατὰ τὴν καλουμένην βελοποιίαν, δι' ἧς οὔτε ἐν εἰρηνικῇ καταστάσει ταραχθήσονταί ποτε ἐχθρῶν καὶ πολεμίων ἐπανόδοις, οὔτε ἐνστάντος πολέμου ταραχθήσονταί ποτε τῇ παραδιδομένῃ ὑπ' αὐτῆς διὰ τῶν

> is concerned with tranquility, about which a great many investigations were made and still now are made by those pursuing wisdom; indeed, I believe that the investigation concerning this [viz. tranquility] will never come to an end through rational discourses. But mechanics, having surpassed the instruction of this [viz. tranquility] through rational discourses, taught all human beings to know how to live tranquilly by means of a single and very small part of this [viz. mechanics], I mean, of course, the one dealing with so-called artillery construction, through which [human beings] will never be disturbed during a state of peace by the assaults of adversaries and enemies, nor when war is upon them will they ever be disturbed, thanks to the philosophy which it [viz. artillery construction] imparts through its instruments. It is therefore necessary at all times to stand calm and to devote every forethought to this part [of mechanics].

Perhaps surprisingly, Hero argues that the best way to achieve tranquility is to build war machines.

First, it is clear that Hero in some way juxtaposes mechanics and a part of mechanics, artillery construction, with philosophical discourses. The interpretation I have advanced is that Hero ascribes two meanings to "philosophy."[39] It is a general field of inquiry, and it is the part of this general field studied by philosophers. In other words, "philosophy" here refers to two categories in a hierarchy of intellectual inquiry. Described at the beginning of the passage, the broadest field of inquiry is the study of philosophy, the greatest part of which is concerned with tranquility, and those individuals who engage in investigations of tranquility pursue wisdom. One may infer, then, that "philosophy" here denotes a very general domain, which one may translate literally as the "love of wisdom." Tranquility – the goal of the largest part of this general area of philosophy – was a keyword in ancient Greek philosophical discourses. Epicureans, Pyrrhonist Skeptics, and Stoics portrayed it as an objective of philosophical inquiry.

Hero proceeds to distinguish two manners by which the individuals who seek wisdom have investigated tranquility: that by means of discourses, and that by means of instruments. These two types of methods imply two sets of individuals, philosophers and mechanicians, and they imply two subspecies of intellectual inquiry, philosophy and mechanics. In other words, Hero does not simply place mechanics in opposition to philosophy. He does not juxtapose mechanics with philosophy as a

ὀργάνων φιλοσοφίᾳ. διὸ τοῦ μέρους τούτου ἐν παντὶ χρόνῳ <ἔμπειρον> καταστῆναι δεῖ καὶ πᾶσαν πρόνοιαν ποιεῖσθαι

39 See Feke (2014).

general field of inquiry, for mechanics, according to Hero, is a type of philosophy; rather, Hero juxtaposes mechanics with the type of philosophy pursued by philosophers. Moreover, he contends that the discourses of philosophers are inadequate. He claims that individuals employing rational discourses will never reach an end in their investigations, but a part of mechanics, artillery construction, already has succeeded in teaching human beings how to live tranquilly. In other words, artillery construction already has achieved the objective that philosophers seek.

Hero also portrays philosophical discourses as inadequate in the introduction to his *Pneumatica* when discussing the theory of micro-voids:[40]

> Now, for those men asserting that there is absolutely no void, it suffices to devise these many dialectical arguments and perhaps they seem more plausible in the discourse since no perceptible demonstration is available; however, if it is shown on the basis of phenomena and by what falls under perception that continuous void is against nature but can come into being, and that void is in accordance with nature but finely interspersed, and that bodies fill up the interspersed voids in accordance with compression, then those men who bring forward the plausible of the discourses concerning these [phenomena] will not have any loophole by which to escape.

Hero again criticizes individuals who rely on discourses rather than some observable phenomena. In the *Belopoeica*, artillery construction provides philosophy through its instruments. In the *Pneumatica*, perceptible demonstrations reveal the true nature of the phenomena: that voids do exist. Moreover, in claiming that philosophical discourses are merely "plausible," Hero criticizes them. The goal of any discourse is not plausibility but certainty, and by calling philosophical discourses "dialectical," rather than "demonstrative," Hero discredits them.

Hero in fact ascribes certainty to mathematical demonstrations alone. He makes this claim in the introduction to the third book of the *Metrica*, where he discusses yet another topic of philosophical inquiry, the just distribution of land:[41]

40 Hero, *Pneumatica* 16.16-26: τοῖς οὖν φαμένοις τὸ καθόλου μηδὲν εἶναι κενὸν ἐκποιεῖ πρὸς ταῦτα πολλὰ εὑρίσκειν ἐπιχειρήματα καὶ τάχα φαίνεσθαι τῷ λόγῳ πιθανωτέρους μηδεμιᾶς παρακειμένης αἰσθητικῆς ἀποδείξεως· ἐὰν μέντοι δειχθῇ ἐπὶ τῶν φαινομένων καὶ ὑπὸ τὴν αἴσθησιν πιπτόντων, ὅτι κενὸν ἄθρουν ἐστὶν παρὰ φύσιν μέντοι γινόμενον, καὶ κατὰ φύσιν μὲν κενόν, κατὰ λεπτὰ δὲ παρεσπαρμένον, καὶ ὅτι κατὰ πίλησιν τὰ σώματα ἀναπληροῖ τὰ παρεσπαρμένα κενά, οὐδεμίαν οὐκέτι παρείσδυσιν ἕξουσιν οἱ τοὺς πιθανοὺς τῶν λόγων περὶ τούτων προφερόμενοι

41 Hero, *Metrica* 140.5-142.2: Οὐ πολὺ ἀπᾴδειν νομίζομεν τὰς τῶν χωρίων διαιρέσεις τῶν γιγνομένων ἐν τοῖς χωρίοις μετρήσεων· καὶ γὰρ τὸ ἀπονεῖμαι χωρίον τοῖς ἴσοις ἴσον καὶ τὸ πλέον τοῖς ἀξίοις κατὰ τὴν ἀναλογίαν πάνυ εὔχρηστον καὶ ἀναγκαῖον θεωρεῖται.

We believe that the divisions of land do not differ much from the measurements produced in the land; for, the distribution of equal land to equal men and proportionally more [land] to the worthy men is considered entirely useful and necessary. Indeed, the entire earth is already divided according to worth by nature itself: for large groups of people have obtained large areas of land dispensed according to this [viz. nature], but some who are few have individually small [areas of land]. No less the individual cities are divided according to worth: on the one hand, to the leaders and the others who are able to lead [are distributed] great [areas of land] also according to proportion, and on the other hand those who are not able to lead find themselves with small places, and to the men with meaner souls are [distributed] villages, farmsteads, and these types of places. But these cases have determined the proportion in a more rough and lazy manner, and if one wished to divide land according to the given ratio, so that not one grain, so to speak, exceeds or falls short of the given ratio, there would be need of only geometry, in which the fit is fair, justice is in the proportion, and the demonstration concerning these is indisputable, which none of the other crafts or sciences professes.

Not only is the just distribution of land proportional – such that larger groups should receive more land than smaller groups, and men with greater souls should receive more land than men with meaner, or smaller, souls – but only one science or craft renders these rewards precisely proportionate, and that science is geometry. The distribution of land, or other goods and honors, in the Greco-Roman world was the domain of the statesman, and Hero reduces the entire enterprise of statesmanship to a simple geometrical problem. Justice is precise proportionality, and geometry is the only science capable of defining precisely proportionate – and therefore just – apportionments of land. Thus, Hero addresses major concerns of philosophical discourses, he places the studies of philosophers and mathematicians in opposition, and he claims that only the mathematical sciences are successful at solving philosophical problems, whether they be how to attain tranquility, whether voids exist, or how to

ἤδη γοῦν καὶ ἡ σύμπασα γῆ διήρηται κατ' ἀξίαν ὑπ' αὐτῆς τῆς φύσεως· νέμεται γὰρ κατ' αὐτὴν ἔθνη μέγιστα μεγάλην λελογχότα χώραν, ἔνια δὲ καὶ ὀλίγην μικρὰ καθ' αὑτὰ ὑπάρχοντα· οὐχ ἧττον δὲ καὶ κατὰ μίαν αἱ πόλεις κατ' ἀξίαν διήρηνται· τοῖς μὲν ἡγεμόσι καὶ τοῖς ἄλλοις τοῖς ἄρχειν δυναμένοις μείζω καὶ κατὰ ἀναλογίαν, τοῖς δὲ μηδὲν τοιοῦτο δυναμένοις δρᾶν μικροὶ κατελείφθησαν τόποι, κῶμαί τε τοῖς μικροψυχοτέροις καὶ ἐποίκια καὶ ὅσα τοιαῦτά ἐστιν· ἀλλὰ τὰ μὲν παχυμερεστέραν πως καὶ ἀργοτέραν εἴληφε τὴν ἀναλογίαν· εἰ δέ τις βούλοιτο κατὰ τὸν δοθέντα λόγον διαιρεῖν τὰ χωρία, ὥστε μηδὲ ὡς εἰπεῖν κέγχρον μίαν τῆς ἀναλογίας ὑπερβάλλειν ἢ ἐλλείπειν τοῦ δοθέντος λόγου, μόνης προσδεήσεται γεωμετρίας· ἐν ᾗ ἐφαρμογὴ μὲν ἴση, τῇ δὲ ἀναλογίᾳ δικαιοσύνη, ἡ δὲ περὶ τούτων ἀπόδειξις ἀναμφισβήτητος, ὅπερ τῶν ἄλλων τεχνῶν ἢ ἐπιστημῶν οὐδεμία ὑπισχνεῖται apportion land justly. The mathematical sciences, then, achieve the most fundamental objectives of philosophical inquiry.

4.4 Archytas of Tarentum

The characterization of mathematical sciences – and, indeed, mathematicians – as achieving the ultimate objectives of philosophy has its roots deep in Greek history, as early as the classical period, when expertise in Greek mathematics emerged. For early evidence, one may look to the fragments of Archytas of Tarentum, a Pythagorean and contemporary of Plato. First, Archytas ascribes to mathematicians a talent for epistemology:[42]

> Those who concern themselves with the mathematical sciences seem to me to make distinctions well, and it is not surprising that they think correctly, about each thing, how they are. For, since they make distinctions well about the nature of the whole, they ought also to see well, about particular things, how they are. And certainly, about the speed of the heavenly bodies, their risings and settings, they have transmitted to us a clear distinction, and so too about geometry and numbers, and especially about music. For these sciences seem to be sisters.

Individuals who study the mathematical sciences are notable for their epistemic abilities. They make distinctions well – one of the most basic of epistemological skills – and they think correctly about the nature of each thing. Archytas does not limit the scope of these abilities; he does not say that these individuals only make distinctions well about mathematical objects. Rather, these individuals are good in general at making distinctions and identifying the natures of wholes and particulars. Carl Huffman (2005, p. 59) has argued that by "wholes" and "particulars" Archytas means to distinguish universals and particular objects or types of objects. Notably, Archytas does not, like Plato, describe the mathematical sciences as merely propaedeutic to some other, higher, metaphysical study. Archytas does not make Plato's distinction between a sensible and an intelligible realm. Instead, knowledge is of what is perceived, including the risings and settings of heavenly bodies.

42 Archytas, DK 47 B1, from Archytas (2016): καλῶς μοι δοκοῦντι τοὶ περὶ τὰ μαθήματα διαγνῶναι καὶ οὐθὲν ἄτοπον ὀρθῶς αὐτοὺς οἷά ἐντι περὶ ἑκάστου θεωρεῖν. περὶ γὰρ τᾶς τῶν ὅλων φύσιος καλῶς διαγνόντες ἔμελλον καὶ περὶ τῶν κατὰ μέρος, οἷά ἐντι, ὄψεσθαι. περί τε δὴ τᾶς τῶν ἄστρων ταχυτᾶτος καὶ ἐπιτολᾶν καὶ δυσίων παρέδωκαν ἁμῖν διάγνωσιν καὶ περὶ γαμετρίας καὶ ἀριθμῶν καὶ οὐχ ἥκιστα περὶ μουσικᾶς. ταῦτα γὰρ τὰ μαθήματα δοκοῦντι ἦμεν ἀδελφεά.

Archytas also ascribes to one particular mathematical science – calculation – success at tackling political matters. After discussing the problem of how one comes to know anything – similar to Meno's paradox, articulated in Plato's *Meno* – Archytas turns to the merits of calculation:[43]

> The discovery of calculation puts an end to civil strife and increases concord. For when this is present, there is no desire to possess more, and there is equality. For it is thanks to this that we are in accord with one another in our mutual relations. Therefore it is thanks to this that the poor receive from the powerful, and that the rich give to the needy, both parties being convinced that they will thereby have equality. And being a rule and a hindrance for unjust men, it stops those who know how to calculate before they commit injustice by persuading them that they will not be able to remain undetected whenever they have recourse to it. As for those who do not know it, having shown them that they are unjust in this area, it prevents them from committing injustice.

The mathematical science of calculation resolves political tensions. It changes what citizens desire. Understanding the basics of proportionality, citizens no longer desire more than their fair share. The rich then give to the poor, the poor receive goods from the rich, and thus calculation produces equality. Calculation also makes individual citizens better. It demonstrates the injustice of certain acts and indicates the proportionate punishment. When employing calculation, individuals who could commit a crime realize the impossibility of going undetected, and so calculation prevents them from performing unjust actions in the first place. Calculation transforms citizens' behavior and, in this way, it makes them better. Calculation, therefore, is key to the good life.

4.5 Conclusion

Thus, Archytas, Hero, and Ptolemy portray mathematical sciences as achieving the most fundamental goals of philosophy. For Archytas, mathematicians have exceptional epistemic abilities, and calculation is the key

[43] Archytas, DK 47 B3, from Archytas (2016): στάσιν μὲν ἔπαυσεν, ὁμόνοιαν δὲ αὔξησεν λογισμὸς εὑρεθείς· πλεονεξία τε γὰρ οὐκ ἔστι τούτου γενομένου καὶ ἰσότας ἔστιν· τούτῳ γὰρ περὶ τῶν συναλλαγμάτων διαλλασσόμεθα. διὰ τοῦτον οὖν οἱ πένητες λαμβάνοντι παρὰ τῶν δυναμένων, οἵ τε πλούσιοι διδόντι τοῖς δεομένοις, πιστεύοντες ἀμφότεροι διὰ τούτων τὸ ἶσον ἕξειν. κανὼν δὲ καὶ κωλυτὴρ τῶν ἀδικούντων <ἐὼν> τοὺς μὲν ἐπισταμένους λογίζεσθαι πρὶν ἀδικεῖν ἔπαυσε, πείσας ὅτι οὐ δυνασοῦνται λαθεῖν, ὅταν ἐπ' αὐτὸν ἔλθωντι, τοὺς δὲ μὴ ἐπισταμένους, ἐν αὐτῷ δηλώσας ἀδικοῦντας, ἐκώλυσεν ἀδικῆσαι.

to engendering political harmony and making citizens just. Hero characterizes mechanics – and, in particular, artillery construction – as having taught human beings how to live tranquilly, pneumatics as revealing that micro-voids exist, and geometry as the basis for distributing land justly. Ptolemy portrays mathematics as superior to physics and theology with respect to epistemology and ethics. Furthermore, both Hero and Ptolemy define mathematical sciences as types, or parts, of philosophy. Hence, for these ancient Greek mathematicians, no distinction held between the mathematical sciences and philosophy. The mathematical sciences were philosophy, and they achieved the most significant objectives of philosophy: epistemological and ethical.

Given this evidence for the characterization of the mathematical sciences as philosophy – from soon after the emergence of high-level mathematics in the Greek world to the imperial period – it is necessary to re-examine the relationship between the mathematical sciences and philosophy in Greek antiquity.[44] By taking into account the historical actors' categories, as well as the substance of their claims, it is possible to overturn a long-standing anachronistic assumption, which pervades multiple academic disciplines. In both the history of mathematics and the history of philosophy, scholars have assumed that the mathematical sciences and philosophy were in Greek antiquity, as they are today, distinct disciplines. In fact, their relationship was far more complex than our histories have assumed and, by putting into doubt the anachronistic assumption that they were simply distinct, we come to a richer and more accurate understanding of the historical relationship between the mathematical sciences and philosophy in Greek antiquity.

References: Primary Sources

Archytas (2016). Fragments. In: *Western Greek Thinkers, part 1. Early Greek Philosophy, vol. 4*, André Laks and Glenn W. Most (eds). Cambridge, MA: Harvard University Press.

Aristotle (1924). Metaphysics. In: *Aristotle's Metaphysics*, 2 volumes, William David Ross (ed). Oxford: Clarendon, Oxford. Corrected edition 1953, reprinted 1970.

Cicero (1922). Academica. In: *Academicorum reliquiae cum Lucullo*, Otto Plasberg (ed). Leipzig: Teubner. Reprinted 1961.

[44] On the chronological beginning of advanced ancient Greek mathematics, see Netz (1999, 271–275).

Cicero (1928–1930). De finibus bonorum et malorum. In: *Cicéron, Des terms extrêmes des biens et des maux*, 2 volumes, Jules Martha (ed). Paris: Les Belles Lettres.

Diogenes Laertius (1964). Lives. In: *Diogenis Laertii vitae philosophorum*, 2 volumes. Oxford: Clarendon. Reprinted 1966.

Euclid (1969–1973). Elements. In: *Euclidis elementa*, 4 volumes, Euangelos S. Stamatis (ed). Leipzig: Teubner.

Hero (1899). Pneumatica. In: *Heronis Alexandrini opera quae supersunt omnia, vol. 1*, Wilhelm Schmidt (ed). Leipzig: Teubner, pp. 2–332.

Hero (1903). Metrica. In: *Heronis Alexandrini opera quae supersunt omnia, vol. 3*, Hermann Schöne (ed). Leipzig: Teubner, pp. 2–184.

Hero (1918). Belopoeica. In: *Herons Belopoiika, vol. 2*: Hermann Diels, Erwin A. Schramm (eds). Abhandlungen der preussischen Akademie der Wissenschaften, Berlin: Reimer, pp. 5–55.

Nicomachus (1866). Introduction to arithmetic. In: *Nicomachi Geraseni Pythagorei introductionis arithmeticae libri ii*, Richard G. Hoche (ed). Leipzig: Teubner.

Nicomachus (1895). Manual of Harmonics. In: *Musici scriptores Graeci*, Karl von Jan (ed). Leipzig: Teubner, pp. 236–265. Reprinted 1962, Olms: Hildesheim.

Plato (1902). Timaeus. In: *Platonis Opera, vol. 4*, John Burnet (ed). Oxford: Clarendon Press. Reprinted 1968.

Plato (2003). Republic. In: *Platonis Rempublicam*, Simon R. Slings (ed). Oxford: Clarendon Press.

Proclus (1873). In primum Euclidis elementorum librum commentarii. In: *Procli Diadochi in primum Euclidis elementorum librum commentarii*, Gottfried Friedlein (ed). Leipzig: Teubner.

Pseudo-Plato (1907). Epinomis. In: *Platonis opera, vol. 5*, John Burnet (ed). Oxford: Clarendon Press. Reprinted 1967.

Ptolemy (1898). Almagest. In: *Syntaxis mathematica. Claudii Ptolemaei opera quae exstant omnia, vols 1.1–1.2*, Johan Ludvig Heiberg (ed). Leipzig: Teubner.

Ptolemy (1961). On the Kritêrion and Hêgemonikon. In: *Claudii Ptolemaei opera quae exstant omnia, vol 3.2*, 2nd ed., Friedrich Lammert (ed). Leipzig: Teubner, pp. 3–25.

Ptolemy (2006). Geography. In: *Klaudios Ptolemaios Handbuch der Geographie, vols 1–2*, Gerd Grasshoff and Alfred Stückelberger (eds). Basel: Schwabe.

Theon (1878). On Mathematics Useful for the Understanding of Plato. In: *Theonis Smyrnaei philosophi Platonici expositio rerum mathematicarum ad legendum Platonem utilium*, Eduard Hiller (ed). Leipzig: Teubner.

References: Secondary Sources

Barnes, Jonathan (ed) (1984). *The Complete Works of Aristotle: The Revised Oxford Translation*, 2 volumes. Princeton, NJ: Princeton University Press.

Bonelli, Maddalena (2016). Proclus et la dialectique scientifique. In: *Logique et Dialectique dans l'Antiquité*, Jean-Baptiste Gourinat and Juliette Lemaire (eds). Paris: Librarie Philosophique J. Vrin, pp. 397–421.

Cuomo, Serafina (2000). *Pappus of Alexandria and the Mathematics of Late Antiquity*. Cambridge: Cambridge University Press.

Cuomo, Serafina (2001). *Ancient Mathematics*. London: Routledge.

Dillon, John (1977). *The Middle Platonists: A Study of Platonism, 80 B.C. to A.D. 220*. London: Duckworth.

Evans, James and Carman, Christiàn Carlos. (2014) Mechanical astronomy: A route to the ancient discovery of epicycles and eccentrics. In: *From Alexandria, through Baghdad: Surveys and Studies in the Ancient Greek and Medieval Islamic Mathematical Sciences in Honor of J.L. Berggren*, Nathan Sidoli and Glen Van Brummelen (eds). Berlin/Heidelberg: Springer, pp. 145–174.

Feke, Jacqueline (2012). Ptolemy's defense of theoretical philosophy. *Apeiron* **45** (1), 61–90.

Feke, Jacqueline (2014). Meta-mathematical rhetoric: Hero and Ptolemy against the philosophers. *Historia Mathematica* **41** (3), 261–76.

Feke, Jacqueline (2018). *Ptolemy's Philosophy: Mathematics as a Way of Life*. Princeton: Princeton University Press.

Guthrie, William K.C. (1962–1981). *A History of Greek Philosophy*, 6 volumes. Cambridge: Cambridge University Press.

Hadot, Pierre (1995). *Qu'est-ce que la philosophie antique?* Paris: Gallimard.

Hadot, Pierre (2002). *What is Ancient Philosophy?* (translated by M. Chase). Cambridge, MA: Belknap.

Huffman, Carl A. (2005). *Archytas of Tarentum: Pythagorean, Philosopher and Mathematician King*. Cambridge: Cambridge University Press.

Inwood, Brad and Gerson, Lloyd P. (eds) (1994). *The Epicurus Reader: Selected Writings and Testimonia*. Indianapolis: Hackett.

Jones, Alexander (2015). Theon of Smyrna and Ptolemy on celestial modelling in two and three dimensions. In: *Mathematizing Space: The Objects of Geometry from Antiquity to the Early Modern Age*, Vincenzo De Risi (ed), Cham: Birkhäuser, pp. 75–103.

Jones, Alexander (2016). Translating Greek astronomy: Theon of Smyrna on the apparent motions of the planets. In: *Translating Writings of Early Scholars in the Ancient Near East, Egypt, Greece and Rome: Methodological Aspects with Examples*, Annette Imhausen and Tanja Pommerening (eds). Berlin: De Gruyter, pp. 465–505.

Netz, Reviel (1997). Classical mathematics in the classical Mediterranean. *Mediterranean Historical Review* **12** (2), 1–24.

Netz, Reviel (1999). *The Shaping of Deduction in Greek Mathematics: A Study in Cognitive History*. Cambridge: Cambridge University Press.

Netz, Reviel (2016). Mathematics. In: *A Companion to Science, Technology, and Medicine in Ancient Greece and Rome*, Georgia L. Irby (ed). Chichester, West Sussex: Wiley–Blackwell, pp. 79–95.

Nicomachus of Gerasa (1926). *Introduction to Arithmetic* (translated by Martin Luther D'Ooge). New York: Macmillan.

O'Meara, Dominic J. (1989). *Pythagoras Revived: Mathematics and Philosophy in Late Antiquity*. Oxford: Clarendon Press.

Petrucci, Federico M. (2018). Theon of Smyrna: Re-thinking Platonic mathematics in middle Platonism. In: *Brill's Companion to the Reception of Plato in Antiquity*, Harold Tarrant, Danielle A. Layne, Dirk Baltzly, François Renaud (eds). Leiden: Brill, 143–155.

Plato (1997). *Complete Works*, John M. Cooper (ed). Indianapolis/Cambridge: Hackett.

Plato (2004). *Republic*, translated and edited by C.D.C. Reeve. Indianapolis/Cambridge: Hackett.

Popper, Karl (1963). *Conjectures and Refutations: The Growth of Scientific Knowledge*. London: Routledge & Kegan Paul.

Sedley, David (1976a). Epicurus and his professional rivals. In: *Études sur l'Épicurisme Antique*, Jean Bollack and André Laks (eds). Lille: Publications de Centre de Recherche Philosophique de l'Université de Lille III, pp. 121–59.

Sedley, David (1976b). Epicurus and the mathematicians of Cyzicus. *Cronache Ercolanesi* **6**, 23–54.

Zhmud, Leonid and Kouprianov, Alexei (2018). Ancient Greek mathêmata from a sociological perspective: a quantitative analysis. *Isis* **109** (3), 445–72.

5

Productive anachronism: on mathematical reconstruction as a historiographical method

Martina R. Schneider
Johannes Gutenberg-Universität, Mainz

Es giebt also (man kann es eigentlich und kühn sagen) im Universum zu Einer Zeit unzählbar-viele Zeiten (Herder, 1799, p. 120f).

Abstract: The spectrum of practices of mathematical reconstruction is explored on the basis of a case study on a partly successful mathematical reconstruction of the Chinese Remainder Theorem. In the nineteenth century Ludwig Matthiessen reconstructed two versions of this theorem on the basis of a corrupted secondary source concerning ancient Chinese mathematics. He identified the more restricted version of the theorem with a Gaussian approach, whereas the other more general one was described as something new surpassing contemporary European mathematical achievements. I identify and compare two different types of mathematical reconstructions in Matthiessen's contributions, and explore their historiographic functions. To capture the relation between mathematical reconstruction and anachronism the time scheme in the case study is analysed and linked to the concept of pluritemporality (Achim Landwehr). This more complex perspective on the category of time in historical research suggests that anachronism should be reconceptualized. It allows for a discussion of some of the conditions under which mathematical reconstructions can be used in a historiographically sensitive way in a different setting. I argue that this kind of historiographically sensitive mathematical reconstruction can be regarded as a productive historiographical method.

[a] This research was partly funded by the European Research Council under the European Union's Seventh Framework Programme (FP7/2007-2013)/ERC Grant Agreement No. 269804 (SAW).

[b] From *Anachronisms in the History of Mathematics: Essays on the Historical Interpretation of Mathematical Texts*, edited by Niccolò Guicciardini © 2021 Cambridge University Press.

5.1 Introduction

In recent years there has been some discussion on the role of anachronism in historiography. Instead of criticizing the use of anachronism in history and calling it a "sin" (L. Febvre), some historians reflect on anachronism historiographically and explore to what extent a kind of modified anachronism (so-called "deliberate" or "controlled" anachronism) can enrich historiography.[1]

More or less the same applies to the practice of mathematical reconstruction that can be seen as a special form of anachronism in the history of mathematics. In 1975 S. Unguru contrasted "logical reconstruction" (in the history of Greek mathematics) with "historical reconstruction." (Mathematical-)Logical reconstruction was attacked as "anachronistic" and as "*non-history*."[2] At that time and in the decades thereafter, during the processes towards the professionalization and institutionalization of the history of mathematics, "mathematical(-logical) reconstruction" became a slogan to distinguish between "good" and "bad" historiography. This can be seen e.g., in a title like "Mathematical reconstructions out, textual studies in: 30 years in the historiography of Greek mathematics."[3] However, its author K. Saito as well as Unguru did not dismiss reconstructions as such. In recent years there have been some historians who, again, point to the fruitfulness of rational reconstruction for the history of science when applied carefully,[4] or who advocate a rational history of mathematics.[5] The different adjectives used in connection with reconstruction in these debates point to a wide spectrum of different types of reconstruction in the history of mathematics and science.

The aim of this chapter is two-fold: First, an initial attempt at exploring the spectrum of mathematical reconstruction will be made. What kind of activities in the historiography of mathematics can be subsumed under mathematical reconstruction? Are there different practices or types of mathematical reconstruction, and if so, how can they be characterized? What means do they use? What assumptions do they rely upon? This

[1] See e.g. Loraux (1993); Rancière (1996), the special issue on anachronism of *Scientia poetica* (8/2004), especially the instructive overview and analysis by Spoerhase (2004), Arni (2007); Landwehr (2013), Rancière's translations into German and French (both 2015). Another earlier and valuable discussion of anachronism with respect to the history of science is Kragh (1987, pp. 88–107).

[2] Unguru (1975, p. 92, emphasis in the original).

[3] Saito (1998).

[4] Jardine (2000); Landwehr (2013); Loison (2016).

[5] Blåsjö (2014).

will provide the basis for a better historiographic understanding of the practice of mathematical reconstruction.

Second, some more general issues addressed in recent debates will be discussed with respect to mathematical reconstructions. What are the functions of mathematical reconstructions in the history of mathematics? In what way is mathematical reconstruction anachronistic? Or rather: What kind of time schemes are at work in mathematical reconstructions, and, more generally, in the research on the history of mathematics? What is the potential of mathematical reconstruction for the history of mathematics? The answers I will develop here should be seen as a starting point and an invitation to explore and reflect further on the historiography of mathematics.

In this chapter the historiographical practice of mathematical reconstruction is approached by analysing an example taken from the history of Chinese mathematics, namely the mathematical reconstruction of the dayan rule – today known as the Chinese Remainder Theorem – by Ludwig Matthiessen in the 1870s and 1880s. The contextualization and analysis of Matthiessen's contributions in Section 5.2 show how sources and contemporary mathematical knowledge and research come together in the reconstruction. Examples and comparisons play vital and diverse roles in Matthiessen's reconstruction and in his historiography. In Section 5.3.1 I will argue that two different types of mathematical reconstruction can be distinguished in the example, and point to the broader spectrum of practices of mathematical reconstruction. Moreover, the example turns out to be partly historiographically successful (Section 5.3.2) and intrinsically linked to anachronism (Section 5.3.3).

Finally in Sections 5.3.4 and 5.4, the focus is on the complex temporal structure of Matthiessen's reconstructions and its implications. This can be captured very well with the concept of "pluritemporality" introduced by Achim Landwehr (2013) in his paper on anachronism.[6] Adapting and extending Landwehr's argument about the productivity of anachronism I will explore how and under what conditions a certain type of mathematical reconstruction can be perceived as a productive method in the historiography of mathematics.

[6] This concept was later integrated in the more general approach to time via "chronoferences" (Landwehr, 2016).

5.2 Reconstructing a Chinese approach to indeterminate analysis

5.2.1 Background: Sources and their reception

During the nineteenth century very little was known about Chinese mathematics and science in Europe that was based on sources. Only very few Chinese mathematical sources were available to European scholars. In 1856 Karl Biernatzki published a paper entitled "The arithmetics of the Chinese" and written in German in the *Journal für reine und angewandte Mathematik* (JRAM). Biernatzki was no mathematician and had never published anything on the history of mathematics before. He had studied theology. In 1856 he was employed as secretary of the Protestant "Central Commission for Home Mission" (Sekretär des Zentralausschusses für Innere Mission) in Berlin. Before that he had been editor of two German missionary journals on Asia for which he regularly translated into German reports published in British and American (missionary) newspapers, journals and books. Biernatzki (1856) was also based on an English newspaper article which he acknowledged in a footnote. However the footnote contained unreliable bibliographical details and no mention of the author's name.

The original paper that Biernatzki referred to was written by the British Protestant missionary Alexander Wylie, who was based in China. It had been published in nine parts in the *North China Herald* – a newspaper with a focus on economics written in English and published in Shanghai – during August and November 1852.[7] Biernatzki's paper was not a straightforward or careful translation of the original. Biernatzki made many changes, omissions and additions. Later parts of Biernatzki's distorted version of Wylie's paper were republished in French by the French mathematicians Olry Terquem and Joseph Bertrand in *Bulletin de bibliographie, d'histoire et biographie mathématiques*, in *Nouvelles annales de mathématiques* and in *Journal des savants*.[8] Again, these French republications differ from faithful translations.[9] The genesis of Wylie's paper, its "translations" into German and French as well as its reception in Europe through Biernatzki, Terquem and Bertrand are currently being investigated in detail by Karine Chemla, Alfred Zhihui Chen and

[7] Wylie (1852).

[8] Terquem (1862a,b); Bertrand (1869a,b).

[9] Matthiessen (1881a, p. 35) mentioned in passing that Terquem had difficulties in translating Biernatzki's text, but without giving any details.

myself.[10] Figure 5.1 gives a schematic overview of the "translations," interpretations and reconstructions in Europe related to Wylie (1852) between 1856 and 1882.

In Germany Biernatzki's paper was taken up rather quickly by historians of mathematics. They focussed – as I will do in this chapter – on the dayan rule.[11] Moritz Cantor (1858) transformed Biernatzki's description into a mathematical formula which, however, did not work in general. And from this lack of generality he concluded that:[12]

> thus especially with respect to the studies on indeterminate analytics the Chinese seemed to be rather behind other contemporary cultural peoples (Cantor, 1858, p. 336, transl. MS).

In Hermann Hankel's posthumously published monograph *Zur Geschichte der Mathematik im Alterthum und Mittelalter* (1874) an appendix is dedicated to Chinese mathematics. Hankel compared the Chinese dayan rule to the Indian kuttaka method. The Indian kuttaka method had been mentioned by Wylie and Biernatzki, too.[13] Hankel claimed that the two methods were the same, and argued that the Chinese had learned it from the Indians. Both, the Chinese and the Indians, had applied it to chronology and constellations, but the Chinese had also used it for divination. For Hankel this example was of high importance as it showed how the history of mathematics could contribute to more general questions concerning the transmission of knowledge and in this particular case to the deconstruction of the myth of Chinese isolation.

Thus, when Matthiessen became interested in the history of indeterminate analysis, "the" Chinese approach was seen neither as a solution of the problem for any kind of numbers nor as originally Chinese in this community of historians of mathematics. It is exactly these interpretations that Matthiessen's reconstruction will challenge.

[10] This research started as a part of the ERC project "Mathematical sciences in the ancient world" (SAW). See Chen (2017) on Wylie (in Chinese). For a first comprehensive analysis see Libbrecht (1973).

[11] The transliteration of Chinese names varies. Biernatzki used "Ta-yen" instead of "dayan." Here I will use today's standard.

[12] "[. . .] so scheinen gerade in Untersuchungen der unbestimmten Analytik die Chinesen hinter anderen gleichzeitgen Culturvölkern eher zurück gewesen zu sein."

[13] Wylie (1852, p. 46, Oct. 23, 1852), Biernatzki (1856, p. 83).

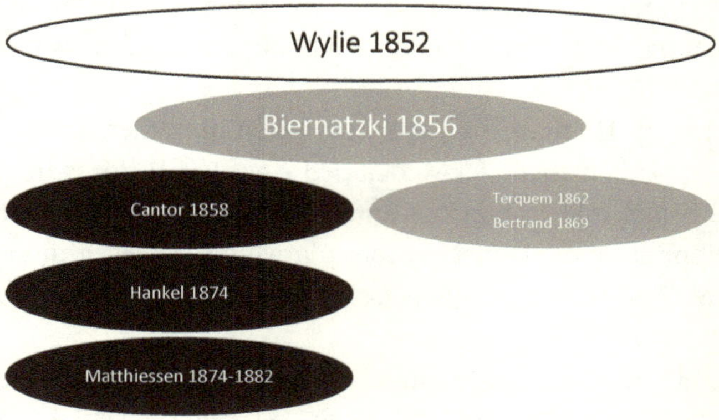

Figure 5.1 Overview of texts related to Wylie (1852) and published in Europe between 1856 and 1882: "translations" of Wylie (1852) or Biernatzki (1856) in light grey, interpretations and reconstructions of Biernatzki (1856) in black.

5.2.2 *On Ludwig Matthiessen's biography*

Heinrich Friedrich Ludwig Matthiessen (1830–1906) was born in Fissau (today part of Eutin in Eastern Holstein, Germany) as the son of a school teacher.[14] He studied mathematics, astronomy and science, in particular physics, at the university of Kiel from 1852 onwards. In 1857 he received his doctoral degree there. Later he habilitated in mathematical physics, and became lecturer (Privatdozent) of physics in Kiel. From 1859 to 1873 he taught mathematics and physics at grammar schools (Gymnasium) in Jever and Husum. During this time, he published a couple of papers on diverse aspects of algebraic equations in the journal *Zeitschrift für Mathematik und Physik* (ZMP) edited by O. Schlömilch and M. Cantor.[15] In 1866 Matthiessen published a short collection of 51 different methods of solving algebraic equations by substitutions. This was motivated by his teaching experience. Among his explicitly expressed aims were to give references to the primary sources containing these solutions and to clarify priority claims.[16] He continued publishing on this topic.

In 1873 he was appointed assistant professor (Extraordinarius) of physics and in the following year as the first full professor (Ordinarius) of physics at the university of Rostock. In 1875 he became director of the

[14] On Matthiessen's biography see Alberti (1868, 1885); Mahnke (1991).

[15] Matthiessen (1861, 1863b,a, 1864, 1865).

[16] Matthiessen (1866, pp. 1f.).

newly established institute of physics, and in 1888 of the astronomical-meteorological observatory. Later he was dean of the department and rector of the university. His research areas concerned algebra, optics, physics and physiology, and in particular the interdisciplinary study of the eye.

In the 1870s Matthiessen continued working on algebraic equations. His research culminated in his extensive, almost 1000-pages long monograph on the history of equations: *Grundzüge der antiken und modernen Algebra der litteralen Gleichungen* (1878).[17] In Section 8 at the end of the book Matthiessen included a list of chronologically ordered sources. He considered this section as particularly precious for historians. The list starts with two pages on ancient Chinese sources as the oldest contributions. Then the list continues with ancient Egyptian, Indian, Greek, Arabic and Persian sources before giving references to almost exclusively Western approaches published after 1600. This historical investigation covering a time span of over 3000 years was new compared to his 1866 publication. However, its presentation in the form of a chronological bibliographical list with very little commentary was rather old-fashioned at the time.

This list can be interpreted as the realization of the motto that Matthiessen had chosen for his studies in the history of equations.[18] The motto is a quotation from Alexander von Humboldt's second volume of the *Kosmos* which is printed in bold and given in context:[19]

> It is the conviction that **the conquered possession is only a very minor portion of what a free humankind will reach through continuous and progressive work and joint formation in the coming centuries** that is more stimulating and more adequate to the idea of a great destiny of our humankind. Everything researched is only a step to something higher in the fatal course of events (von Humboldt, 1847, p. 399, Matthiessen's motto in boldface, transl. MS).

It shows that Matthiessen believed in a history of continuous progress of humankind. The bibliographical list in Section 8 of (Matthiessen, 1878c) on the history of equations can be seen as a demonstration as well as a result of this belief.

[17] Matthiessen (1878c).

[18] Matthiessen (1866, 1878c).

[19] "Belebender und der Idee von der grossen Bestimmung unseres Geschlechtes angemessener ist die Ueberzeugung, dass der eroberte Besitz nur ein sehr unbeträchtlicher Theil von dem ist, was bei fortschreitender Thätigkeit und gemeinsamer Ausbildung die freie Menschheit in den kommenden Jahrhunderten erringen wird. Jedes Erforschte ist nur eine Stufe zu etwas Höherem in dem verhängnissvollen Laufe der Dinge." In fact Matthiessen had chosen this motto already (Matthiessen, 1866).

Some of his research in the history of equations concerned the Chinese *dayan* rule of indeterminate analysis. In addition to a few pages in his monograph,[20] Matthiessen devoted a total of nine publications to this topic in various German journals and one French one dedicated to mathematics, science, or to the teaching of these disciplines and published between 1874 and 1882.[21] In the following I will outline how he reconstructed the dayan rule without taking any Chinese sources into account. I will identify two different types or practices of mathematical reconstruction that are at work in Matthiessen's approach.

5.2.3 Matthiessen's reconstructions of the dayan rule

The dayan rule

The dayan rule is introduced by Wylie by giving and solving an explicit problem:

> Given an unknown number, which when divided by 3, leaves a remainder of 2; when divided by 5, it leaves 3; and when divided by 7, it leaves 2; what is the number? Answer: 23 (Wylie, 1852, Oct. 16, p. 43).

In modern notation the problem can be described as: Find a number $n \in \mathbb{N}$ such that

$$\begin{aligned} n &\equiv 2 \mod 3 \\ n &\equiv 3 \mod 5 \\ n &\equiv 2 \mod 7. \end{aligned}$$

Wylie went on to explain where this problem can be found in ancient Chinese texts, how it was solved by different scholars and gave some more examples of application. This is reflected in the following structure deduced from Wylie's text and also present in Biernatzki's, but introduced by those of us working on the project:

Part A based on the *Mathematical Classic by Master Sun*[22] the problem can be solved by "a special rule [. . .] concise and elliptical" followed by "a more general note" (Wylie, 1852, p. 43, sp. 1, Oct. 16, 1852), (Biernatzki, 1856, pp. 77f).

[20] Matthiessen (1878c, p. 291, pp. 964f.).

[21] Matthiessen (1874, 1876a,b, 1878c, 1879, 1881c,b,a, 1882) – some of these are just reprints. Two more publications touch upon the topic indirectly (Matthiessen, 1878b,a).

[22] Wylie: "Sun Tsze's Arithmetical Classic."

Part B the *Mathematical Writings in Nine Chapters* (1247) by Qin Jiushao[23] "put us in full possession of the principle, and enable us to unravel the meaning of the above mysterious assemblage of numerals" (Wylie, 1852, p. 43, sp. 1, Oct. 16, 1852), (Biernatzki, 1856, pp. 78f).

Part C "various applications of this theory" (9 problems; based on Qin, who took up some from Yi Xing[24] according to Wylie; Biernatzki wrongly attributed all of the examples to Yi Xing) (Wylie, 1852, p. 43, sp. 1f, Oct. 16, 1852; p. 46, sp. 4f, Oct. 23, 1852), (Biernatzki, 1856, pp. 79–82).

Wylie made clear that Part B explained Part A. Part B contained a procedure called "Finding Unity," that went back to Qin. Wylie (1852, p. 43, Oct. 16) mentioned the name of the procedure, but did not include it. So it is not clear what the procedure actually does.

Biernatzki did not mention the name of the unclear procedure. Instead he gave an explanation of the procedure in Part B in modern mathematical notation. It is this explanation which Matthiessen proved to be wrong (see Phase 1). Biernatzki also modified Part C considerably.

Phase 1: Spotting and correcting a mistake, and various comparisons (1874–1878)

In 1874 Matthiessen wrote directly to Cantor to tell him that there was a mistake in Biernatzki's article and that hence Cantor's interpretation of the Chinese achievements in arithmetics in Cantor (1858) was wrong.[25] Cantor had some parts of this letter published under the heading "On the algebra of the Chinese" in the ZMP. Matthiessen (1874) argued as follows: first he gave a general reason for the Chinese capability to calculate. It concerned their practical astronomical knowledge:[26]

> [...] how would it have been possible that for many centuries the Chinese arithmeticians had been applying rules for calculating their cycles if those had not been generally valid (Matthiessen, 1874, p. 271, transl. MS).

Then he blamed Biernatzki for wrongly translating from the English original. His argument was not based on a comparison between Wylie's

[23] Wylie: "*Nine sections of the art of numbers* by Tsin Keu chaou."

[24] Wylie: "Yih Hing"; Biernatzki: "Yih King."

[25] Unfortunately the original letter is not contained in Cantor's Nachlass in Heidelberg.

[26] "[...] wie wäre es es auch möglich, dass die chinesischen Arithmetiker bei der Berechnung ihrer Cykeln Regeln angewendet hätten viele Jahrhunderte hindurch, wenn sie nicht allgemeine Giltigkeit hatten."

Wir haben hier offenbar die Gauss'sche Methode vor uns und die Darstellung in ihren Zeichen ist demnach:

$$x \left\{ \begin{array}{l} \equiv n_1 \ (mod\, a_1) \\ \equiv n_2 \ (mod\, a_2). \\ \equiv n_3 \ (mod\, a_3) \end{array} \right.$$

Man bilde drei Hilfszahlen (chinesisch *tsching su*, das ist Multiplicator) h_1, h_2, h_3 in der Weise, dass

$$h_1 a_2 a_3 \equiv 1 \ (mod\, a_1),$$
$$h_2 a_1 a_3 \equiv 1 \ (mod\, a_2),$$
$$h_3 a_1 a_2 \equiv 1 \ (mod\, a_3),$$

alsdann ist

$$x = h_1 n_1 a_2 a_3 + h_2 n_2 a_1 a_3 + h_3 n_3 a_1 a_2.$$

Setzen wir die Zahlen ein, so ist

$$h_1 . 5 . 7 \equiv 1 \ (mod\, 3),$$

und dies drücken Sun tsze's Worte aus: „für 1 durch 3 gewonnen, setze 70"; der Multiplicator h_1 ist nämlich 2. Da aber zufälligerweise der Rest (*ki*) von $\frac{35}{3}$ auch 2 ist, so hat Biernatzki diese verwechselt.

Figure 5.2 Excerpt from Matthiessen (1874, p. 271).

original paper and Biernatzki's translation of it, let alone a comparison with original Chinese sources, instead it pointed to the lack of consistency of Biernatzki's text. Matthiessen got his clue from what Biernatzki had called "the aphoristic note" (and Wylie "the more general note," part A), namely

> For 1 obtained by 3, set down 70; for 1 obtained by 5 set down 21; for 1 obtained by 7, set down 15; when the sum is 106 or above, subtract 105 from it, and the remainder is the number required (Biernatzki, 1856, p. 78; this is Wylie's original English version).

He recognized in this the Gaussian method as explained in *Disquisitiones Arithmeticae* (§32–36) in 1801.[27] Matthiessen gave the solution of the general and the concrete problem in contemporary (Gaussian) notation and style (see Figure 5.2). He identified what Gauss had called an "auxiliary number" (Hilfszahl) with the Chinese concept that according to Biernatzki was called the "multiplicator" ("Multiplicator") or "tsching su."[28] Then he showed how the Gaussian construction of the solution was in line with "the aphoristic note" quoted above.

[27] For the rather slow reception of (Gauss, 1801) see (Goldstein and Schappacher, 2007a,b). Matthiessen (1876a,b) also referred to a corresponding section (§25) in Richard Dedekind's first edition of Dirichlet's lectures on number theory (Dedekind, 1863). Matthiessen (1881a,c) mentioned this, too, but without specifying the edition. Section §25 was revised in the second edition (Dedekind, 1871).

[28] Wylie used "Ching suh" or "Multiplying term."

Matthiessen identified Biernatzi's mistake. He made clear that for the given numerical example Biernatzki's approach to the calculation of the multiplicator gave the numerically correct answer, but this answer was based, as Matthiessen (as well as Cantor and Hankel) pointed out, on a wrong procedure. Thus, Matthiessen claimed, this was the reason why Biernatzki got mixed up about the procedure.

After showing how the Gaussian method was actually the Chinese method, Matthiessen stated that Sun tsze (Master Sun) was the discoverer of indeterminate analysis, and dated him around 200 CE. Then he claimed that the Indians began dealing with that subject more than a century later (than Master Sun) and that the Indian kuttaka method "coincided" with an alternative method by Euler. Matthiessen concluded that the Indian kuttaka method was different from the Chinese dayan method. He pointed out that "the method of the Chinese as well as their exercises are original" (Matthiessen, 1874, p. 271, transl. MS). Thus, he disagreed with Hankel's interpretation that identified the dayan rule with the kuttaka (see p. 135) without mentioning Hankel.

After this short note, Matthiessen gave several talks on the topics and published more comprehensive papers in journals related to teaching. In Matthiessen (1876a)[29] he gave a little bit more mathematical, historical and historiographical information. He had written to Alexander Wylie in Shanghai who had answered him in October 1874. This is the first time Wylie's name is mentioned in connection with Biernatzki's paper.[30] Apparently Matthiessen (1876a, p. 126) had asked him for his original paper as well as more translations of ancient Chinese sources and more historical studies. However, Wylie could not help him.[31]

Matthiessen's exposition of the historiography is more detailed in that Matthiessen more closely followed part B of (Biernatzki, 1856), i.e. Qin's *Mathematical Writings*, introducing more Chinese terminology. In Matthiessen (1876a, p. 128) he explicitly stated the conditions underlying the Gaussian method, that were explicitly mentioned by Gauss as well as by P.G. Lejeune Dirichlet (in R. Dedekind's edition): namely that the moduli be relatively prime to each other. He then claimed that

29 (Matthiessen, 1876b) is an almost identical publication. It contains everything mentioned in this paragraph, too.

30 It is unclear how Matthiessen found out Wylie's name. He could have written to Biernatzki.

31 Wylie's 1852 paper was reprinted in Europe several times (Wylie, 1864b,a,c, 1882, 1897). It is unclear if, and if so when, Matthiessen got hold of one of these reprints. Like Libbrecht (1973, p. 313, fn. 21; p. 314, fn. 25) we believe Matthiessen did not come across any of the reprints while publishing on the topic.

the Chinese were also dealing with problems that did not satisfy this condition.[32] However, he did not go into it any further at the time.

Phase 2: Reconstructing the general rule and the transformation into a modern research object (1879–1882)

In 1879 Matthiessen returned to his claim (about non-relatively prime moduli) and published a paper entitled "On an ancient solution to the so-called remainder problem in modern presentation" (Matthiessen, 1879) in the *Zeitschrift für mathematischen und naturwissenschaftlichen Unterricht* (ZMNU) – a journal for teachers that occasionally published papers concerning the history of science. His explicit intention was to show that contemporary papers that deal with the problem of simultaneous congruences of the first degree (like the one by von Schäwen, 1878a)[33] "accorded"[34] with the ancient dayan rule. Matthiessen thought that any differences were only due to either the introduction of modern symbolism or the choice of the starting point of the calculation.[35]

Matthiessen's second aim was to "deduce"[36] a generalized dayan rule dealing with non-relatively prime moduli and going back to ancient Chinese mathematicians. He remarked that the original deduction of the dayan rule by Chinese mathematicians and astronomers was unknown. According to Matthiessen the reason for this was that "a scientific treatment of mathematical problems was completely alien to them."[37] He chose to restrict historical remarks and references almost exclusively to the introduction and the footnotes,[38] and to use exclusively modern notation and concepts for his exposition.

In the following the main ideas of Matthiessen's deduction are out-

[32] Matthiessen (1876a, p. 128). In Matthiessen (1876b, p. 80) Matthiessen altered this remark slightly by evaluating Yi Xing's (Yih-Hing) approach as difficult to understand ("schwer verständlich").

[33] There are also others, e.g. by F. Reuschle jun. (1874).

[34] "übereinstimmen" (Matthiessen, 1879, p. 106).

[35] "[Recent methods of solutions] only differ from the dayan rule either through the introduction of modern symbolism or through the choice of a special starting point of the applied calculus; however, the root form is always the same" (Matthiessen, 1879, p. 106, transl. MS). German original: "[Auflösungsmethoden des Restproblemes in neuerer Zeit] unterscheiden sich von der Tayen–Regel entweder nur durch die Einführung einer modernen Symbolik oder in der Wahl eines besonderen Ausgangspunktes des angewandten Calcüls; die Wurzelform dagegen ist überall die gleiche."

[36] "herleiten" (Matthiessen, 1879, p. 107).

[37] "[. . .] da eine wissenschaftliche Behandlung mathematischer Probleme denselben vollständig fremd war" (Matthiessen, 1879, p. 107).

[38] Matthiessen referred the reader to his previous papers (Matthiessen, 1874, 1876b) and references therein. In a footnote Matthiessen (1879, p. 106) included additional historical references, to Biot (1839) and Schäfer (1831).

lined. Starting with the general problem

$$N = m_1 x_1 + r_1 = m_2 x_2 + r_2 = \cdots$$

Matthiessen distinguished two cases: the moduli m_i being pairwise prime, and non-relatively prime moduli. He claimed that both cases were treated by "both old Chinese and modern authors."[39]

In the first case Matthiessen transformed the simultaneous congruences $N \equiv r_i (\text{mod } m_i)$ into one single congruence with respect to the module $A = m_1 m_2 m_3 \cdots m_u$. This was done by multiplying each equation first with A/m_i and then by integers k_i that were to be determined later. Finally the different equations were added up to form one single congruence

$$N \sum_1^u k_z \frac{A}{m_z} \equiv \sum_1^u r_z k_z \frac{A}{m_z} \quad (\text{mod } A).$$

If $\sum_1^u k_z \frac{A}{m_z} \equiv 1 (\text{mod } A)$ for some integer numbers k_z, then N can be determined basically by solving systems of linear equations by use of determinants. The condition, which is equivalent to the one reconstructed by Matthiessen previously,[40] and the way of solution appear, as Matthiessen pointed out, in von Schäwen (1878a, p. 116).

The second case of non-relatively prime moduli worked similarly. Instead of using the product of the moduli m_i as the new modulus A common to all (transformed) equations, the least common multiple of the m_i, designated by B, was used. An analogous congruence and condition (with B instead of A) were the result of solving the problem.

At the end of his paper Matthiessen illustrated the second case with a concrete numerical example that seems to be new. In this example the determination of the k_z was not straightforward and required one further case distinction: (i) k_z that were smallest in absolute term, and (ii) the smallest non-negative solutions. For both cases Matthiessen determined the solution N to the problem using the same method he had outlined before. The smaller positive number of the two solutions was the solution sought.

It is in this paper that for the first time the case of non-relatively prime moduli was treated extensively by Matthiessen.[41] Matthiessen is

39 Matthiessen (1879, p. 107).

40 However, there is a misprint in the reconstructed condition: instead of A the correct modulus is m_z for the congruences $k_z \frac{A}{m_z} \equiv 1$.

41 Matthiessen (1879, pp. 108ff.).

explicit neither about how he arrived at this "deduction" that he ascribed to the ancients nor about his sources. The link to (Biernatzki, 1856) is extremely weak since Matthiessen's approach does not much resemble the calculations in example 1 (part C) presented by Biernatzki. Perhaps Matthiessen recognized that the least common multiple of the moduli was calculated and used in this example – a concept that was used explicitly by Gauss (1801, §32) and Dedekind (1863, §25, p. 58).[42] In addition to that there is a close structural similarity with Gauss's approach: Gauss (1801, §32) pointed out that the general case (with an arbitrary number of congruences of arbitrary moduli) could be reduced to solving one congruence modulo the least common multiple of the original moduli. So this time Matthiessen tried to model the Chinese solution on the Gaussian approach, but neither did he make this explicit nor did he identify the two. Other contemporary influences are also obvious in Matthiessen's treatment of the example (negative numbers, von Schäwen's determinant method).

This seems to indicate that at that time Matthiessen was more interested in a solution of the general problem than in "the" ancient Chinese approach. This attitude is in line with his monograph on solving equations (Matthiessen, 1878c) and might be seen as a first step towards his second reconstruction of the generalized dayan rule.

Matthiessen continued to work on the problem. In April 1880 he wrote to Cantor that he had resolved the problem completely:[43]

> This dark passage in Crelle Journ. [JRAM] No. 52 [. . .] paper Biernatzkie [sic!] gave me terribly many headaches. – but now it is clear and unquestionable.

In the following two years Matthiessen published four papers on it: a more historical one in ZMP (Matthiessen, 1881a), a mathematical one in the JRAM (Matthiessen, 1881c), a short mathematical one in *Comptes Rendus hebdomaires des séances de l'Académie des sciences* (Matthiessen, 1881b), and finally an improved mathematical summary with some historical introduction in the ZMNU (Matthiessen, 1882). In

[42] Matthiessen used the term "kleinster Dividuus" (Matthiessen, 1879, p. 108), and later "kleinster gemeinschaftlicher Dividuus" [Matthiessen (1881c, p. 260), Matthiessen (1881a, p. 35)] to refer to the least common multiple. (Gauss, 1801, §32, p. 22) used "minimos communes dividuos," Dedekind (1863, §25, p. 58) "kleinste gemeinschaftliche Multiplum."

[43] "Diese dunkle Stelle in Crelle Journ. No. 52 [. . .] Art. Biernatzkie [sic!] hat mir entsetzlich viel Kopfzerbrechen gemacht. – nun ist sie aber klar und unzweifelhaft." Nachlass Moritz Benedikt Cantor, Universitäsbibliothek Heidelberg, Heid. Hs. 4028, 277, Matthiessen an Cantor, Rostock, 20.4.1880.

the course of his publishing of the papers Matthiessen's mathematical reconstruction of the generalized dayan rule as presented in 1879 was completely reworked.

- Matthiessen added a condition for the generalized problem being solvable: The given congruences

$$N \equiv r_1 (\text{mod } m_1) \equiv r_2 (\text{mod } m_2) \equiv r_3 (\text{mod } m_3) \quad \text{etc.}$$

  have to fulfill

$$r_p \equiv r_q \qquad \text{mod } \delta(m_p, m_q)$$

  for all p, q with $p \neq q$ where $\delta(m_p, m_q)$ is the greatest common divisor of m_p and m_q.[44] This insight is a generalization of Dirichlet's condition for two congruences being solvable to $n > 2$ (Dedekind, 1863, p. 57).
- Matthiessen adapted its structure so as to resemble more closely the one found in Biernatzki. He split the least common multiple (of the original moduli), now called m, into factors of prime powers and, if necessary, extended it by adjoining factors of unity (i.e. moduli equal to 1)

$$m = 1^p \cdot 1^q \cdots 2^r \cdot 3^s \cdot 5^t \cdots = \mu_1 \mu_2 \mu_3 \cdots,$$

  and

$$m_z \equiv 0 \qquad \text{mod } \mu_z$$

  for all z, so that the number of congruences stayed the same and the new moduli μ_i were relatively prime to each other.[45] This meant that Matthiessen reduced the general problem to one of relatively prime moduli which could be solved.
- For choosing a number k_z – the multiplicator – such that

$$k_z \frac{m}{\mu_z} \equiv 1 \qquad \text{mod } \mu_z$$

  (m, μ_z as above) Matthiessen (1882, 189) opted for the smallest solution in absolute value in his last paper.
- Matthiessen asked the mathematicians for a proof of the reconstructed procedure for non-relatively prime moduli, which he ascribed to Yi Xing.[46]

44 Matthiessen (1881c, p. 261), Matthiessen (1882, p. 189).
45 Matthiessen (1881c, p. 260), Matthiessen (1882, p. 189). Of course, $\mu_i = 1$ is special.
46 Matthiessen (1881b, p. 294), Matthiessen (1882, p. 189) – see Figure 5.3.

2. Die Methode von Yih-hing für nicht teilfremde Moduln. Es sei wiederum

$$N \equiv r_1 \pmod{m_1} \equiv r_2 \pmod{m_2} \equiv r_3 \pmod{m_3} \equiv \ldots$$

und der kleinste gemeinschaftliche Dividuus der beliebigen Moduln

$$m = 1^p \cdot 1^q \ldots 2^r \cdot 3^s \cdot 5 \ldots = \mu_1 \mu_2 \mu_3 \ldots,$$

wo also $\mu_1, \mu_2, \ldots$ Potenzen von Primzahlen bedeuten und derart auf die vorgelegten Kongruenzen zu verteilen sind, dafs sie folgenden Kongruenzen genügen:

$$m_1 \equiv 0 \pmod{\mu_1},\quad m_2 \equiv 0 \pmod{\mu_2},\quad m_3 \equiv 0 \pmod{\mu_3}, \text{ u. s. f.}$$

Man bestimme die kleinsten (positiven und negativen) Wurzeln k der Kongruenzen

$$k_1 \frac{m}{\mu_1} \equiv 1 \pmod{\mu_1},\quad k_2 \frac{m}{\mu_2} \equiv 1 \pmod{\mu_2}, \text{ u. s. f.}$$

Dann ist die gesuchte Zahl

$$N = r_1 k_1 \frac{m}{\mu_1} + r_2 k_2 \frac{m}{\mu_2} + r_3 k_3 \frac{m}{\mu_3} + \ldots - mn.$$

Noch allgemeiner ist nach Yih-hing

$$N = \Sigma r_1 k_1 \frac{m}{\mu_1} (1 + m_1 - \mu_1) - mn,$$

wo n eine beliebige ganze Zahl bedeutet.

Die Richtigkeit dieses Verfahrens entbehrt noch eines strengen Beweises, dessen Auffindung ich den Herren Fachgenossen hiermit als Problem vorgelegt haben möchte. Es giebt Fälle, in denen

Figure 5.3 Excerpt from Matthiessen (1882, p. 189).

Matthiessen shaped two of his texts to explain the calculations of the first example of Part C in Biernatzki (1856, pp. 79f).[47] The analysis of this example was a key factor for his new reconstruction. Matthiessen calculated more numerical examples with this new approach. These examples were either invented by Matthiessen himself, like the one used in 1879, or taken (and sometimes modified) from other sources. For instance, Matthiessen modified two examples of (Biernatzki, 1856, 81), namely the third example of different parties building a dam (Matthiessen, 1881c, 259f) and the fourth example of an amount of money diminished by a certain quantity each day (Matthiessen, 1882, 190), by introducing concrete numbers. He also took over a contemporary example by von Schäwen (1878a, 117).

47 Matthiessen (1881c, pp. 258f), Matthiessen (1881b, pp. 292f).

Historiographically, Matthiessen judged the dayan rule as a great Chinese achievement:[48]

> The dayan rule is one of the best achievements of the Chinese in the field of number theory (Matthiessen, 1881c, p. 254, transl. MS).

Neither Gauss nor Dirichlet/Dedekind had dealt with its generalization to non-relatively prime moduli according to Matthiessen.[49] Matthiessen attributed to Yi Xing the solution of the generalized dayan rule and stressed the diversity of realms of application that he had come up with. In Matthiessen's view, Biernatzki had wrongly narrowed down this diversity to applications in the field of divination. It is this misrepresentation that Matthiessen identified as "one of the greatest stupidities ever committed in the historiography of mathematics."[50] Matthiessen hence relativized some of the achievements of his contemporaries in number theory that Dedekind, in his preface to Dirichlet's lectures on number theory, called "the immortal fruit of a truly noble competition of the European peoples."[51]

Matthiessen did not exclusively address historians of mathematics when publishing on the reconstruction of the generalized dayan rule. His places of publication, the style and contemporary notation as well as his request for a proof show that Matthiessen was also addressing working mathematicians and teachers. In doing so, he had turned the reconstructed generalized dayan rule into an object of contemporary mathematical research.

To sum up Matthiessen's reconstruction activities around the Chinese dayan rule: Matthiessen corrected a wrong algorithm for relatively prime moduli in Biernatzki's paper and reconstructed the structure of the rule for non-relatively prime numbers. This was done on the basis of a close analysis of Biernatzki's paper, by a careful analysis of its examples and by linking it to contemporary approaches (Gauss, Dirichlet/Dedekind) – see Figure 5.4 for Matthiessen's resources from diverse fields, in particular the history of mathematics, didactics, and contemporary math-

48 "Diese Tayen [. . .] ist jedenfalls eine der besten Leistungen der Chinesen auf dem Gebiete der Zahlentheorie."

49 Matthiessen (1881a, p. 34). However, both indicate a kind of stepwise procedure to deal with non-relatively prime moduli. Matthiessen seems neither to have realized this nor to have known of V.-A. Le Besgue's solution (Le Besgue, 1859).

50 "gehören unter die grössten Thorheiten, welche je in der Geschichtsschreibung der Mathematik begangen worden sind" (Matthiessen, 1881a, pp. 34f).

51 "die unvergängliche Frucht eines wahrhaft edelen Wettkampfes der europäischen Völker" (Dedekind, 1871, p. x, transl. MS).

ematics. Contemporary mathematical notation, style (formulation of a theorem) and concepts (moduli, prime numbers, etc.) were used, and hence the dayan rule was incorporated into contemporary mathematics. Matthiessen also turned it into a contemporary research object by discussing criteria of solvability of the general problem (remainders, negative numbers) and by asking for a proof of the general procedure that he set out in the form of a theorem.[52]

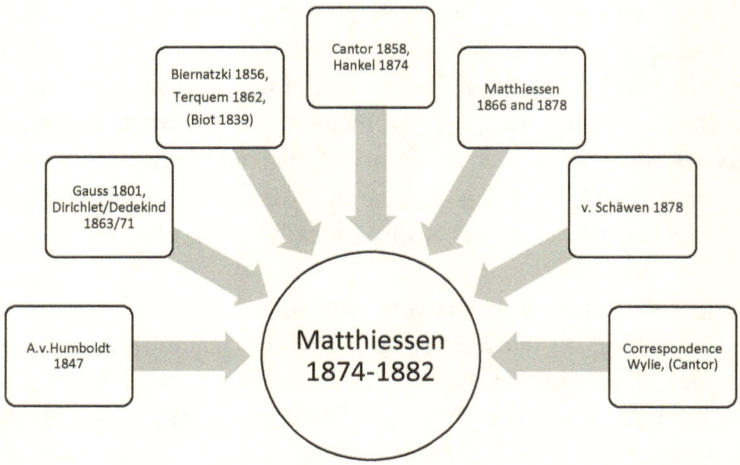

Figure 5.4 Resources of Matthiessen's study of the dayan rule.

On the historiographic level, Matthiessen used this to contradict the standard interpretation of Chinese mathematics in Germany at the time. He claimed that the dayan rule worked, that it was originally Chinese, distinct from the Indian kuttaka method and "the same" as Gauss's approach in his *Disquisitiones arithmeticae*. In addition to that, he discovered that the Chinese were able to solve a more general problem, i.e. where the moduli are not relatively prime. Matthiessen pointed out that this (reconstructed) generalization to non-relatively prime moduli surpassed the contributions by Gauss as well as by Dirichlet/Dedekind.

[52] Matthiessen (1881c,b). Proofs had been given before by V.-A. Le Besgue (1859, pp. 56–58) and later by Stieltjes (1890, pp. 28–33) and Mahler (1958).

5.3 Discussion

In the following I will discuss four issues related to the analysis of the example raised above, to the questions raised in the introduction and to the theme of this book. The first two issues discussed concern the concept of mathematical reconstruction. What type(s) of mathematical reconstruction can be identified in the above example? The third one discusses the example's link to the book's theme of anachronism. Finally, Matthiessen's example is analyzed through the perspective of pluritemporality. This change of perspective allows us to capture in a coherent way many of the complex spatio-temporal features of the example exhibited in Section 5.2. Moreover it shows that these features are in a certain way typical of any kind of historical research.

5.3.1 Different types of mathematical reconstruction

After the above analysis of Matthiessen's approach one might hesitate and wonder whether it really exemplifies mathematical reconstruction. He himself did not call it a mathematical reconstruction. At first glance one might rather perceive him as dealing mathematically with the contemporary work of a contemporary scholar. Matthiessen's contributions might be seen as a follow-up to Biernatzki's work and as a part of contemporary mathematical discussions.[53] This perspective, however, omits two essential aspects: (1) Biernatzki's (or rather Wylie's) paper is about ancient Chinese mathematics and is based on ancient Chinese sources; (2) Matthiessen attributed his findings to ancient Chinese scholars (rather than to himself). There are further good reasons to perceive Matthiessen's research as a mathematical reconstruction of an ancient Chinese mathematical procedure.

Matthiessen reconstructed something for which he had no access to the relevant primary sources. The sources were neither available to scholars in Europe at the time nor would Matthiessen have been able to read them. Instead he worked with a secondary source (Biernatzki, 1856). By carefully studying this text, in particular the translated Chinese quotations, and the examples given in it and by comparing the Chinese approach to the problem with (contemporary) approaches known to him he developed procedures that made sense to contemporary mathematicians.

[53] von Schäwen (1878a); Matthiessen (1878b); von Schäwen (1878b); Matthiessen (1878a).

He ascribed these procedures to the Chinese scholars (Master Sun, Yi Xing) and dated them. He used contemporary notation, concepts and argumentation to clarify and communicate the dayan rule. His "making sense" of Biernatzki's text led to a new historiographic perspective that contradicted the standard evaluation of Chinese mathematics in Germany at the time. It opened the way to a much more positive assessment of ancient Chinese contributions to mathematics.

If attention is paid to the details of Matthiessen's reconstructing activities, then the two different phases of Matthiessen's work on the dayan rule outlined above not only differ in time, but also correspond to a qualitative change. They show two different types of mathematical reconstruction present in Matthiessen's approach.

Type 1: Conflating ancient with contemporary mathematics

In the first phase Matthiessen's mathematical reconstruction can be characterized as conflating an ancient mathematical rule with a modern one: due to his expertise in the history of equations Matthiessen was able to detect in a short quotation (the "aphoristic note" translated from an ancient Chinese source in part A) a contemporary method, namely that of Gauss. He re-wrote the ancient Chinese procedure by using the contemporary Gaussian notation and by identifying a Chinese concept with a Gaussian one ("tsching su," "Multiplicator," "Hilfszahl"). Thus, the ancient procedure is presented as the Gaussian one in contemporary mathematical notation garnished by some Chinese terminology. Matthiessen's form of presentation suggests that the two procedures are essentially the same. As a consequence or special bonus of this sameness, the correctness of the dayan rule as well as the progressiveness of ancient Chinese mathematics follow almost automatically. In addition to that, new questions, such as those concerning the realm of applicability of the dayan rule, emerge.

The sameness is produced at the expense of the ancient approaches. Most features of the ancient Chinese approach described by Biernatzki get lost. The sameness is an effect of this type of mathematical reconstruction. At the same time this type of mathematical reconstruction allows for the creation of differences and makes comparisons possible: Matthiessen's identification of the Indian kuttaka method with an approach by Euler and later with one by Bachet de Méziriac[54] allowed Matthiessen to claim that the kuttaka method and the dayan rule were different.

As Goldstein (2019) points out, the "art" of producing sameness (and differences) is historiographically closely connected to the writing of long-term histories of mathematics. This type of mathematical reconstruction relies not only on a profound knowledge of mathematics, but also seems to be based on two beliefs: first, the belief that ancient mathematical practices are mathematically interesting. Matthiessen believed in the effectiveness of Chinese astronomy and hence in the effectiveness of the mathematics behind it. Second, the belief in a mathematical core that is universal and transcends and unifies mathematical practices over time and space. Matthiessen identified the Gaussian method as the core of the Chinese approaches despite huge discrepancies between the two mathematical practices. These differences were of no significance to him, hence he made almost all of them vanish in his presentation. What mattered historically and mathematically to Matthiessen was identifying different ways of solving a problem, not different mathematical practices.

Type 2: Creating ancient mathematical knowledge from ancient material

The second phase of Matthiessen's research on the dayan rule shows a different type of practice of mathematical reconstruction, namely the creation of ancient mathematical knowledge (the generalized dayan rule) with the help of ancient material. In this type of mathematical reconstruction both old and contemporary knowledge of mathematics are used to produce new mathematical knowledge about ancient mathematics.

Matthiessen's starting point was an example in Biernatzki's text (example 1 in part C) that did not deal with pairwise prime moduli, but with non-relatively prime moduli. It provided him with a problem that was new to him. From the analysis of the example Matthiessen came to the conviction that the Chinese had been in possession of a generalized dayan rule, which also worked for non-relatively prime moduli. Matthiessen formulated such a generalized dayan rule in modern notation and style and put forward two different solutions. These solutions – the second more so than the first – were based on elements from contemporary

54 Matthiessen (1876b, p. 79). See the analysis by Libbrecht (1973, Chapter 18) on the comparison between the dayan rule and the kuttaka.

contributions: the first was modelled mainly on the structures outlined in Gauss (1801, §32), the second on features of Biernatzki's text that Matthiessen identified as significant. For example, Matthiessen identified the concept of the least common multiple in the example and its factorization as key elements for the solution. Moreover he employed the dayan rule that he had reconstructed before in the solution. However, the solutions Matthiessen came up with cannot be found in Biernatzki's text nor in ancient Chinese sources.[55] They were at least partly Matthiessen's creations.

Matthiessen's presentation of the generalized dayan rule as a theorem looked like a contribution to contemporary mathematics. Matthiessen provided conditions for the solvability of the generalized dayan rule, applied the rule to various problems (mostly not related to Biernatzki's text) and asked mathematicians for its proof. Nevertheless he claimed that the solution to the problem was old, created by an ancient Chinese scholar (Yi Xing). Historiographically this was a new claim about Chinese arithmetics.

This type of mathematical reconstruction relies on the same kind of beliefs about mathematics as the first type. Both types of mathematical reconstruction rest on the contemporary knowledge of and developments in mathematics. Hence, they are themselves a product of their time and can be subjected to historical investigation.

Other types

There are, however, also other types of mathematical reconstruction that do not necessarily exhibit such a close connection to contemporary mathematics. For example, J. Hogendijk's explanation of Sharaf al-Dīn al-Ṭūsī's work on the number of positive roots of cubic equations is a contemporary example of this kind.[56] Hogendijk shows how al-Tūsī's results can be summarized and analyzed on the basis of ancient and medieval mathematical methods that were known at the time of al-Tūsī in the twelfth century and how they fit into a geometrically motivated approach to algebra. Hogendijk's mathematical reconstruction relies crucially on his knowledge of the history of ancient mathematics, and not so much on his knowledge of contemporary mathematics. Hogendijk contrasts his

[55] See Section 5.3.2 for more details.
[56] Hogendijk (1989).

approach to R. Rashed's – two different reconstructions compete with each other.[57]

By now reconstruction has become an actor's category. However, Hogendijk avoids calling his reconstruction a "mathematical" one. This again points to the negative connotation associated with "mathematical reconstruction" at the time as alluded to in the introduction. Hogendijk's reconstruction might be seen as an example of – what I will call in Section 5.4 – a "historiographically sensitive" mathematical reconstruction.

The different types of mathematical reconstruction sketched here are the result of a first analysis of the practices of mathematical reconstruction. I am sure a deeper and broader analysis of the history of the practices of mathematical reconstruction will yield a more refined spectrum of types. It may even be necessary to introduce sub-types.[58] Some examples will show a combination of different types of mathematical reconstruction at work. The differences between mathematical reconstruction and interpretation as well as their links are also worthy of further research. A better understanding of the practices of mathematical reconstruction, their bases and their consequences will shed light on the complex relationship between mathematics and its historiography.

5.3.2 A historiographic success or failure?

General criteria to judge the adequacy of a reconstruction are not easy to provide since reconstructions, like historical interpretation in general, go, by their very nature, beyond the sources available at the time.[59] However, in Matthiessen's case some of the primary sources are available to us today, so that we may compare some of Matthiessen's claims with source material. This is rather a special situation. Libbrecht (1973) pointed out that Matthiessen's claim that the Buddhist priest Yi Xing[60]

57 Rashed, according to Hogendijk, claims that the concept of a derivative of a function or a polynomial is implicit in al-Tūsī's work. Rashed's claim rests on yet another type of mathematical reconstruction that attributes implicit knowledge of later concepts to ancient scholars. Hogendijk's mathematical reconstruction gives an alternative explanation that avoids this concept that was developed much later.

58 For example one could further differentiate according to what it is that is reconstructed (theorem, method, concept) or to its attached epistemological status (implicit vs. explicit knowledge).

59 E. Robson (2001, p. 176) has developed six criteria to judge interpretations of the cuneiform tablets: historical sensitivity, cultural consistency, calculational plausibility, physical reality, textual completeness, tabular order. Some of them can be applied more generally, i.e. outside of Mesopotamian mathematics, see Section 5.4.

60 Libbrecht's transcription is "I-hsing." See Libbrecht (1973, pp. 280–282; pp. 313f., fn.21) for more details.

had invented the generalized dayan rule in the eighth century was wrong. It was Qin. However, Libbrecht also suggested that Matthiessen's reconstruction was correct as far as the mathematical procedure in its modern transformation was concerned. K. Chemla's subtle analysis of available ancient Chinese sources suggests that it was only the structure of the Chinese procedure that Matthiessen reconstructed correctly. However, in Matthiessen's reconstruction two procedures in the text of Qin were missing.

First, in his mathematical reconstruction of the first phase Matthiessen discovered that it was important to find numbers (chenglü) that are 1 (modulo some product). This is the key idea behind an ancient Chinese procedure called "Finding unity." The procedure "Finding unity" is not mentioned in Biernatzki's text (and only briefly in Wylie's paper). It involves a numerical "algorithm" to calculate numbers that solve the above problem.[61] Matthiessen did not restore this algorithm. Instead he left it to the reader to find numbers that solve the problem. Only having access to Biernatzki's text that did not even mention the name of the procedure, Matthiessen had no chance of correctly reconstructing the algorithm that was actually given in the Chinese sources.

Second, in his reconstruction of the generalized rule Matthiessen also omitted a procedure that reduced non-relatively prime moduli to relatively prime ones. It was contained in the sources, but only cryptically in Biernatzki's text. Instead of this procedure Matthiessen created an approach using the least common multiple. Here, Matthiessen did get some clues from Biernatzki's text. However, he did not closely follow the procedure outlined in the first example of part C, but rather set it out differently. Comparing Matthiessen's approach to Qin's reveals huge discrepancies between them.[62]

So in fact Matthiessen apparently correctly reconstructed some mathematical parts of the Chinese approach, but not all. As Libbrecht (1973, p. 314, fn. 21) put it:

> I [Libbrecht] cannot but hold Matthiessen in esteem; he was incontestably a very clever mathematician, as he was able to understand the *ta-yen* rule with such inaccurate data as Biernatzki's translation provided him.

Matthiessen's reconstruction opened the way to a more positive assessment of Chinese mathematics.

[61] The procedure of "Finding unity" is first discussed by Mikami (1913, pp. 66f).
[62] Libbrecht (1973, Chapter 22).

As for the reception of Matthiessen's paper, Libbrecht (1973, pp. 317–327) showed that it was mixed: in Germany Cantor incorporated Matthiessen's results in the three editions of the first volume of his very successful *Lectures on the history of mathematics* (Vorlesungen über Geschichte der Mathematik) (Cantor, 1880, 1894, 1907). T.J. Stieltjes (1890, p. 47) mentioned (Matthiessen, 1881c) in a bibliographical study on number theory. Leonhard E. Dickson (1920, pp. 57–64) dedicated a chapter to the "Chinese problem of remainders" in the second volume called "Diophantine analysis" of his *History of the Theory of Numbers*.[63] The chapter was based on Matthiessen's reconstruction, but also attempted to give an overview of the different historical solutions to the problem up to the nineteenth century. Matthiessen's mistake about Yi Xing as the person who developed the dayan rule was replicated by all of them.

However, as Libbrecht (1973) has shown, there were also other historians of mathematics publishing on ancient Chinese mathematics who did not refer to Matthiessen's contribution (like H.G. Zeuthen (1893), D.E. Smith (1912), Y. Mikami (1913), L. van Hee).[64] Some of them, especially the Jesuit Van Hee, expressed a very low opinion of Chinese mathematics.

To sum up, Matthiessen's paper was neither ignored nor highly influential. It influenced the historiography of mathematics probably mostly via Cantor's *Lectures*. A more detailed study of its reception is needed to better understand the dynamics. So, taking everything into account, Matthiessen's reconstruction was partly a successful one both historiographically as well as historically.

However, the two criteria for historiographic success given above – adequateness with respect to sources and positive reception – are not as straightforward as it seems at a first glance. Firstly, a positive wide reception of an historical paper does not prove its adequateness. It rather shows that it fits very well into the historiographic framework of (large or dominating parts of) the community of historians of mathematics at the period. Secondly, judging a mathematical reconstruction (or an interpretation) with the original sources at hand presupposes that the sources can speak for themselves and do not need an interpretation to be

[63] This could be the origin of calling (versions of) the theorem "Chinese Remainder Theorem" nowadays.

[64] It is not clear to Libbrecht whether G. Vacca referred to Matthiessen.

understood. This view can, of course, be questioned – especially with regard to ancient sources. For example, the famous cuneiform tablet Plimpton 322 does not speak for itself, as the different interpretations show.[65]

5.3.3 The dayan rule and anachronism

There is a very close link between the dayan rule and the book's topic of anachronism. The term "anachronism" is, as its history reveals,[66] an Italian product ("anacronismo" derived from the Greek) of the late sixteenth century. During the seventeenth century it appeared in French and English, in the early eighteenth century in German. In the beginning the term meant a mistake in synchronizing different calendrical systems, i.e. a mistake in the field of chronology. The term was coined during a time when different Western calendars and also Chinese calendars were compared, synchronized and put into something that might be called a universal global time. It was only later in eighteenth-century England that the meaning of anachronism shifted to "late in time," and from that it developed to our present day understanding as something "out-dated." Landwehr (2013) points out that as a Greek neologism it was itself an anachronism (in the latter sense) in the time of the rise of vernacular languages.

As Qin pointed out, the dayan rule was used by calendar makers. In an example, Qin calculated conjunctions of the solar year and the lunar month in the Old (Chinese) calendar with respect to a given conjunction.[67] So it means synchronizing two different calendars – the lunar and the solar one – and hence this example belongs to the field of chronology. We find this example as the second example in part C of Biernatzki (1856, p. 81). Biernatzki classified it as an astronomical calculation. In his *Disquistiones Arithmeticae* (1801) Gauss also mentioned the use of the rule on indeterminate analysis for chronology as an application.[68] As shown above Matthiessen (1874, p. 271) argued that due to its application in Chinese astronomy he believed that the Chinese dayan rule could not be wrong.

So the dayan rule has been linked to chronology for a long time. It is

65 See Robson (2001); Britton *et al.* (2011); Mansfield and Wildberger (2017).
66 See for example Landwehr (2013, pp. 11–14).
67 Libbrecht (1973, Chapter 20).
68 Gauss (1801, §36).

in fact a means to prevent anachronisms in the original meaning of the word.

5.3.4 On the example's complex time structure

The careful analysis of Matthiessen's reconstruction in Section 5.2 as well as the discussion in 5.3.1 and 5.3.2 show that the temporal structure of the example is very complex. On the one hand, the past is absent – Matthiessen had no access to relevant ancient Chinese sources – but it is also present to him – in the form of Biernatzki's text. This present past accessible to Matthiessen, however, is a distorted one due to Biernatzki's (and Wylie's) mistakes and misunderstandings.

On the other hand, Matthiessen used the contemporary knowledge of mathematics to understand what the Chinese scholars had done in the past. Thus, he dealt with the past in terms of the present, or rather, as argued before, he dealt with a distorted present past. He corrected parts of it, but at the same time added mistakes.

Finally, by asking for a proof of the reconstructed generalized theorem Matthiessen set a research agenda. Thus, he tried to influence future research. Hence the present might also influence the future.

This kind of complex temporal structure in which past, present and future are interwoven in a multi-layered and intricate relationship, a kind of time knot, is what Landwehr (2013) calls "pluritemporality." The structure also has a spatial component in that time should be localized. A feature that the example analyzed above also exhibits is that the primary sources like Qin's *Mathematical Writings*, published in 1247 and critically edited by Song Jingchang in 1842 (as well as Wylie's paper), were available in Shanghai in the 1850s, but not in Europe. So the primary sources are present and absent at the same time.

According to Landwehr (2013), pluritemporality is not the exception, but rather the norm. The historian is confronted with a more complex structure of time than the simplistic idea of a linear global time that chronology (and indeed also the concept of anachronism) is based upon. This has been pointed out before, e.g. prominently in Reinhart Koselleck's extensive *Begriffsgeschichte* (conceptual history) of time (Koselleck, 2017). Landwehr argues that historical research is based upon the pluritemporality of time. Doing history means drawing upon, dealing with, and creating this complex time structure. Pluritemporality is at the heart of the practice.

Mathematical reconstructions embody typical features of this temporal structure. They bring together a contemporary understanding of mathematics (Matthiessen) or of its history (Hogendijk) at a given place and time with sources and objects from the past that come with their own histories (of transmission and interpretation) in order to say something about the mathematics of the past. The result of mathematical reconstructions is based to some extent on sources, but also transcends them.

The types of mathematical reconstruction that are based on contemporary mathematical knowledge are mostly applied by persons trained in mathematics. Their research and publications can be easily grasped by the peer-group. One might say that these types of mathematical reconstruction offer the opportunity of a dialogue of present mathematicians with past and present mathematicians – a dialogue that not only many mathematicians enjoy, but that also shapes the past and the contemporary image of mathematics in certain ways.[69] For example, the first two types of mathematical reconstructions mentioned in Section 5.3.1 rely on and co-produce the image of a universal mathematics.

5.4 Concluding remarks on historiographically sensitive mathematical reconstruction

Some types of mathematical reconstruction have been criticized as resting on universal logico-mathematical reasoning, and neglecting past ways of reasoning that are different.[70] However, taking into account the more complex time scheme of pluritemporality, the concept of anachronism can be rethought. Landwehr (2013) has done so and comes to the conclusion that anachronism can play a productive role in historical research. Following and adapting his analysis I will argue that mathematical reconstruction may also be seen as a productive tool in the history of mathematics for generating new hypotheses and interpretations.[71]

[69] For a more refined overview of how mathematicians can relate to past mathematics see Fried (2018).

[70] Saito (1998, 134f.).

[71] Indeed, this is exactly the function that Saito (1998) has pointed out for mathematical reconstructions that integrate textual studies.

To use mathematical reconstruction productively three typical pitfalls have to be avoided:

(i) Scholarly mistakes have to be avoided, i.e. one has to point out on what sources and assumptions the reconstruction rests.

(ii) Conceptual mistakes have to be avoided, like projecting contemporary concepts back onto the past or assuming that if the name of a concept is the same, the concept has not changed;[72] this does not only pertain to the "body of knowledge," but also to the "image of knowledge."[73] Both change over time and should be studied "locally." For example, the style of writing mathematics, e.g. interpreting rules as formulae (Cantor) or theorems (Matthiessen),[74] or values like generality (Cantor, Matthiessen), proofs (Matthiessen) or abstractness,[75] are not universal and therefore should not be projected back either.

(iii) Discipline-based mistakes should be avoided, e.g. projecting contemporary understanding of disciplines[76] and ways of practising them onto the past and onto different cultures (Matthiessen). For a historical understanding of the disciplinary context one might have to take into account not only mathematical texts and objects (together with their material features and historical sediments), but also, if possible, social features: Who practised mathematics? For what reasons? In what contexts? What was the status of the people doing mathematics and of mathematical practices in a particular society? etc.

Of course, avoiding these mistakes needs careful distinction between past and contemporary practices of mathematics. However, this very distinction is challenged by the concept of pluritemporality. The borders between past and present (and future) become blurred. This means that as historians we have to be all the more aware of a kind of temporal fragility of our interpretational concepts and frames as well as of the texts and objects under investigation and of the texts we write. One way to answer this challenge might be to be aware of and to lay open our

72 See K. Plofker's contribution to this volume.

73 See e.g. Corry (1996) for this distinction between the content of mathematics and metamathematical issues (based on Y. Elkana).

74 See also K. Chemla's contribution to this volume.

75 See also J. Gray's contribution to this volume.

76 See the contributions by J. Feke and by J. Lorenat in this book.

own personal backgrounds, motivations, and goals that go along with our historical research in our publications.

Taking the above issues as well as the available primary sources (and their different (mis-)interpretations) seriously into consideration means that the constructing activity is far from being arbitrary. It is, on the contrary, restricted by a lot of constraints. It should live up to criteria like historical sensitivity, cultural consistency or calculation plausibility, to name a few.[77] If this is the case, one might call this type of mathematical reconstruction a "historiographically sensitive" mathematical reconstruction.

The benefits of using such historiographically sensitive mathematical reconstructions in the history of mathematics are the following:

- The juxtapositioning of contemporary forms of mathematical knowledge and practices with old forms allows us to show the specific form of mathematical knowledge at a specific place and time. It highlights the historicity of mathematical knowledge.
- Historiographically such reconstructions are a tool to put forward new hypotheses and/or to provide different interpretations of the historical material. These can then be compared, discussed and their appropriateness assessed. Indeed, Matthiessen's and Hogendijk's mathematical reconstructions challenge older ones (by Cantor, Hankel, and Rashed). Eventually this might lead to a different (if not better) understanding of the past. Thus, reconstructions are historiographically productive.
- Based partly on contemporary knowledge mathematical reconstructions change over time. Contemporary concepts might better capture some aspects of past mathematical practices, like the algorithmic approach to Mesopotamian (Knuth, Proust) or Egyptian (Imhausen/Ritter) mathematics.[78] Thus, the historicity of writing the history of mathematics becomes apparent as well.

Thus, in the dynamic process of historical interpretation historiographically sensitive mathematical reconstructions might lead to a deeper understanding of the mathematical traces of the past. Of course, some of these benefits are not exclusively achieved by mathematical reconstructions, but they can be the result of careful historical and historiographical

[77] Robson (2001).

[78] See K. Chemla's contribution to this book.

research, too. With regard to the concept of pluritemporality the aim of mathematical reconstruction can be seen as twofold: destroying the differences between today and the past (in order to see similarities) and at the same time highlighting the very differences.

The example of Matthiessen's reconstruction described above neither avoids all the pitfalls nor does it exhibit all the benefits. Yet I have argued in Section 5.3.2 that it gave rise to a new and in some respects better interpretation of Chinese mathematics at the time. Thus Matthiessen's study was productive despite its shortcomings. This seemingly paradoxical result can be linked to the epistemic status of mathematics – either as an effect of or as contributing to the impression of a universal mathematics. Moreover, my analysis in Section 5.3.4 shows that the concept of pluritemporality captures very well the complex spatio-temporal structure of Matthiessen's reconstruction.

The above historiographical study of Matthiessen's reconstruction can be seen as a first step towards a history of the practices of mathematical reconstruction. Such a history would provide an excellent basis for the further study of uses, functions and underlying assumptions of mathematical reconstruction as a tool in the historiography of mathematics and of its consequences for the image of mathematics. It would not only contribute to a better understanding of mathematical reconstruction, its various types and their roles in the historiography of mathematics, but also of the special role mathematicians have had in the writing of the history of their discipline.

Acknowledgements I would like to thank all my colleagues who gave me feedback on various versions of my talk on this topic and on drafts of this article, especially K. Chemla, N. Guicciardini, T. Hoff Kjeldsen, M. Remenyi, and E. Scholz. The final version has benefitted considerably from our discussions.

References

Alberti, Eduard, (Ed.) (1868). *Lexikon der Schleswig-Holstein-Lauenburgischen und Eutinischen Schriftsteller von 1829 bis Mitte 1866*, volume 2. Kiel: Akademische Buchhandlung.

Alberti, Eduard, (Ed.) (1885). *Lexikon der Schleswig-Holstein-Lauenburgischen und Eutinischen Schriftsteller von 1866–1882*, volume 2. Kiel: Karl Biernatzki.

Arni, Caroline (2007). Zeitlichkeit, Anachronismus und Anachronien. Gegenwart und Transformationen der Geschlechtergeschichte aus geschichtstheoretischer Perspektive. *L'Homme* **18**(2), 53–76.

Bertrand, Joseph (1869a). Les mathématiques en Chine. Die Arithmetik der Chinesen, von Herrn Dr. K.L. Biernatzki, zu Berlin. 1er article. *Journal des savants*, Juin, 317–329.

Bertrand, Joseph (1869b). Les mathématiques en Chine. Die Arithmetik der Chinesen, von Herrn Dr. K.L. Biernatzki, zu Berlin. 2ième article. *Journal des savants*, Août, 464–477.

Biernatzki, Karl Leonhard (1856). Die Arithmetik der Chinesen. *Journal für die reine und angewandte Mathematik* **52**, 59–94.

Biot, Edouard (1839). Table générale d'un ouvrage chinois intitulé Souan-fa-tong-tsong, ou Traité complet d'art de compter (Fourmont no. 350), traduite et analysée par M. Éd. Biot. *Journal Asiatique* **7**, 193–217 (transl. and analyzed by Éd. Biot).

Blåsjö, Viktor (2014). A critique of the modern consensus in the historiography of mathematics. *Journal of Humanistic Mathematics* **4**(2), 113–123.

Britton, John P., Proust, Christine, and Shnider, Steve (2011). Plimpton 322: a review and a different perspective. *Archive for History of Exact Sciences* **65**(5), 519–566.

Cantor, Moritz (1858). Zur Geschichte der Zahlzeichen. *Zeitschrift für Mathematik und Physik*, 3:325–341.

Cantor, Moritz (1880). *Vorlesungen über Geschichte der Mathematik. Von den ältesten Zeiten bis zum Jahre 1200 n. Chr.*, volume 1. First edition. Leipzig: Teubner.

Cantor, Moritz (1894). *Vorlesungen über Geschichte der Mathematik. Von den ältesten Zeiten bis zum Jahre 1200 n. Chr.*, volume 1. Second edition. Leipzig: Teubner.

Cantor, Moritz (1907). *Vorlesungen über Geschichte der Mathematik. Von den ältesten Zeiten bis zum Jahre 1200 n. Chr.*, volume 1. Third edition. Leipzig: Teubner.

Chen, Zhihui A. (2017). Alexander Wylie's Comparisons of Algebra between China and Europe, and Its Historiographical Origin. *Studies in the History of Natural Sciences* **36**(4), 502–518.

Corry, Leo (1996). *Modern Algebra and the Rise of Mathematical Structures*. Basel, Boston: Birkhäuser.

Dedekind, Richard (Ed.) (1863). *Vorlesungen über Zahlentheorie von P.G. Lejeune-Dirichlet*. First edition. Braunschweig: Vieweg.

Dedekind, Richard (Ed.) (1871). *Vorlesungen über Zahlentheorie von P.G. Lejeune-Dirichlet*. Second revised and enlarged edition. Braunschweig: Vieweg.

Dickson, Leonard Eugene (1920). *History of the Theory of Numbers*, volume 2. Washington: Carnegie Institution.

Fried, Michael N. (2018). Ways of relating to the mathematics of the past. *Journal of Humanistic Mathematics* **8**(1), 3–23.

Gauss, Carl Friedrich (1801). *Disqusitiones Arithmeticae*. Leipzig: Fleischer. Reprints: 1863 (vol. 1), 1870 (vol. 1); translations: 1806 (French by A.-C.-M. Poullet-Deslisle), 1889 (German by H. Maser), 1965 (English by Arthur A. Clarke).

Goldstein, Catherine (2019). Long-term history and ephemeral configurations. In: *Proceedings of the International Congress of Mathematicians 2018, Rio de Janeiro*, volume 1, B. Sirakov, P. N. de Souza, and M. Viana (eds.). Singapore: World Scientific, pp. 487–522.

Goldstein, Catherine and Schappacher, Norbert (2007a). A book in search of a discipline (1801–1860). In *The Shaping of Arithmetic after C.F. Gauss's Disquisitiones Arithmeticae*, C. Goldstein, N. Schappacher, and J. Schwermer (eds). Berlin, Heidelberg: Springer, pp. 3–65.

Goldstein, Catherine and Schappacher, Norbert (2007b). Several disciplines and a book (1860–1901). In *The Shaping of Arithmetic after C.F. Gauss's Disquisitiones Arithmeticae*, C. Goldstein, N. Schappacher, and J. Schwermer (eds). Berlin, Heidelberg: Springer, 66–103.

Hankel, Hermann (1874). *Zur Geschichte der Mathematik im Alterthum und Mittelalter*. Leipzig: Teubner.

Herder, Johann Gottfried (1799). *Verstand und Erfahrung. Eine Metakritik zur Kritik der reinen Vernunft. Erster Theil*. Leipzig: Hartknoch.

Hogendijk, Jan P. (1989). Sharaf al-Dīn al-Tūsī on the Number of Positive Roots of Cubic Equations. *Historia Mathematica* **16**, 69–85.

Jardine, Nicholas (2000). Uses and abuses of anachronism in the history of the sciences. *History of Science* **38**, 251–270.

Koselleck, Reinhart (2017). *Vergangene Zukunft. Zur Semantik geschichtlicher Zeiten*. Tenth edition. Frankfurt a.M.: Suhrkamp.

Kragh, Helge S. (1987). *An Introduction to the Historiography of Science*. Cambridge: Cambridge University Press.

Landwehr, Achim (2013). Über den Anachronismus. *Zeitschrift für Geschichtswissenschaft* **61**(1), 5–29.

Landwehr, Achim (2016). *Die anwesende Abwesenheit der Vergangenheit*. Frankfurt a.M.: Fischer.

Le Besgue, Victor-Amédée (1859). *Exercices d'Analyse Numerique*. Paris: Leiber et Farguet.

Libbrecht, Ulrich (1973). *Chinese Mathematics in the Thirteenth Century*. Cambridge, Mass., London: MIT Press.

Loison, Laurent (2016). Forms of presentism in the history of science. Rethinking the project of historical epistemology. *Studies in History and Philosophy of Science Part A* **60**(1), 23–37.

Loraux, Nicole (1993). Eloge de l'anachronisme en histoire. *Le genre humain* **27**(1), 23–39.

Mahler, Kurt (1958). On the Chinese Remainder Theorem. *Mathematische Nachrichten* **18**(1–6), 120–122.

Mahnke, Reinhard (1991). Ludwig Matthiessen – erster ordentlicher Professor der Physik an der Universität Rostock. *Beiträge zur Geschichte der Universität Rostock* **17**, 19–33.

Mansfield, Daniel F. and Wildberger, Norman J. (2017). Plimpton 322 is Babylonian exact sexagesimal trigonometry. *Historia mathematica* **44**, 395–419.

Matthiessen, Ludwig H.F. (1861). Anwendung der oscillierenden Kettenbrüche zur gleichzeitigen Bestimmung zweier Wurzelwerthe einer Gleichung. *Zeitschrift für Mathematik und Physik* **6**, 51–59.

Matthiessen, Ludwig H.F. (1863a). Eine neue Auflösung der biquadratischen Gleichungen. *Zeitschrift für Mathematik und Physik* **8**, 140–142.

Matthiessen, Ludwig H.F. (1863b). Neue Auflösungen der quadratischen, cubischen und biquadratischen Gleichungen. *Zeitschrift für Mathematik und Physik* **8**, 133–140.

Matthiessen, Ludwig H.F. (1864). Über einen Zusammenhang der Seiten eines Kreisvierecks mit den Wurzeln einer biquadratischen Gleichung. *Zeitschrift für Mathematik und Physik* **9**, 453–454.

Matthiessen, Ludwig H.F. (1865). Über eine Beziehung der Seiten und der Diagonalen eines Kreisvierecks zu den Wurzeln einer biquadratischen Gleichung und ihrer Resolvente. *Zeitschrift für Mathematik und Physik* **9** 331–332.

Matthiessen, Ludwig H.F. (1866). *Die algebraischen Methoden der Auflösung der litteralen quadratischen, kubischen und biquadratischen Gleichungen nach ihren Prinzipien und ihrem inneren Zusammenhang dargestellt.* Leipzig: Teubner.

Matthiessen, Ludwig H.F. (1874). Zur Algebra der Chinesen. Auszug aus einem Briefe an M. Cantor. *Zeitschrift für Mathematik und Physik* **19**, 270–271.

Matthiessen, Ludwig H.F. (1876a). Vergleichung der indischen Cuttuca und der chinesischen Tayen-Regel, unbestimmte Gleichungen und Congruenzen ersten Grades aufzulösen. In *Verhandlungen der Dreissigsten Versammlung Deutscher Philologen und Schulmänner in Rostock vom 28. September bis 1. October 1875*, H. E. Bindseil, (ed). Leipzig: Teubner, pp. 125–129.

Matthiessen, Ludwig H.F. (1876b). Vergleichung der indischen Cuttuca und der chinesischen Tayen-Regel, unbestimmte Gleichungen und Congruenzen ersten Grades aufzulösen. *Zeitschrift für den Mathematischen und Naturwissenschaftlichen Unterricht* **7**, 77–81.

Matthiessen, Ludwig H.F. (1878a). Antwort. *Zeitschrift für Mathematischen und Naturwissenschaftlichen Unterricht* **9**, 368f.

Matthiessen, Ludwig H.F. (1878b). Die Diophantischen Gleichungen ersten Grades. Eine Entgegnung von Prof. Ludwig Matthiessen aus Rostock.

Zeitschrift für Mathematischen und Naturwissenschaftlichen Unterricht **9**, 194–197.

Matthiessen, Ludwig H.F. (1878c). *Grundzüge der Antiken und Modernen Algebra der Litteralen Gleichungen.* Leipzig: Teubner.

Matthiessen, Ludwig H.F. (1879). Ueber eine antike Auflösung des sogenannten Restproblemes in moderner Darstellung I. *Zeitschrift für Mathematischen und Naturwissenschaftlichen Unterricht* **10**, 106–110.

Matthiessen, Ludwig H.F. (1881a). Die Methode Tá jàn im Suán-king von Sun-tsè und ihre Verallgemeinerung durch Yih-hing im I. Abschnitte des Tá jàn li schu. *Zeitschrift für Mathematik und Physik* **26**, 33–37.

Matthiessen, Ludwig H.F. (1881b). Le problème des restes dans L'Ouvrage chinois Swan-king de Sun-tsze et dans L'Ouvrage Tayen lei schu de Yih-hing. *Comptes Rendus Hebdomaires des Séances de l'Académie des Sciences* **92** (Janvier–Juin), 291–294.

Matthiessen, Ludwig H.F. (1881c). Ueber das sogenannte Restproblem in den chinesischen Werken Swan-king von Sun-tsze und Tayen lei schu von Yih-hing. *Journal für die Reine und Angewandte Mathematik* **91**, 254–261.

Matthiessen, Ludwig H.F. (1882). Ueber eine antike Auflösung des sogenannten Restproblemes in moderner Darstellung II. *Zeitschrift für den mathematischen und naturwissenschaftlichen Unterricht* **13**, 187–190.

Mikami, Yoshio (1913). *The Development of Mathematics in China and Japan.* Volume 30 of *Abhandlungen zur Geschichte der mathematischen Wissenschaften mit Einschluss ihrer Anwendungen.* Leipzig: Teubner.

Rancière, Jacques (1996). Le concept d'anachronisme et la vérité de l'historien. *L'Inactuel* **6**(1), 53–68. (Engl transl. by Noel Fitzpatrick and Tim Stott: "The concept of anachronism and the Historian's Truth", InPrint(2015) **3**(1), 21–52; German transl. by Ronald Voullié n *Kunstgeschichtlichkeit* Eva Kernbauer (ed). Historizität und Anachronie in der Gegenwartskunst. München: Wilhelm Fink, (2015), pp. 33–50.)

Reuschle jun., Carl (1874). Ueber die unbestimmten Gleichungen ersten Grades. *Zeitschrift für Mathematik und Physik* **19**(3), 272.

Robson, Eleanor (2001). Neither Sherlock Holmes nor Babylon: a reassessment of Plimpton 322. *Historia Mathematica* **28**, 167–206.

Saito, Ken (1998). Mathematical reconstructions out, textual studies in: 30 years in the historiography of Greek mathematics. *Revue d'Histoire des Mathématiques* **4**, 131–142.

Schäfer, Johann Christoph (1831). *Die Wunder der Rechenkunst.* Weimar: Voigt.

Spoerhase, Carlos (2004). Zwischen den Zeiten: Anachronismus und Präsentismus in der Methodologie der historischen Wissenschaften. *Scientia Poetica* **8**, 169–240.

Stieltjes, Thomas-Joannes (1890). Étude bibliographique. Sur la théorie des nombres. *Annales de la Faculté des Sciences de Toulouse* **4**(3), 1–103.

Terquem, Olry (1862a). K.-L. Biernatzki, arithmétique et algèbre des Chinois. *Bulletin de Bibliographie, d'Histoire et Biographie Mathématiques* **8**, 35–44.

Terquem, Olry (1862b). K.-L. Biernatzki, arithmétique et algèbre des Chinois (fin) (voir Bulletin Mathématique, t. VIII). *Nouvelles Annales de Mathématiques* **2**, 529–540.

Unguru, Sabetai (1975). On the need to rewrite the history of Greek mathematics. *Archive for History of Exact Sciences* **15**(1), 67–114.

von Humboldt, Alexander (1847). *Kosmos. Entwurf einer physischen Weltbeschreibung*, volume 2. Stuttgart, Tübingen: Cotta'scher Verlag.

von Schäwen, Paul (1878a). Die Diophantischen Gleichungen ersten Grades. *Zeitschrift für den Mathematischen und Naturwissenschaftlichen Unterricht* **9**, 105–118.

von Schäwen, Paul (1878b). Nochmals die Diophantischen Gleichungen 1. Grades. Erwiderung auf Herrn Prof. Matthiessens Entgegnung. *Zeitschrift für den Mathematischen und Naturwissenschaftlichen Unterricht* **9**, 367f.

Wylie, Alexander (1852). Jottings on the science of Chinese arithmetics. *North China Herald* **108**, 10–11, Aug. 21, 1852; **111**, 23, Sept. 11, 1852; **112**, 26, Sept. 18, 1852; **113**, 30–31, Sept. 25, 1852; **116**, 42–43, Oct. 16, 1852; **117**, 46, Oct. 23, 1852; **119**, 55, Nov. 6, 1852; **120**, 59–60, Nov. 13, 1852; **121**, 62–63, Nov. 20, 1852.

Wylie, Alexander (1864a). The science of China. Algebra. *The Chinese and Japanese Repository of Facts and Events in Science, History, and Art Relating to Eastern Asia* **2**, (Aug. 12, 1864) (13), 22–25.

Wylie, A. (1864b). The science of the Chinese. Arithmetics. *The Chinese and Japanese Repository of Facts and Events in Science, History, and Art Relating to Eastern Asia* **1**, (April 3, 1864; May 3, 1864; June 3, 1864)(10–12), 411–417; 448–457; 494–500.

Wylie, Alexander (1864c). The science of the Chinese. Arithmetics. *The Chinese and Japanese Repository of Facts and Events in Science, History, and Art Relating to Eastern Asia* **2** (Sept. 12, 1864)(14), 69–73.

Wylie, Alexander (1882). Jottings on the science of the Chinese. *Copernicus* **2**, 169–195.

Wylie, Alexander (1897). Jottings on the science of the Chinese arithmetics. In *Chinese Researches by Alexander Wylie*, J. Thomas (ed). Shanghai, pp. 159–194.

6

Anachronism in the Renaissance historiography of mathematics

Robert Goulding
University of Notre Dame

Abstract: The historiography of mathematics in the Renaissance involves two kinds of anachronism. First, the tendency to anachronism in the authors themselves, whose understanding of broad historical structures and of the development of mathematics may be colored by concerns of their own time. And second, our anachronism in reading these histories of mathematics as if they were attempting precisely the same thing as modern historians of mathematics. This article focuses on the author of the first modern work dedicated to the history of mathematics, Petrus Ramus (1515–1572), singling out three occasions in which his historical account seems to diverge widely from that of modern historians, and examining them in the light of both types of anachronism. First, in his account of the development of mathematics in the early Platonic Academy; second, his assessment of Eudoxus of Cnidus; and finally in his dating of the Neoplatonic philosopher and commentator on Euclid, Proclus.

6.1 Introduction

The historiography of mathematics before Jean-Étienne Montucla's eighteenth-century pioneering work can, at first sight, seem to be a "mere jumble of names, dates, and titles," as one writer in the *Dictionary of Scientific Biography* put it (Scriba *et al.*, 2002, p. 112). Imitating ancient and medieval doxographies and chronicles, Renaissance authors seemed to do no more than collect the names of mathematical authors and practitioners, gathering under each chronologically ordered heading a list of works (extant or recorded in sources such as Diogenes Laertius or *Suidas*), some second-hand accounts of their discoveries, and a few improving anecdotes about their lives. Motivated by nothing more than antiquarian concerns, their attempts at a history of mathematics fell far short of the accounts of the development of ideas themselves that characterized the great Enlightenment histories from Montucla onwards. Moreover, the historical details they preserved were often wrong, and (to a modern eye) *obviously* wrong.

As I have argued at length elsewhere, while there is some truth to this assessment, it misses entirely the purpose and methodology of the history of mathematics in this period.[1] Renaissance scholars wrote on the history of mathematics precisely because they were interested in mathematical ideas, their epistemological status and origin, and the correct way of teaching them. Moreover, the historical errors themselves grew out of their preexisting convictions about the nature of mathematics or of philosophy. Some may have started out as simple blunders, but they were propagated because they seemed plausible given certain preconceptions that we no longer share.

Here, my focus will be on Petrus Ramus (or Pierre de la Ramée), the French logician and educational reformer at the University of Paris whose *Prooemium Mathematicum* of 1567 was the first great history of mathematics. He first published it in three substantial books, and then in 1569 republished it with minor variations as the first part of his 31-book *Scholae Mathematicae*, the extra 28 books of which were devoted to a systematic critique of Greek mathematics (especially Euclid), also incorporating a great deal of historical reasoning. Out of this vast work, for the purposes of this volume, I will consider three subjects where Ramus seems to embrace some kind of anachronism. The first two are idiosyncratically his own. First, his construction of Plato and the early Academy as sober-minded, anti-metaphysical materialists, a reading of the past quite out of step with the contemporary Neoplatonic revival – and an essential component of his historiography of mathematics. Next, and briefly, his assessment of Eudoxus of Cnidus, which seems so bizarrely different from our own. The third may seem the strangest of all; a literal anachronism, in which a major historical figure, the philosopher and mathematician Proclus, was misdated hundreds of years out of his

[1] See Goulding (2010), from which some of the argument of this chapter is taken.

own time. This error, however, was not Ramus's fault at all, but was a mere fragment of a larger mistaken narrative about the history of later Platonism – a fitting complement, in its way, to Ramus's own distortion of the early Academy. In this last case, it may seem scarcely credible to the modern scholar that such a patently false historical narrative should have been so widely accepted for so long. But given the imperfect and ambiguous sources at their disposal, as well as their very different understanding of the nature of philosophy, it is not entirely surprising that it seemed plausible at the time.

6.2 Ramus on the early Academy

From the beginning of his published career as a logician and reformer at the University of Paris, Ramus portrayed himself as a Platonist. His first publication, the *Dialecticae Institutiones* of 1543, ended with something he called "third judgement," a quasi-mystical intuition of the complete logical structure of the arts, which evoked the philosopher's escape from the cave in Plato's *Republic*. Ramus edited third judgement out of subsequent editions of his dialectic (Ong, 1958, pp. 189–190). But throughout his career he continued to insist that his famous "method" was nothing more than an elaboration of a passage from the *Phaedrus* (which he quoted in every edition), filled out with the dialectic of the *Philebus* and *Sophist*. And, to an extraordinary degree, Ramus saw himself not only as a faithful student of Socrates but, in a sense, as himself a Socrates sent to provoke the University of Paris – and subject to almost all the same hostility as his ancient teacher.[2]

But to Ramus, Plato looked quite different from the Plato of, say, Marsilio Ficino. Ramus's Plato was skeptical, undogmatic, practical and distrustful of metaphysics; he was, in other words, an inheritor and imitator of Ramus's philosophical hero, Socrates. Ramus's Platonism was, in some ways, in the same vein as the Academic skepticism of his close collaborator Omer Talon, who prefaced his important editions of Cicero's *Academica* in 1547 and 1550 with an essay lauding the skeptical

[2] See the autobiographical passage in the *Scholae dialecticae* V.4 (Ramus, 1569a, col. 155). At the end of his philosophical and dialectical self-education, he finally decided that the logic of the University was profoundly mistaken, and "I began to think to myself, 'Well, why shouldn't I socratize a bit?'" ("Coepi egomet mecum . . . sic cogitare, Hem? Quid vetat paulisper sôcratizein . . . ?"). Ramus compared his temporary ban from teaching philosophy, following his 1543 criticism of Aristotle, as analogous to Socrates's death sentence (Ramus, 1569b, p. 77).

or zetetic philosophy of the Academy as the only guarantee of a *libertas philosophandi*. The Academic philosophy, far from being a denial of knowledge itself, was a method of remaining open-minded before the competing claims of the schools. It amounted to withholding payment for philosophical goods until they had proved their worth (Talon, 1550, pp. 16–20). Yet Ramus was not a skeptic, and he did *not* identify Plato with the skeptical Academy of antiquity. Plato's positive guidance on methodology, and assurance (in Ramus's favorite passages) that an objective truth could be found, were central to Ramus's own epistemology. Nevertheless, he imagined a Platonism that would hold back from metaphysical speculation, that was epistemologically modest (though not actually skeptical), and that would accept only that which could be experienced, or *done*.

Two of Ramus's most interesting works in this context are his very little studied lectures on Aristotle's *Physics* and *Metaphysics*, published in 1565 and 1566. These works, the *Scholae Physicae* and *Scholae Metaphysicae*, consist of chapter by chapter refutations of Aristotle's works. They have become the focus of my next research project, because within them, between Ramus's spirited and frequently scurrilous abuse of Aristotle, he develops his own natural philosophy. Nature, he thinks, is atomistic. And it is also mathematical – but not, perhaps, for the reasons we might imagine. Ramus argues that *geometry too* is atomistic – lines are made up of minute, indivisible lines – and, for that reason, at a deep level physics and mathematics are the same[3] It was a doctrine that, Ramus argued, had been taught by Plato himself – and for Ramus, it fit perfectly with his conviction that Plato valued practice above all else, and shunned metaphysical subtleties. As he wrote in his *Scholae Physicae*,

> Geometry is not some empty and imaginary theoretical science. Like the other arts, it is something concrete and definite. The art of measurement does not measure magnitudes that can only be thought about and that never actually exist; rather, it measures sensible, real magnitudes. Nor is the geometer some kind of inert contemplator, but a busy and productive protagonist. He does not only conceive in his mind sections, productions and equalities of magnitudes and, from them, of figures, as well as the dimensions of lengths, breadths and heights; but he actually measures them in reality. He uses the rule, plumb and square. He does this not by the power of magnitude, from the potential of

[3] Goulding (2018). Ramus's atomistic mathematics and natural philosophy will be the subject of a forthcoming study of mine.

the geometer's soul, but from the nature of quantity, innate and intrinsic to magnitude.[4]

In other words, geometry ran up against the same practical limits as physical measurement. And since measurement would eventually come to an absolute physical minimum, so too would geometry. Plato's embrace of finite line elements was, to Ramus, the clearest sign that he rejected metaphysical speculation, whether in philosophy or mathematics.

Of course, in the general atmosphere of Renaissance Platonism, it was difficult to maintain Ramus's vision of a non-metaphysical Plato. It posed an additional challenge when he engaged in a systematic refutation of Aristotle: both because Aristotle *does* depict a metaphysically engaged Plato, and because Ramus was indebted, for many of his anti-Aristotelian arguments, to the Aristotelian commentators of the late Academy, whose reading of Plato (and often of Aristotle too) was thoroughly Neoplatonic.

When he wrote his *scholae* on the thirteenth and fourteenth books of the *Metaphysics*, for example, Ramus relied on the work of Proclus's teacher and predecessor as Platonic *diadoche*, Syrianus, who presented difficult challenges to him. The last two books of Aristotle's *Metaphysics* constituted a systematic attack on the theories of numbers and ideas of Plato and the early Academy (including, incidentally, the theory of indivisible lines); in his defense of Plato, Syrianus abandoned the usual even-handedness of the late Academy, and its conviction that Aristotle's work could be assimilated into the Platonic scheme, and instead paid Aristotle back like for like. It is a bruising commentary, in which Syrianus not only dismantled all of Aristotle's arguments, but taunted the Philosopher, belittled him sarcastically, and addressed him as a barely competent schoolboy.[5] In short, its *style* was much the same as that of Ramus's own *scholae*, and at times Ramus could barely contain his glee

[4] (Ramus, 1569a, col. 710): "Neque tamen geometria, inanis quaedam et phantastica contemplatio est, sed ars artium reliquarum similis est solida quaedam res et expressa; et metiendi denique ars, non eas magnitudines quae tantum cogitentur, nusquam autem sint, sed eas, quae sensibus percipiantur et revera sint. Neque geometres, est contemplator quidam iners, sed actor navus et industrius: neque tantum concipit animo sectiones, productiones, aequationes magnitudinum, et ex iis, figurarum, dimensiones denique longitudinum, latitudinum, altitudinum quaslibet, sed reipsa ac veritate metitur: regula denique perpendiculo, norma utitur: neque omnino potentia magnitudinis, ex potentia animi geometrici, sed ex quantitatis natura, magnitudini insita et ingenerata est."

[5] See Longo (2010, pp. 625–29); the author notes Syrianus's change of style in the commentary on books M and N, addressing Aristotle in the second person and with an intensity reflecting that, on this point, there was no reconciling Platonic and Aristotelian teachings. For the text of Syrianus's commentary, see Syrianus (2006).

at encountering a kindred spirit; he stole some of Syrianus's better lines and wove them back into his own *scholae*.

Yet, however much he liked Syrianus as a fellow enemy of Aristotle, the high Neoplatonic metaphysics of his commentary was quite uncongenial to Ramus. Worse still, Syrianus was concerned to defend a complicated, unworldly metaphysics of abstract ideas and numbers *that he attributed to Plato* – something that Ramus could hardly tolerate, given his own construction of a modest and pragmatic Plato. Ramus dealt with this quandary, as he did so often in his various *scholae*, by constructing a useful history. The particular narrative he reached for, he had in fact constructed at the very beginning of his *Metaphysics* lectures. At the end of book 1 of the *Metaphysics*, having surveyed earlier accounts of first principles, Aristotle turned to the theory of forms of his own teacher Plato – a theory, needless to say, that bore no resemblance to the undogmatic Socratism that Ramus preferred. Ramus explained how Aristotle came to attribute such doctrines to Plato:

> What's that? What was Plato's philosophy like? As Aristotle said, it came along after all those other philosophies [i.e., those of the Pythagoreans, Empedocles, etc.], but it diverged in different ways, according to the different minds of his followers. Plato's accounts of Ideas are varied and complex. He sometimes seems to mean some sort of separated exemplars of things, as in the *Phaedrus* and *Republic* X. Elsewhere, he purposefully undermines such fictions (*commenta*), as in the *Parmenides* – the book from which Plato's opinion about Ideas really ought to have been sought out. And for this reason, Plato's Academy divided according to contradictory teachings about the philosophy of Ideas. Xenocrates (and the members of the Academy after him), defended those crabbed fantasies of exemplary Ideas; Aristotle led a more sophisticated school, and one that hewed closer to logical truth.[6]

One should note here Ramus's use of a favorite critical term, *commenta* (or sometimes *commentitia*). In his historical account of the earliest philosophy, Ramus believed that the original philosophers hewed closely to nature itself; it was the later vanity of men such as Aristotle who replaced a philosophy that accurately mirrored nature with their own

[6] Ramus (1569a, col. 847): "Quid? Platonis philosophia qualis est? Ea successit (ait Aristoteles) post illas philosophias omnes; sed pro variis discipulorum ingeniis varie dissecta. Platonis sermones sunt varii de ideis et multiplices: videtur interdum separata quaedam rerum exemplaria dicere, ut in Phaedro, in 10 de Republica; alias de industria commenta illa labefactare, ut in Parmenide, quo in libro, sententia Platonis de ideis exquirenda fuit. Hinc Platonis academia contrariis haeresibus de idearum philosophia discissa est. Xenocrates, et ab eo Academici, spinosiora illa exemplarium idearum figmenta tutati sunt; Aristoteles sectam elegantiorem et logicae veritati germaniorem sectatus est."

contrived fantasies, out of a desire to impress others and build up their own personal prestige.[7] The force of the term here, then, implies not just a philosophical error but a deliberate obfuscation of reality which, in Ramus's reconstruction of history, was the catalyst to split the Platonic school in two:

Let us also recall from so many of my *scholae* against Aristotle, that Aristotle was of a most litigious and argumentative mind, so that he made it his business to fight against the Ideas in so many passages, and with so many words – to declaim against them, even, as if to have some reason to show off his ingenuity. On the other hand, we ought not entirely blame Aristotle's zeal for so varied an attack [on the Ideas]. As I said, the legacy of the Academy was split up between the sects of Xenocrates and Aristotle. And both the Xenocratics and the Aristotelians insisted that they were following Plato's philosophy of Ideas. And so Aristotle's criticism of the Ideas was principally directed against Xenocrates and his exemplars.[8]

Thus it was Xenocrates (and sometimes the Pythagoreans) who was really the object of Aristotle's arguments, both in *Metaphysics* I, and in the final two books. And Syrianus, the admirable critic of Aristotle, had to be treated carefully, as Ramus would later say, since he was a *patronus* of the Xenocratic *commenta* and Pythagorean *somnia* (Ramus, 1569a, col. 973). Yet even Syrianus, partisan Xenocratic that he was, seemed to Ramus to reveal the continuing existence of a schism in the Academy, almost every time he cited his predecessors' opinions on any matter. To give one more example:

These are Aristotle's arguments against the [theory of] mathematical abstraction of the Pythagoreans and Xenocratics, [a concept] which some of the Pythagoreans and Platonists, such as Severus the commentator on Plato, in fact understand according to this very doctrine of Aristotle. Others, however, cling

7 See Meerhoff (2001) in particular for an account of "nature" in Ramus's conception of the arts; (Goulding, 2009, pp. 73–4) on *commentitia* in philosophy and mathematics. (Ong, 1958, pp. 36–41) argued that Ramus never defended for his master's degree the thesis "Whatever was said by Aristotle is a fiction" ("*Quaecumque ab Aristotele dicta essent, commentitia esse*"), as was claimed posthumously of him; but even if he did not, the statement was entirely in his spirit.

8 Ramus (1569a, cols. 851–2): "Recordemur autem (quod Aristoteles antea jam nos admonuit) Platonis ideas a logica consideratione profectas, non aliud esse, quam in logica genera et species. Recordemur etiam e tam multis adversus Aristotelem scholis nostris, Aristotelis ingenium litigiosum valde, et contentiosum fuisse, ut adversus ideas, tot locis et tam multis verbis decertandum, et velut exercendi ingenii caussa, declamandum sibi proponeret; neque tamen tam multiplicis concertationis caussam totam Aristotelis alacritati tribuamus. Fuit enim Academiae patrimonium Xenocraticis et Aristoteleis sectis divisum, uti dictum est. Atque utrique tum Xenocratici, tum Aristotelei, Platonis philosophiam de ideis sectari se profitebantur. Quapropter Aristotelis concertatio de ideis adversus Xenocratem et Xenocratica idearum exemplaria, praecipue declamatur."

stubbornly to those imaginary entities of mathematical abstraction: men such as Plotinus, Iamblichus, Nicomachus, Numenius, Cronius, Amelius, and similar men cited by Syrianus. Imbued more with the scholastic wrangling of the Academy than the solid Platonic philosophy, they trifle away their time on such trifles.[9]

On the one hand, then, there were commentators like Severus (a Middle Platonist, as we would now call him),[10] who (Ramus believed) supported a commonsense philosophy of mathematics much like Aristotle's; on the other hand, there were many – mostly Neoplatonists, as we would say – who replaced the solid philosophy of Plato (that is, the Plato of Ramus's imagination) with fantastical metaphysics, driven by a desire for victory in scholastic argumentation. What Ramus was actually observing in the pages of Syrianus was an account of the development of the philosophy of mathematics, from Aristotle's theory of abstraction, to the mathematical realism of late Platonism, especially Proclus. He *interpreted* this development, however, as if in fact there were two schools that had coexisted since the time of Plato, battling not only over mathematical objects, but over the very meaning of Plato's philosophy itself.

No modern historian of philosophy would understand the history of the Platonic tradition in this way: Ramus was "wrong," and we might say that he was also *anachronistically* wrong. For, Ramus was encouraged in this reading of the past because he could only view the history of ancient philosophy through the filter of an institution like the University of Paris. Throughout his lectures on the *Metaphysics*, Ramus portrayed the variety of opinions over long periods of time, as though they were squabbles between rival professors and their students at his own institution (squabbles that he himself spent the whole of his career embroiled in). At times, he made the analogy quite explicit. At the end of the

9 Ramus (1569a, col. 974): "Haec Aristotelis argumenta sunt adversus Pythagoreorum et Xenocraticorum abstractionem mathematicam, quam Pythagoreorum et Platonicorum nonnulli ex praesenti Aristotelis sententia interpretantur, ut Severus Platonis interpres; alii autem mathematum abstractorum phantasticas illas essentias mordicus tuentur, ut Plotinus, Jamblichus, Nicomachus, Numenius, Cronius, Amelius et similes homines a Syriano citati, qui scholasticis Academiae contentionibus potius, quam solida Platonis philosophia imbuti, multa de his nugis nugantur."

10 Dillon (1996, pp. 262–4). In a passage cited by Dillon from the *Metaphysics* commentary, Syrianus criticized Severus for "misusing mathematical concepts in the explanation of physical questions," which Dillon takes to mean his theory of the soul and extension (p. 264). Ramus seems to have interpreted Syrianus to mean that Severus was a physicalist geometer after his own heart.

passage cited above on the division of the Platonic school between the followers of Xenocrates and those of Aristotle, Ramus added:

> This whole controversy in the schools of Greece provoked much uproar, much like that which the University of Paris had to put up with last century from the nominalists and realists, when one group of them argued that words were found in Aristotle's categories, and the other that things were found there; and they argued not just with words and disputations, but with blood, slaughter, and nigh on civil war and sedition, so that the unrest had to be put down by royal authority.[11]

Ramus was recalling here the edict of Louis XI in 1473, suppressing nominalism at the University of Paris.[12] Nothing of the sort took place in the history of ancient philosophy, of course. But Ramus's recollection of this event emphasized the contemporary relevance of the divisions he discerned in the Platonic school. In particular, this account reflected his own struggle to introduce into the university a mathematics founded in experience and practice, a mathematics which he believed had its roots in the deepest antiquity. His reading of history, then, was both shaped by his convictions about the nature of mathematics, and corroborated them. At the same time, as mistaken as his narrative might seem in its details, he drew upon historical texts that were at least capable of this interpretation. One of the most important things that he missed was the development of ideas over a long period; I will suggest later a reason why that may have seemed much less obvious to him than it is to us.

6.3 Mathematics in the early Academy

Ramus's lectures on physics and metaphysics were written in the early to mid-1560s, at the high point of his confidence in both his atomic natural philosophy, and in his vision of a practically minded Plato – a vision, as we see, that was bolstered by an account of the history of the Academy. His works of the late 1560s show a distinct change of tone. In 1566,

[11] Ramus (1569a, col. 847): "Haec vero controversia in scholis Graeciae maximas turbas excitavit, simillimas iis quas superiore saeculo Parisiensis Academia a Nominalibus et Realibus pertulit, cum alii nomina, alii res in Aristotelis categoriis et Praedicamentis esse contenderent, et quidem contenderent, non verbis tantum et argumentis, sed sanguine, caede, civili prope discordia et seditione, ut Regum authoritate fuerit opus ad tantos motus componendum."

[12] See Normore (2017). Ramus himself would have leaned towards the nominalist camp, though (as Normore notes) the terms even by this time had a strictly historical flavor to them, and were hardly ever used by Ramus. After this very passage, in fact, he stated that the whole of logic (not just the categories) was about words *and* things, so the quarrel was pointless. In that assessment, he was in the mainstream of humanist logic.

Ramus lost a protracted court battle, over the teaching of mathematics at the university, to his long-time rival Jacques Charpentier. Charpentier was both an Aristotelian and a Platonist, who had published several books on both philosophers, and on the harmony in their philosophy. In speeches and pamphlets, Charpentier attacked Ramus's philosophy mercilessly, including his atomism and his perverse reading of Plato.[13] As a result, Ramus abandoned his atomism entirely, and began to turn away from Plato as a champion of practical, non-metaphysical philosophy and mathematics, and instead to reconstruct him as an addled metaphysician, who led mathematics away from the world and into the realm of abstract speculation.

In the historical sections of the *Scholae Mathematicae*, Ramus drew on the ancient traditions, according to which Plato had learned mathematics in his travels around Greece, Italy, and Egypt. These were, as Ramus interpreted them, the three traditions of mathematics that had diverged since the time of the Fall, and which Plato had thus been able to reunite in his own person (Goulding, 2010, p. 43). The Academy, then, was an immensely important place in the history of mathematics, where mathematics was taught at a level last seen in the Garden of Eden itself. But it was also (like the Garden) a place where pride overcame goodness: Plato distorted his legacy of reunited mathematics, and urged his students to move away from practice to speculation.

To see how Ramus put this historical model to work, it is interesting to consider his account of the greatest mathematician of Plato's Academy, Eudoxus of Cnidus. In his biographical account in the first book of the mathematical lectures, Ramus began with Eudoxus's most famous and (from the point of view of modern mathematics) most lasting achievement: the fifth book of the *Elements* and the new theory of proportion developed there. Ramus dismissed it briefly: "In its own place," he says, "I will talk about those sophistical demonstrations."[14] He was more impressed by his astronomical models, his measurement of the length of the year, and above all the effort that he and Archytas of Tarentum applied to developing new instruments of practical mathematics. Ramus took this last detail from Plutarch's *Life of Marcellus*, which went on to say that Plato was annoyed with his two mathematicians for vulgar-

[13] For this court case, see Girot (1998); Loget (2004); Goulding (2010, pp. 50–56); for the ensuing turn away from atomism, see Goulding (2018, pp. 375–81).

[14] Ramus (1569b, p. 17): "Totum in Elementis quintum librum de analogiis invenit, de quarum sophisticis demonstrationibus dicetur suo loco."

izing mathematics and revealing secrets to the masses that should have been reserved for philosophers. This incident provoked one of Ramus's most passionate tirades against his erstwhile hero, Plato, accusing the philosopher of "womanish jealousy," and a fastidiousness about practical mathematics that had marked the beginning of its long decline to its present state: an obscure, abstract subject, neglected because of its uselessness and difficulty.[15]

And obscurity, abstraction, and difficulty were what Ramus found when he turned to the Eudoxan theory of proportion "in its own place," the thirteenth book of the *Scholae Mathematicae*.[16] Ramus found in it the typical sign of a mathematical theory that had been perverted from nature: *metabasis*, or the violation of his rule *de omni* that forbade one to go outside of an art. As he so often put it, "arithmetic must be treated arithmetically, geometry geometrically." In the Eudoxan theory, he believed he found an unholy mixture of arithmetic, geometry, and logic. The sense of magnitudes being in proportion was natural to all human beings and to nature itself; Eudoxus's definition of proportionality buried this natural sense under a mountain of sophisticated cleverness. As such, Eudoxus was one of many examples in the *Scholae Mathematicae* of a poor teacher, preening over his own contra-natural invention while the simple truth remained untaught.

But equally, he believed that Eudoxus's theory was simply *wrong*. And here we come up against Ramus's serious limitations as a mathematician: his enthusiasm for the subject and his erudition as a historian always greatly outpaced his mathematical abilities. His rejection of the Eudoxan theory shows an almost comical level of misunderstanding. In order to show Eudoxus's error, Ramus chose four numbers that, as he said, were definitely *not* in proportion: 4,3,5,4. And then he tried to show that, according to the Eudoxan definition, they actually *were* in proportion. He did this by offering two examples of equimultiples, in which the equality or inequality between the pairs of numbers was the same. If

15 Ramus (1569b, p. 18): "... ista pene muliebris zelotypia ..."

16 Ramus was concerned with the famous definition of equality of ratios, or proportionality, which is now listed as the fifth definition (numberings vary in Renaissance editions) of book V: "Magnitudes are said to be *in the same ratio*, the first to the second and the third to the fourth, when, if any equimultiples whatever are taken of the first and third, and any equimultiples whatever of the second and fourth, the former equimultiples alike exceed, are alike equal to, or alike fall short of, the latter equimultiples respectively taken in corresponding order." See Euclid (1956, 2, pp. 120–129) on this definition. For the challenges it presented to medieval and Renaissance mathematicians, see Murdoch (1963), Sylla (2008).

the first and third terms are multiplied by 6, and the second and fourth by 9, we have 24 < 27 and 30 < 36. And if the multipliers are instead 6 and 7, we have 24 > 21 and 30 > 28. Therefore, concluded Ramus, the pairs alike exceed or fall short of each other, and so by Eudoxus's criterion these numbers must be proportional, which they clearly are not. The flawed definition, he added, should be removed from geometry all together, one more casualty in his radical renovation of the whole discipline.[17] It was convenient to Ramus that he found (or imagined he found) such a factual error in the Eudoxan theory, since the theory itself could not be retained within Ramus's conception of a mathematics founded on nature and the experience of nature.

6.4 Proclus unstuck in time

The reform of Euclid's *Elements* was central to Ramus's mathematical program for the University; and in his lectures on the history of mathematics, he laid the basis for revising, or even replacing, the text of the *Elements* as the authoritative text on geometry. In particular, he raised doubts about the authorship of the text – not just whether it should be attributed to Euclid, but whether *anyone* had really overseen its composition. And he offered a historical source that, he believed, cast doubt on the accepted authorship of the work:

> Theon, it seems, far surpassed Euclid, and was the last "elementator." Indeed, the *Elements* of mathematics which are popularly attributed to Euclid should, it seems, be attributed to Theon. For, among his praises of Euclid, Proclus does not mention the discovery of a single proposition, but only the more careful construction of proofs. I have found solid confirmation of this by comparing the demonstrations in the first book of the *Elements* found in Proclus with any proof by Theon. Proclus, who lived before Theon, could neither have seen nor known about the later Theon. Proclus lived in the second century after Christ, Theon almost in the fourth. Proclus had Euclid's genuine proofs, in which he sometimes calls Euclid the "Elementator" par excellence, sometimes the Geometer, and sometimes by his name, Euclid.[18]

[17] Ramus (1569b, pp. 221–2). The problem, of course, it that Ramus did not understand that Eudoxus's definition requires the condition to hold for *all* equimultiples, not just two, and we can easily find a counterexample in this case. With the multipliers 3 and 4, we find that 12 = 12, but 15 < 16, so (by Eudoxus's definition) the four numbers are indeed not proportional.

[18] Ramus (1569b, p. 39): "Theon videtur Euclidem longissime superasse, et *stoikheiôtês* ultimus fuisse. Etenim mathematica elementa, quae Euclidi vulgo tribuuntur, videntur Theoni tribuenda. Nec enim ullius propositionis inventio inter Euclidis laudes a Proclo numeratur, sed demonstrationum accuratior explicatio. Cuius rei fidem amplissimam nactus sum, comparandis primo

Let us leave aside, for the moment, the details about which parts of the *Elements* are to be attributed to whom. What is most peculiar here is Ramus's chronological confusion. Theon of Alexandria lived squarely in the fourth century, not "almost in the fourth," as Ramus had it. But this is a very minor error compared to his confusion over Proclus, who flourished some hundred years *after* Theon, in the fifth century, not two hundred years *before*, as Ramus asserted.

The textual argument founded on the misdating of Proclus was crucial to Ramus's extended critique of the *Elements*. Proclus's *Commentary on the First Book of Euclid's Elements* was Ramus's single most important source for the philosophy and history of mathematics; and, misplaced some three hundred years out of his historical setting, he appeared to Ramus also to preserve, in his quotations of the text, a more primitive, more genuinely "Euclidean" version of the *Elements* than the text that had come down to Ramus's own age. On the foundation of this historical pseudo-fact, Ramus built his criticism of the *Elements*. It was, he claimed, a text that had been laid down haphazardly over the centuries. Accordingly, he analyzed it into Pythagorean, Euclidean, and Theonine strata. Built as it was by a historical process, a cohesive logical structure could no longer be perceived, he claimed. This unplanned text, argued Ramus, had no place in the mathematics classroom (and should be replaced by his own *Geometry* and *Arithmetic*, works built on natural, commonsense reasoning and practical application).[19]

Ramus's false dating of Proclus seems so obviously absurd: for one thing, it would make this late Neoplatonist earlier than the *founder* of the Neoplatonic school, Plotinus. But, as we shall see, in this case Ramus was not indulging his own historical imagination: the pseudo-facts about Proclus were widely accepted in the sixteenth century. While Ramus may be guilty of not thinking through the consequences of this date for Proclus, he was not alone in his negligence. There is an interesting, and tangled, story behind the origin of Ramus's error, and that story will be the main focus of the remainder of this article, in which I will examine both the evidence that contributed to the error, and how it gained

Elementorum libro Procli demonstrationibus cum Theonis qualibet demonstratione. Proclus aetate maior Theonem minorem neque videre, neque nosse potuit. Proclus floruit proximo post Christum seculo, Theon fere quarto. Proclus veras Euclidis demonstrationes habuit, in quibus appellatur Euclides per excellentiam modo *stoikheiôtês*, modo *geômetrês*, interdum suo nomine Euclides appellatur."

19 On Ramus's attacks on the *Elements*, see Goulding (2010, pp. 49–50; 170–72).

currency by opening up the possibility of subjecting the *Elements* to a rudimentary kind of source criticism.

To begin, then, with Euclid. As the quote from his mathematical lectures suggests, Ramus was interested in the date of Proclus's life because it might help to distinguish between the work of Euclid himself and that of his commentator Theon in the *Elements*. His concern with this problem reflected the general conviction in the sixteenth century that the *Elements* had been the work of *both* Euclid and Theon; most editors of and commentators on the *Elements* were convinced that Euclid was responsible only for the bare propositions and their arrangement, while Theon had provided all the demonstrations. This division of labor had been very recently arrived at. Ancient commentators, such as Proclus himself, thought of the work quite differently: the *Elements* was an account of mathematics that was entirely unified, both in its overall structure, and in the cohesion between the propositions and demonstrations. In the brief history of mathematics which he prefaced to the *Commentary*, Proclus wrote:[20]

> Not long after these men came Euclid, who brought together the *Elements*, systematizing many of the theorems of Eudoxus, perfecting many of Theaetetus, and putting in irrefutable demonstrable form propositions that had been rather loosely established by his predecessors.

And just in case there was any ambiguity about whether the demonstrations were an important part of Euclid's contribution to the *Elements*, Proclus went on to leave no doubt about the matter:[21]

> He also included reasonings of all sorts, both proofs founded on causes and proofs based on signs, but all of them impeccable, exact and appropriate to science. Besides these, the book contains all the dialectical methods: the method of division for finding kinds, definitions for making statements of essential

[20] Proclus (1873, p. 68; 1992, p. 56): "οὐ πόλυ δὲ τούτων νεώτερός ἐστιν Εὐκλείδης ὁ τὰ στοιχεῖα συναγαγὼν καὶ πολλὰ μὲν τῶν Εὐδόξου συντάξας, πολλὰ δὲ τῶν Θεαιτήτου τελεωσάμενος, ἔτι δὲ τὰ μαλακώτερον δεικνύμενα τοῖς ἔμπροσθεν εἰς ἀνελέγκτους ἀποδείξεις ἀναγαγών."

[21] Proclus (1873, pp. 69–70; 1992, pp. 57–58): "...ἔτι δὲ τοὺς τῶν συλλογισμῶν παντοίους τρόπους, τοὺς μὲν ἀπὸ τῶν αἰτίων λαμβάνοντας τὴν πίστιν, τοὺς δὲ ἀπὸ τεκμηρίων ὡρμημένους, πάντας δὲ ἀνελέκτους καὶ ἀκριβεῖς καὶ πρὸς ἐπιστήμην οἰκείους, πρὸς δὲ τούτοις τὰς μεθόδους ἁπάσας τὰς διαλεκτικάς, τὴν διαιρετικὴν ἐν ταῖς εὑρέσεσι τῶν εἰδῶν, τὴν δὲ ὁριστικὴν ἐν τοῖς οὐσιώδεσι λόγοις, τὴν δὲ ἀποδεικτικὴν ἐν τοῖς ἀπὸ τῶν ἀρχῶν εἰς τὰ ζητούμενα μεταβάσεσι, τὴν δὲ ἀναλυτικὴν ἐν ταῖς ἀπὸ τῶν ζητουμένων ἐπὶ τὰς ἀρχὰς ἀναστροφαῖς ... ἔτι δὲ λέγομεν τὴν συνέχειαν τῶν εὑρέσεων, τὴν οἰκονομίαν καὶ τὴν τάξιν τῶν τε προηγουμένων καὶ τῶν ἑπομένων, τὴν δύναμιν, μεθ' ἧς ἕκαστα παραδίδωσιν. ἢ καὶ τὸ τυχὸν προσθεὶς ἢ ἀφελὼν οὐκ ἐπιστήμης λανθάνεις ἀποπεσὼν καὶ εἰς τὸ ἐναντίον ψεῦδος καὶ τὴν ἄγνοιαν ὑπενεχθείς;"

properties, demonstrations for preceding from premises to conclusions, and analysis for passing in the reverse direction from conclusions to principles. [. . .] We mark also the coherence of its results, the economy and orderliness in its arrangements of primary and corollary propositions, and the cogency with which all the several parts are presented. Indeed, if you add or take away any detail whatever, are you not inadvertently leaving the way of science and being led down the opposite path of error and ignorance?

One of the very earliest readers of Proclus's *Commentary* in the Renaissance was the Venetian humanist Bartolomeo Zamberti. His preface to his 1505 Latin translation of the *Elements* praised the *Elements* as a work of incomparable unity, and a jewel of Platonic philosophy. In the same preface, ironically, he also put into currency the notion that Euclid had not himself written the demonstrations, a conclusion he reached while comparing the text of Euclid found in Byzantine manuscripts, with the common and already printed Latin version made by Johannes Campanus in the thirteenth century. Campanus's edition, as we now know, was a highly eclectic, composite text. The enunciations and order of the propositions were taken, for the most part, from various Arabo-Latin versions already in circulation; and the proofs were freely adapted from the same texts, or composed anew by Campanus, drawing on contemporary mathematical works such as Jordanus de Nemore's *Arithmetica*, Boethius, and other medieval commentary material. It would be quite accurate to say that Campanus did indeed replace Euclid's demonstrations with his own commentaries; indeed, the printer of the 1482 edition described the work he was publishing as "the *Elements* of Euclid of Megara, together with the commentaries on the geometrical art by the most perceptive Campanus."[22]

Zamberti intended his new version, made directly from the Greek, to replace this standard translation. On the title page, he described *his* edition as "the thirteen books of Euclid's *Elements* with the exposition of the great mathematician Theon. Many things that were missing in the translation of Campanus have been added from the Greek text, and many things that were disordered and absurd have been returned to order and corrected."[23] Zamberti assumed that the novel (and, he thought, better)

[22] Euclid (1482, colophon): "Opus elementorum euclidis megarensis in geometriam artem, in id quoque Campani perspicacissimi Commentationes finiunt."

[23] Euclid (1505): ". . . elementorum libros xiii cum expositione Theonis insignis mathematici, quibus multa quae deerant ex lectione graeca sumpta addita su[nt], necnon plurima subversa et prepostere [con]voluta in Campani interpretatione, ordinata, digesta, et castigata sunt."

demonstrations he found in the Greek text bore the same relation to the propositions as Campanus's commentaries did to his Latin version of the propositions: they were commentaries, and not integral parts of the text. And he attributed the authorship of these "commentaries" to Theon, whose name he found at the head of most Euclid manuscripts, where it was stated that the text was taken ἐκ τῶν συνουσιῶν Θέωνος ("from the lectures of Theon") – a phrase whose precise meaning would be controversial. In the *catena* of sources about Euclid and the *Elements* that Zamberti included in his edition, he quoted in full the passage from Proclus on the unity of the *Elements* and the excellence of his demonstrations. But in his own preface, Zamberti attributed only the statement and arrangement of the propositions to Euclid, praising Theon at much greater length:

> And you should see how great was the perception, skill and learning of the commentator, Theon, who explains the sublime sense of the problems and theorems in quite marvellous order, and makes them clear through his investigations. Through the preliminary specification, he sets out what the questions demand. Through the construction, he constructs and builds up marvellously that which is said [in words]. Then, in the proof, he proves the question, laying it out to the senses. And finally, he closes with a conclusion that is both valid and most stable, tying it up so tightly that one would hardly dare to deny what has been proposed and then proven.[24]

Zamberti's edition, together with its prefatory essays and collections of texts, enjoyed great authority over the next century, with the result that the demonstrations of the *Elements* became closely associated with the name of Theon – an identification that persisted into the eighteenth century. It became commonplace in sixteenth-century editions to alter the text of Theon quite drastically (but *not* that of Euclid), improving the proofs, replacing them with new ones, or simply omitting them altogether. One of the few voices questioning the attribution of the proofs to Theon was the mathematician Jean Borrel (or Johannes Buteo). In an appendix to his 1559 work on the quadrature of the circle he attacked modern editors for their cavalier approach to the text, replacing the original

24 Euclid (1505, fol. 6): "Videasque quanta sit acuitas, quantum sit ingenium, quantaque doctrina Theonis ipsius interpretis, qui miro quoddam ordine sublimes problematum et theorematum sensus explicat, magnaque indagine patefacit, per prodiorismum nanque ea quae in quaestionibus posita sunt proponit; per constructionem ea quae dicuntur construit et mirabiliter aedificat; inde per demonstrationem comprobat sensui subiiciens, postmodum conclusione firmissima et valida claudit, et astringit adeo ut ea quae proposita et comprobata sunt minime negare audeamus." In dividing the parts of the demonstration in this way, Zamberti ironically echoes Proclus's analysis of *Euclid's* proofs.

proofs with ones of their own devising that frequently did not even make mathematical sense. The very assumption on which they had rejected the original proofs of the *Elements*, was, he said, misguided: it was simply an error that Theon of Alexandria had written the demonstrations of the *Elements*. The mistake arose, he wrote, from a misunderstanding of the title found in the Greek codices, ἐκ τῶν συνουσιῶν Θέωνος, a phrase which, he argued, should be translated "from the lectures" or "from the expositions," and not, as these translators had apparently understood it, "from the demonstrations," a phrase that would be quite different in Greek. Borrel insisted that it would be unheard of for any ancient geometer to publish his theorems without proofs. Moreover, ancient authors unanimously attributed the *Elements* to Euclid alone, and Proclus in particular testified that Euclid was the author of the theorems and proofs alike (Borrel, 1559, pp. 209–11).

Borrel thought that the "lectures" of Theon were now lost; all that had survived was the text of Euclid that was once read alongside it. But he assumed that the lectures must have been much like Proclus's commentary – and, like the commentary, quite distinct from the text of the *Elements* itself: [25]

> I would not deny that Theon did some work on the demonstrations in that work . . . but I do insist that he did this separately and distinctly, in the course of commenting on some passages. This is just what Proclus did with the first book of the *Elements*. He brought in demonstrations of his own and of others everywhere, but distinguished them from those that we now have in the Greek text by mentioning the author [i.e., Euclid], whom he most often calls the elementator or the geometer, and sometimes calls by his actual name.

Borrel's point was that Proclus quoted demonstrations *as they were now extant in the Greek text*, and attributed them explicitly to Euclid; other material, not found in the received text of the *Elements*, he was careful to attribute to other authors, or to himself. This was, in its own way, a momentous observation. Up to this point, Proclus has been read by sixteenth-century mathematicians as an authority who had important things to say *about* Euclid; Borrel recognized that Proclus was also a witness to the text as it stood in antiquity – and later readers of Proclus

[25] Borrel (1559, p. 210): "Non autem negaverim Theonem aliquid demonstrationum in eo opere fecisse . . . Hoc autem dico factum separatim atque distincte inter exponendum locis quibusdam. Quemadmodum et fecit Proclus in primum Elementorum. Nam suas et aliorum demonstrationes passim adducens, ab his quas habemus in Graecis libris authoris mensione distinguit quem vel stoikheiotên, vel geômetrên saepius appellat, interdum etiam nomine proprio."

would be looking as much to his quotations from Euclid, as to his commentaries upon them. But if Proclus provided a snapshot of the text, when was that snapshot taken? Obviously his quotations from Euclid were a much older witness than the extant Greek manuscripts of Euclid themselves – but how much older? When, precisely, did Proclus live? And this is where the story begins to become very odd.

One might expect that Marsilio Ficino, who had translated and commented upon such vast amounts of Proclus's philosophy, would be a trustworthy guide to the chronology of the Platonic school, and the date of Proclus's life and writings. He wrote, for instance, a historical discursus in his *Platonic Theology*, greatly expanding upon the account of the Platonic succession that Proclus had inserted into his own *Platonic Theology*, and placing the history of the Academy within the history of the Christian revelation (Allen, 1998, pp. 70–75). In fact, though, his accounts here and elsewhere of the Platonic succession were just that: a connected chain of teachers and students, correct in almost all of its details, but referring to no external historical events except the birth of Christ. The subsequent authors that I will examine all knew Ficino's order of succession, either directly or indirectly, and often drew on it to place their historical references in the right order: he provided a dependable *relative* history. But Ficino gave them no help whatsoever in discovering the *absolute* place in history of their philosophers. The text that sixteenth-century authors consulted on the date of Proclus and other Platonic philosophers was of altogether lesser philosophical and intellectual calibre.

In 1506, Raffaele Maffei published a humanist encyclopedia, entitled the *Commentaria Urbana*. A substantial part of the work was devoted to an enormous collection of biographies of philosophers and writers, arranged alphabetically, detailing the extant and lost writings of almost every known ancient author. The work was reprinted many times, and this list was habitually consulted, though seldom acknowledged by later writers: the Wikipedia of the early sixteenth century, and often much less reliable. And here we find the seeds of almost all the subsequent confusion over Proclus.

For the entries that are relevant to our problem, Maffei relied largely on the Byzantine encyclopedia *Suda*; from his own wider reading and by inference, he made connections between authors and added dates (which *Suda* itself rarely did). It was in this attempt at reconciliation between sources that the limitations of Maffei's scholarship became apparent: often he made non-existent connections on the basis of mere similarity of names – without, moreover, much historical consistency across separate entries.

To begin, then, with his entry on Proclus. He began simply by translating the *Suda* entry on Proclus of Lycia. This entry tells us, among much else, that Proclus was a Platonic philosopher – the head, in fact, of the Athenian Academy. He was a student of Syrianus, and the teacher of Marinus of Neapolis. *Suda* mentions his commentaries on Homer and on Plato's *Republic*; Maffei confirmed that this Proclus was the author of all the commentaries on Platonic dialogues that appear under the name Proclus (neither he nor Suda mentions the commentary on Euclid). Suda and Maffei both went on to record that he was the second author, after Porphyry, to attack the Christians, and was subsequently attacked by John Philoponus. Maffei then added a sentence that would create no end of subsequent trouble:[26]

> He was also the tutor of M. Antoninus [that is, Emperor Marcus Aurelius], who raised him to the consulship, as Spartianus tells us.

Maffei's careless reference here is to the *Historia Augusta*, the notoriously unreliable lives of the later emperors, supposedly written by Aelius Spartianus and several other authors. The passage Maffei is thinking of comes from the life of Marcus Aurelius (attributed in fact to Julius Capitolinus, not Spartianus), where we read:[27]

> Besides these, his teachers in grammar were the Greek Alexander of Cotiaeum, and the Latins Trosius Aper, Pollio, and Eutychius Proculus of Sicca. ... he advanced Proculus to a proconsulship, though assumed the [financial] burdens of the office himself.

This is, without a doubt, the passage Maffei had in mind. Yet the incompatibility of this passage with the Greek philosopher Proclus seems evident. We are told here of a philosopher called Eutychius Proculus. Proculus is the Latin version of the Greek Proclus, it is true; but the text takes pains to distinguish him explicitly from the Greek philosophers attending the emperor, identifying him as a Latin, and states his origin as

26 Maffei (1506, fol. 259v): "Praeceptor etiam M. Antonini, quem ad consulatum usque provexit, ut autor Spartianus."

27 [Julius Capitolinus], *Vita M. Antonini* (*Historia Augusta*), 2: "Vsus praeterea grammaticis, Graeco Alexandro Cotiaeensi, Latinis Trosio Apro et Pollione et Eutychio Proculo Siccensi. ... Proculum vero usque ad proconsulatum provexit oneribus in se receptis."

the town of Sicca [Veneria] in North Africa, not Lycia in Asia Minor, the home of the Platonic philosopher. Still, Maffei does not hesitate to identify the two men, and on the slenderest of evidence, Proclus, the Platonic philosopher of the fifth century, became tutor to the philosopher-emperor Marcus Aurelius (who ruled 161–180).

If we now cross-reference the other figures mentioned in Maffei's biography, we find the confusions multiplying. Looking up Proclus's student and biographer Marinus, we read that he was a philosopher and orator, well versed in Greek learning, who wrote a life of Proclus "in verse and prose." Thus far from *Suda*, from which Maffei also reported that he succeeded Proclus as head of the Academy "under the Emperor Hadrian" – the last words were, of course, added by Maffei. This would mean that Marinus succeeded Proclus in the period 117–38, at least 23 years before Proclus was imperial tutor (not chronologically impossible, though it implies that Proclus returned to philosophical work long after his retirement). The source of Maffei's confusion here was the existence of *another* Marinus, a geographer whom Ptolemy cited in his *Geography*, and who indeed did flourish under the early Antonines.

The case of Syrianus, the teacher of Proclus, whose commentary on the *Metaphysics* so engaged Ramus, is rather more complicated. Maffei did not have an entry for Syrianus himself, but he did have one for Syrianus's teacher, Plutarch. In his entry for the latter, he managed to distinguish (quite uncharacteristically) between the Plutarch of Athens who taught Syrianus and the other more famous Platonist, Plutarch of Chaeronea, who wrote the *Moralia* and the *Parallel Lives*. The philosopher and head of the Academy, he wrote, flourished under Julian the Apostate (in other words, the late fourth century), while the biographer lived during the reign of Trajan, in the second century.[28] If Maffei had joined all the dots here, he would have ended up with an approximately correct date for Proclus himself: if his teacher Syrianus flourished in the late fourth century, then clearly Proclus himself must have lived in the fifth century.

But confusion once again reigns when we begin to search out other members of the late Platonic Academy. If we examine the entry for "Hermes," for instance, alongside the expected Hermes Trismegistus,

[28] Maffei (1506, fol. 255v–256): "Plutarchus Cheronaeus Traiani temporibus, a quo in Illyricum missus consulari potestate . . . Scripsit complura praeter vitas parallelas iam vulgo notas. Plutarchus Nestorii F. Atheniensis sophista, praeceptor Syriani, qui scholae praesidebat Atheniensi sub Iuliano principe. Autor utriusque Suidas."

we are told there was another Hermes, this one the pupil of Syrianus and thus a "condiscipulus" of Proclus – "under Hadrian," added Maffei helpfully.[29] Although in another entry he seemed to realize correctly that Syrianus lived in the fourth century, he now has pushed him back in the second century. What has happened? It seems that, while writing the entry for the "other Hermes," Maffei realized that he must have been a fellow student with Proclus – and thus, according to his own entries for Proclus and Marinus, a contemporary of Hadrian. Syrianus, therefore, was also dragged back into the second century, despite Maffei's earlier and correct realization that he flourished under Julian.

This is as far as I shall take Maffei, although the confusions and inconsistencies multiply far beyond this, as one follows the thread of each member of the Platonic school. It is enough to see that checking and cross-checking a few entries in the *Commentaria* could give the reader the impression that Proclus lived in the second century, with the whole Platonic school arrayed around him, in the correct order, at least. And it seems that several later scholars made precisely this inference. The German mathematician Johannes Stöffler published in 1534 an edition of the Latin version of the *Sphere*, an elementary astronomical work falsely attributed to Proclus. In his preface, he set out a brief biography of the putative author, writing:[30]

> Proclus of the Lycian nation, pupil of Syrianus, was a Platonic philosopher, the head of the school at Athens. Syrianus of Alexandria, a Platonic philosopher, taught at Athens. His pupil and successor was our Proclus. Marinus of Naples, a philosopher and orator, and well-versed in Greek learning, was Proclus's pupil and successor. He wrote, in the time of Hadrian, about Proclus' life and works, in verse and prose. Therefore, by conjecture, Proclus lived in the time of Trajan or thereabouts.

Stöffler's indebtedness to Maffei is obvious; his information on Marinus, in particular, he copied *verbatim* from the *Commentaria*. But he can be forgiven for not recognizing the error; Stöffler was no Greek scholar, as the rest of his preface demonstrates. He puzzled over Proclus's cognomen

29 Maffei (1506, fol. 214): "Hermes alter item, philosophus Aegyptius, auditor Syriani sophistae, condiscipulus Procli sub Hadriano principe. . . . Haec Suidas."

30 Stöffler (1534, fol. 1): "Quartus et est noster Proclus, natione Lycius, discipulus Syriani; philosophus Platonicus, praefuit scholae Atheniensi. Syrianus Alexandrinus philosophus Platonicus, docuit Athenis. Huius discipulus et successor fuit noster Proclus. Marinus Neapolitanus philosophus et orator, graece eruditus, Procli discipulus et successor, sub Adriano scripsit ipsius Procli et vitam et dissertationes, versibus et soluta oratione. Quare iuxta coniecturam, Proclus floruit sub Traiani temporibus, aut circiter."

"Diadochus" (which, of course, means "successor" to the head of the Academy). Stöffler was convinced, however, that it was a reference to a variety of beryl mentioned by Pliny called *diadochos*, and concluded that, if Proclus were German, he would have a surname that referred to hard stones or mountains, such as Herttenfelser, Trakkenfelser, or Gryffenfelser!

It is interesting to note that, after Maffei, the false dating of Proclus is found only in introductions to *mathematical* works attributed to the philosopher, despite the fact that Maffei says nothing about Proclus's mathematical writings in his entry on Proclus. To give another example, in 1553, some 20 years after Stöffler, Martin Hopper published in Antwerp a translation of the pseudo-Proclan *Sphere*, in the preface to which he quoted *verbatim* (though without credit), the whole of Maffei's entry on Proclus. Eight years later, he published a deluxe edition of the same text, in Basel, with copious commentary by the Austrian mathematical humanist Erasmus Oswald Schreckenfuchs. He handed over to Schreckenfuchs also the task of writing a new life of Proclus. The biography shows a much greater knowledge of the range of Proclus's writings than any previous account of his life and work; but, at heart, it preserves the same chronological confusion, deriving directly or indirectly from Maffei (one notes that he preserves Stöffler's absurd detail that Diadochus was Proclus's "family name"):[31]

> As far as can be determined, Proclus was of the Lycian nation, and from his family or ancestry had the surname Diadochus; a student of Syrianus, Platonist philosopher, he was the head of the Athenian school. This Proclus wrote many extraordinary works of philosophy, mathematics, and grammar, commentaries on the whole of Homer. Certain of his writings have reached us, namely his commentaries on Plato, outlines of astronomy, construction of the astrolabe, and this little handbook on the celestial circles. He also wrote many things against the Christians. He lived and flourished in the age of Trajan.

None of these authors was a true Greek scholar; they were mathematicians and editors, republishing a pseudo-Proclan text that had originally been translated into Latin by Thomas Linacre (who, incidentally, in *his*

[31] Proclus (1561, pp. 2–3): "[Proclus], quatenus constat, natione fuit Lycius, Diadochus a familia seu prosapia cognominatus, discipulus Syriani, Philosophus Platonicus, Scholae Atheniensi praefectus. Hic, inquam, Proclus scripsit multa & egregia in Philosophia, mathematica & grammatica, commentarios in totum Homerum; ex cuius scriptis, haec ad nostros devenerunt, scilicet Commentarii in Platonem, Hypotyposes Astronomicorum, fabrica Astrolabii, & hoc compendiolum de circulis coelestibus. Scripsit etiam multa contra Christianos. Vixit & floruit Traiani temporibus."

editions of the text says nothing about the identity of Proclus whatsoever). It is perhaps forgivable that they rifled through Maffei to fill out their introductions to Proclus's *Sphere*. It is much more surprising to find a scholar of the stature of Francesco Barozzi relying on Maffei and committing the same chronological error. Barozzi's 1560 Latin translation of Proclus's *Commentary* on Euclid was a masterpiece of scholarship. Rejecting the very poor Greek *editio princeps* of the work, he established his own text from several manuscripts. Although he did not publish his Greek text, modern editors have used his Latin translation to infer Barozzi's excellent readings of the Greek. Yet his scholarship seemed to desert him in his preface, where he wrote:[32]

> I would first like you to know that, although there were several Procluses, one was the most famous, who had the cognomen "Diadochus," that is, successor. He was from Lycia, a Platonic philosopher and extraordinary mathematician, who (if we are to believe *Suda*) was a student of the great Syrianus. When he became head of the Athenian school, he had many students. A notable one was Marinus of Neapolis, and another was M. Antoninus, by whom he was raised to the consulate, as Spartianus records.

Barozzi lifted all of this directly from Maffei; the knowing, but inaccurate reference to "Spartianus" is a dead giveaway. Like a sly student cribbing from Wikipedia, however, he added a few flourishes here and there to make it *look* like original research: the arch "if we are to believe Suda," for example. Ramus used this translation of the *Commentary* in preference to the Greek throughout his lectures on mathematics; and here, no doubt, in Barozzi's confident survey of evidence he had never apparently examined himself, was the source of Ramus's own misdating of Proclus to the second century.

To return, then, to Ramus. His dating of Proclus was the one part of his argument, we now see, that was *not* controversial. Everywhere he might have gone to check, he would have been told that the author of the *Commentary* on the first book of Euclid lived well before Theon, who was the final editor of the *Elements* and, as Ramus repeatedly insisted, the last mathematician of antiquity. Equipped, then, with Proclus's more primitive, pre-Theonine version of the *Elements*, he set out to dissect the

[32] (Proclus, 1560, sig. **2): "Primum itaque te scire velim praeter alios multos Proclos, unum Clarissimum omnium fuisse, cognomine Diadochum, hoc est successorem, patria Lycium, Platonicum philosophum, Mathematicumque praestantissimum, qui (si Suidae credendum est) magni Syriani fuit discipulus, cumque Atheniensi Scholae praefuisset, alios ipse discipulos habuit, e quorum numero unus, insignisque fuit Marinus Neapolitanus eius successor; alter M. Antoninus, a quo etiam (ut refert Spartianus) ad consulatum usque provectus fuit."

text historically. He was able to discover, in fact, very few substantial differences between the text of Euclid cited by Proclus, and the received Greek text. Rather, there were many, very small changes, some of which did indeed reflect editorial changes by Theon, but most of which were nothing but trivial variants. But those differences that he *did* find, he magnified and returned to obsessively through his *Scholae Mathematicae*. He insisted that alterations had been made, and many, but he gathered from them as a whole that Theon had not made any *drastic* changes to the *Elements*. But he had, nevertheless, made small incremental, almost random changes, of exactly the sort that Ramus believed had been the way that the *Elements* had grown. It was a sedimentary text, laid down over centuries, of which Theon was as much the author as Euclid had ever been. Or, to put it another way, as *little* as Euclid had been; the *Elements* was a text without an author, an unplanned, and unsatisfactory product of history, not of reasoning men. "Now nothing is left to Euclid but an empty name," he concluded triumphantly.[33]

Ramus's *Scholae Mathematicae*, the first ambitious history of mathematics ever written, were widely read in Europe, and helped to propagate the false dating of Proclus more widely, at least for a short time. It made its way into Ramist textbooks, like Johannes Freig's *Quaestiones* on Euclid and Ramus.[34] Elsewhere, I have examined at length the lectures on the history of mathematics by the young Oxford scholar Henry Savile, who based himself solidly on Ramus's *Scholae* (though disagreeing philosophically with them), and also accepted the false date of Proclus as a means to examining the textual history of the *Elements* (Goulding, 2010, pp. 173–4). But eventually, the false dating of Proclus faded with the sixteenth century like a fever dream. In 1592, Johann Jakob Fries published his *Bibliotheca philosophorum classicorum*, in which he provided without any fanfare or explanation an accurate enough chronology of Proclus – and, indeed, of all of the members of the Platonic school (Fries, 1592, p. 41). It marked the beginning of the modern historical understanding of late-antique philosophy; and after its publication, the second-century Proclus was never mentioned again. Indeed, it is completely forgotten today that anyone in the Renaissance held this bizarre belief.

33 Ramus (1569b, p. 39): "ut Euclidi praeter inane nomen nihil admodum relinquatur."

34 Freig (1583, sig. α8v). In the catalogue of mathematicians and astronomers that Freig extracted from Ramus's history, Proclus is located *before* Theon, who (as Ramus often stated in the *Scholae*) was the last Greek mathematician.

There were circumstances that encouraged humanist scholars to accept a wildly anachronistic date for Proclus. First, there was simply the repetition of the "fact" in encyclopedias, handbooks, and introductions; for all their professed desires to return to the clear springs of Greek and Latin texts, humanist writers were addicted to compendia and shortcuts. An idea that made it into the standard references took on a life of its own. Moreover, there were aspects of the false story that were attractive to humanists. The fact that Proclus was not only the head of a school, but also beloved by an emperor and raised to the consulship, no doubt flattered their own sense of the closeness of scholarship to power, and the possibilities of patronage. For much the same reason, they retold *ad nauseam* the story of Euclid trading witticisms with the Ptolemaic ruler of Egypt, or Aristotle advising Alexander.

6.5 Conclusion

Renaissance historians of mathematics and philosophy inherited from their ancient models (such as Diogenes Laertius and Proclus himself) a pair of complementary structuring patterns for the history of learning: there were competing schools of thought into which everyone, one way or another, must be fitted; and there was a chain of masters and pupils passing on the teachings of the school. Even though Proclus himself used these models,[35] and the "golden chain" of Platonic teachers reaching back to Plato was proverbial, it was not in fact a model that fitted Platonism very well. For periods of its history at least, Platonism was a disordered and syncretistic tradition, rather than a well-ordered "school." And, as an institution, it had a broken history; the original Academy (if it is even possible to speak of such a thing)[36] came to an end in the time of Sulla, and the school which Proclus headed was a very recent foundation by Plutarch of Athens. It was simply not possible to draw an unbroken line, master to student, across the thousand years or more from Plato to the last scholarch Damascius. It was not even possible to complete the chain over the 250 years from Plotinus to Proclus.

It is for this reason, I believe, that the compressed history of Platonism entailed by an early Proclus was so plausible. Over a millennium, it was impossible to draw continuous lines of succession, while the persever-

35 In, for example, the opening of his *Platonic Theology*.

36 See Dillon (1983) on anachronistic notions of the early Academy as an institution.

ance of an unchanged tradition (as Renaissance Platonists believed it to be, following Proclus) over such a vast stretch of time must have seemed unlikely. But over 500 years, a school could be sustained, and a chain of unbroken learning from historically misplaced teacher to historically misplaced student could be traced. A Proclus who lived in the second century made it possible to knit together the Platonic school and the succession within it. One can now begin to see perhaps why Ramus, in his account of warring schools in the early Academy, was so oblivious to philosophical development. He, too, was committed to the idea of all but unchanging philosophical schools; and the much shorter time-frame in which ancient philosophy had to unfold only encouraged him to see the tradition as more or less static, but divided. Any change that he could *detect* within this truncated period of time he assumed was for the worse – examples of poor pedagogy obscuring natural simplicity, as attested by his understanding (or misunderstanding) of the Eudoxan theory of proportion.

To return to the theme of this volume, the Renaissance historiography of mathematics involves two parallel anachronisms. The first consists of the anachronistic errors one can easily uncover in any Renaissance work on the history of mathematics. Ramus's peculiar account of the early Academy and his dismissal of Eudoxus as a mathematician both stem in their own way from his attempt to impose his contemporary ideas of institutional structure and university pedagogy onto the ancient mathematical past. The false dating of Proclus, a very literal anachronism shared by many besides Ramus, was very likely encouraged by a conception of a tightly knit and neatly organized Platonic school which is lost to us now, and which bears no resemblance to the modern historical understanding of ancient Platonism.

But if we criticize Renaissance historiography for making these errors, for jumbling together names, dates, and titles, we ourselves commit a second, parallel anachronism. We forget, for instance, how recently the modern understandings of the history of mathematics and of philosophy have emerged. The broadly well-established narratives of the history of Platonism, or the transformations in ancient mathematics brought about by the discovery of incommensurables, are creations only of the nineteenth and twentieth centuries, and of course are still subject to constant revision. We have only very recently become "modern" in our understanding of the philosophical and mathematical past. And our current understanding is almost certainly conditioned by preconceptions about the nature of the ancient world that we are not even aware of. When we dismiss the past's conception of the past, we risk missing the peculiar logic that held the narrative together. One may still conclude that the accounts Ramus and his contemporaries wrote about ancient mathematics and philosophy tell us little useful about antiquity; but one cannot deny that they speak eloquently about the Renaissance itself. Perhaps our distant successors will come to a similar determination about the intellectual histories we write too.

References

Allen, Michael J.B. (1998). *Synoptic Art: Marsilio Ficino on the History of Platonic Interpretation*. Vol. 40. Istituto Nazionale di Studi sul Rinascimento: Studi e Testi. Florence: Leo S. Olschki.

Borrel, Jean (1559). *De quadratura circuli libri duo, ubi multorum quadraturae confutantur, et ab omni impugnatione defenditur Archimedes*. Lyon: G. Rouillé.

Dillon, John M. (1983). What happened to Plato's garden? *Hermathena* **134**, 51–9.

Dillon, John M. (1996). *The Middle Platonists, 80BC to 220AD*. Ithaca, NY: Cornell University Press.

Euclid (1482). *Preclarissimus liber elementorum Euclidis perspicacissimi*. Venice: Erhard Ratdolt.

Euclid (1505). *Euclidis megarensis philosophi platonici mathematicarum disciplinarum janitoris ... Elementorum libros xiij habent*. Edited by Bartolomeo Zamberti. Translated by Bartolomeo Zamberti. Venice.

Euclid (1956). *Elements, Vol. 2*, Second Edition. Edited by T.L. Heath in 3 volumes. New York: Dover Publications.

Freig, Johannes Thomas (1583). *Quaestiones geometricae et stereometricae in Euclidis et Rami* στοιχείωσιν. Basel: per Sebastianum Henricpetri.

Fries, Johann Jakob (1592). *Bibliotheca philosophorum classicorum authorum chronologica*. Zurich: Apud Ioannem VVolphium typis Frosch.

Girot, Jean-Eudes (1998). La notion de lecteur royal: le cas de René Guillon (1500–1570). In: *Les Origines du Collège de France (1500–1560): actes du colloque international (Paris, décembre 1995)*, Marc Fumaroli (ed). Paris: Klincksieck, pp. 43–108.

Goulding, Robert (2009). Pythagoras in Paris: Petrus Ramus imagines the prehistory of mathematics. *Configurations* **17**, 51–86.

Goulding, Robert (2010). *Defending Hypatia: Ramus, Savile, and the Renaissance Rediscovery of Mathematical History*. New York: Springer.

Goulding, Robert (2018). Five versions of Ramus's geometry. In: *Et Amicorum: Essays on Renaissance Humanism and Philosophy in Honour of Jill Kraye*, Margaret Meserve and Anthony Ossa-Richardson (eds). Leiden: Brill, pp. 355–387.

Loget, François (2004). Héritage et réforme du quadrivium au XVIeme siècle. *Sciences et Techniques en Perspective* **8**, 211–30.

Longo, Angela (2010). Syrianus. In: *The Cambridge History of Philosophy in Late Antiquity Vol. 2*, Lloyd P. Gerson (ed). Cambridge: Cambridge University Press, pp. 616–29.

Maffei, Raffaele (1506). *Commentariorum urbanorum libri XXVIII*. Rome.

Meerhoff, Kees (2001). "Beauty and the Beast": Nature, logic and literature in Ramus. In: *The Influence of Petrus Ramus: Studies in Sixteenth and Seventeenth Century Philosophy and Science*, Mordechai Feingold, Joseph S. Freedman, and Wolfgang Rother (eds). Schwabe Philosophica 1. Basel: Schwabe and Co. AG, pp. 200–14.

Murdoch, John E. (1963). The medieval language of proportions: Elements of the interaction with Greek foundations and the development of new mathematical techniques. In: *Scientific Change: Historical Studies in the Intellectual, Social and Technical Conditions for Scientific Discovery and Technical Invention*, A.C. Crombie (ed). London: Heinemann, pp. 237–71; pp. 334–43.

Normore, Calvin G. (2017). Nominalism. In: *Routledge Companion to Sixteenth Century Philosophy*, Henrik Lagerlund and Benjamin Hill (eds). New York: Routledge, pp. 135–50.

Ong, Walter J. (1958). *Ramus, Method and the Decay of Dialogue: From the Art of Discourse to the Art of Reason*. Cambridge MA: Harvard University Press.

Proclus (1560). *In primum Euclydis elementorum librum commentariorum ... libri IV*. Translated and edited by Francesco Barozzi. Padua.

Proclus (1561). *De sphaera*. Edited by Erasmus Oswald Schreckenfuchs. Basel.

Proclus (1873). *In primum Euclidis Elementorum librum commentarii*. Edited by G. Friedlein. Leipzig: Teubner.

Proclus (1992). *A Commentary on the First Book of Euclid's Elements*. Translated and edited by G.R. Morrow. Princeton: Princeton University Press.

Ramus, Petrus (1569a). *Scholae in liberales artes*. Basel.

Ramus, Petrus (1569b). *Scholarum mathematicarum libri unus et triginta*. Basel.

Scriba, Christoph J., Menso Folkerts, and Hans Wussing (2002). Germany. In: *Writing the History of Mathematics: Its Historical Development*, Joseph W. Dauben and Christoph J. Scriba (eds). Science Networks. Basel: Birkhäuser, pp. 109–49.

Stöffler, Johannes (1534). *In Procli sphaeram*. Tübingen.

Sylla, Edith Dudley (2008). The origin and fate of Thomas Bradwardine's *De proportionibus velocitatum in motibus* in relation to the history of mathematics. In: *Mechanics and Natural Philosophy Before the Scientific*

Revolution, Walter R. Laird and Sophie Roux (eds). Boston Studies in the Philosophy of Science. Dordrecht: Springer, 67–119.

Syrianus (2006). *On Aristotle, Metaphysics 13–14*. Translated by J.M. Dillon and D.J. O'Meara. Ancient Commentators on Aristotle. Cornell: Cornell University Press.

Talon, Omer (1550). *Academia*. Paris.

7

Deceptive familiarity: differential equations in Leibniz and the Leibnizian school (1689–1736)

Niccolò Guicciardini
University of Milan

It would be quite anachronistic to read every equation of general and celestial mechanics in the eighteenth century as they are read today, that is, as equations in which both sides are numerically equal (Roche, 1998, p. 126).

Abstract: The differential equations as written by Leibniz and by his immediate followers look very similar to the ones in use nowadays. They are familiar to our students of mathematics and physics. Yet, in order to make them fully compatible with the conventions adopted in our textbooks, we only need to change a few symbols. Such "domesticating" renderings, however, generate a remarkable shift in meaning, making those very equations – when thus reformulated – unacceptable for their early-modern authors. They would have considered our equations, as we write them, wrong and would have corrected them back, for they explicitly adopted tasks and criteria different from ours. In this chapter, focusing on a differential equation formulated by Johann Bernoulli in 1710, I evaluate the advantages and risks inherent in these anachronistic renderings.

7.1 The "birth of analytical mechanics": qualifying a historiographical category

The eighteenth century was long defined as the "century of Newton." There is no doubt that the Newtonian philosophy of nature – the theory

[a] From *Anachronisms in the History of Mathematics: Essays on the Historical Interpretation of Mathematical Texts*, edited by Niccolò Guicciardini

[b] This research was funded by the Department of Philosophy "Piero Martinetti" of the University of Milan under the Project "Departments of Excellence 2018–2022" awarded by the Ministry of Education, University and Research (MIUR).

of universal gravitation, the theory of colours, the "Regulae" of his *Principia*, the "Queries" of his *Opticks*, and so on – had an enormous influence on the culture of the century of the Enlightenment. Particularly thanks to the fundamental works by Michel Blay and Clifford Truesdell, historians of mathematics are now aware of the fact that this image is the result of gross oversimplification. Eighteenth-century mathematics and mechanics flourished thanks to the contributions of mathematicians like Johann Bernoulli, Pierre Varignon, Alexis-Claude Clairaut, Jean le Rond d'Alembert, Leonhard Euler, Joseph-Louis Lagrange, and Pierre-Simon de Laplace, who relied on mathematical methods and on physical principles that can only in the slightest part be ascribed to the Newtonian tradition.[1]

The thesis of a "crisis of Newtonian mathematics" that took place precisely in the century of Newton has brought numerous advantages. Blay and Truesdell succeeded in making it canonical in the history of mathematics thanks to works exemplary for their mathematical depth and historical accuracy. It cannot be doubted that it is especially thanks to these two scholars that today we may view the eighteenth century as something more than a period of advancement of "normal science" along the trajectory traced by Isaac Newton's *Principia*. We no longer consider the eighteenth century to be one in which the mathematical discoveries of the *Principia* were simply generalized and refined with no substantial conceptual changes.

According to Blay and Truesdell, the birth of "analytical mechanics" led to the rewriting of the theory of Newtonian gravitation in a new mathematical and physical language. The new calculus and the new dynamics of the eighteenth century were made possible thanks to concepts, theories, and principles that realized Leibnizian intuitions, while going beyond both Isaac Newton's and Gottfried Wilhelm Leibniz's works: the abstract concept of function and its generalization to the concept of multivariable functions, the theories of partial differential equations and the calculus of variations, and minimum principles for dynamics which provided an alternative to the mechanics based on the three Newtonian "laws or axioms of motion."

As often takes place with historiographical theses, especially when

[1] See Truesdell (1954, 1960a,b); Blay (1992). Their theses, moreover, are compatible with the results obtained by other highly reliable and competent historians, like Pierre Costabel and David Speiser, to take two names it might be appropriate to cite here (Costabel, 1967; Speiser, 1996).

they become so influential as to be adopted somewhat sight unseen, one may at times notice the limits, increased rigidity, and simplifications to which this indiscriminate adoption of Blay's and Truesdell's theses may lead. It is certainly true that one of the most significant innovations in eighteenth-century mechanics was the abandonment of the geometrical methods of Galileo Galilei, Christiaan Huygens, and Newton. However, as I shall show in this chapter, a strong interest in geometry and in mechanical constructions survived – or better, played a vital role – in the work of those, including Leibniz, who saw themselves as the first advocates of analytical mechanics, the theory that was to rid Newtonian mechanics of its geometric dead weight.

Indeed, if we carefully read the differential equations that Leibniz and his acolytes employed, we discover that these equations were written with a geometrical interpretation in mind, and that their solutions were often sought in terms of mechanical constructions. By reading the differential equations of the first proponents of analytical mechanics as dealing with functions and their derivatives – a reading that neither Blay nor Truesdell would endorse – we project onto them a modernity that was not envisaged by their authors, who still shared a deep concern for geometrical interpretation and for mechanical construction with older generations of mathematicians. It is here that a form of anachronism in textual interpretation emerges, one due to what might be called "deceptive familiarity."

7.2 Deceptive familiarity

It is useful to focus our attention on the most important field of investigation for the Leibnizian school active in the decades between 1690 and 1730: ordinary differential equations. Indeed, research on the calculus by Leibniz, Varignon, Jacob Hermann, and the Bernoulli brothers – that is, of the first generation of mathematicians skilled in calculus, who were responsible for the birth of analytical mechanics – finds its maximum realization in the study of the solutions to ordinary differential equations applied to increasingly difficult problems in mechanics, optics, and geometry. It was in the final decade of the seventeenth century and the first decades of the eighteenth that problems were dealt with – such as those of the brachistochrone, the *elastica*, the catenary, central force motion, and resisted motion – that led to the formulation and solution of differential equations using methods that were to enjoy enormous success.

These methods, in fact, are still presented in the first pages of integral calculus textbooks. By way of example, I might mention the integrating factor method and the method of separation of variables. A mathematics student in the first two years of study who has the desire to read an article concerning differential equations published in the *Acta Eruditorum* at the turn of the eighteenth century will find a notation (the Leibnizian notation for differentials and integrals), equations (e.g. the equation for motion in a central force field), and methods of solution (e.g. by substitution of variables) that will appear wholly familiar to him. For sure, our student, even when exceptionally gifted, will need a historian by his side, to help with the difficulties inherent in neo-Latin, typographically awkward notations, and a style of writing in which prose predominates on formulas to an extent that is unusual today.

This familiarity is a notable fact: the seventeenth and early eighteenth century is the period in which certain mathematical texts (such as Descartes's *Géométrie* and the papers on differential equations written by the Leibnizians) began to be written in a way that makes them immediately understandable by a present-day mathematician. And it is in fact this familiarity – what I choose to call "deceptive familiarity" – that raises problems of historical interpretation. As I am about to show, the deceptive familiarity of Leibnizian calculus results in a rather insidious form of anachronism. Under an anachronistic lens, whose aberration is due to the deceptive familiarity of the objects it is focussed on, Leibniz's methods and those of the early Leibnizians acquire a modernity not intended by their authors. The anachronistic interpretation is fostered precisely by a similarity between the language and methods of the first Leibnizians, and those in use today – a similarity that depends on the fact that that language and those methods had a profound influence on eighteenth-century mathematics and were quite successfully adopted by subsequent generations, until they came down to us as they are written in our textbooks. These methods, as they are to be found in present-day textbooks, are similar to their originals: that is why we are deluded into not seeing the distance that separates them. It is a phenomenon of deceptive familiarity that overshadows the distance, the "unbridgeable distance or a nearness that is merely apparent," as Paolo Rossi would

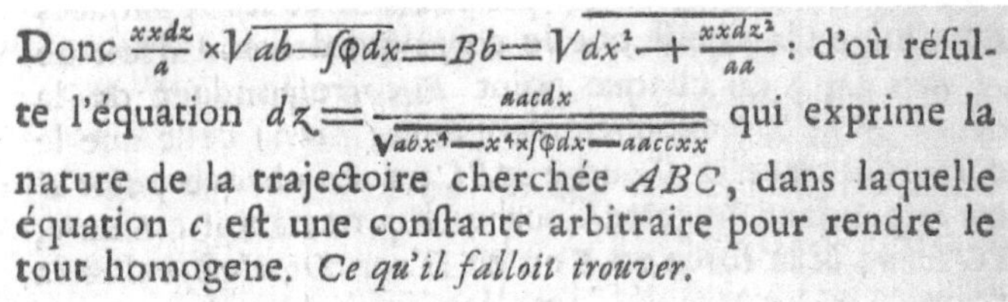

Donc $\frac{xxdz}{a} \times \sqrt{ab - \int \phi dx} = Bb = \sqrt{dx^2 + \frac{xxdz^2}{aa}}$: d'où réſulte l'équation $dz = \frac{aacdx}{\sqrt{abx^4 - x^4 \int \phi dx - aaccxx}}$ qui exprime la nature de la trajectoire cherchée *ABC*, dans laquelle équation *c* eſt une conſtante arbitraire pour rendre le tout homogene. *Ce qu'il falloit trouver.*

Figure 7.1 Bernoulli's differential equation for central force motion. Source: Bernoulli (1710, p. 526).

say, that lies between the texts of Leibniz's times and those in use today (Rossi, 1999, p. 28).

7.3 Bernoulli's equation for central force motion

At this juncture, in order to illustrate the "familiarity," as well as the "distance," of the works on differential equations produced by the Leibnizians, I will focus on a specific example. Several of the differential equations written by Leibniz, Jacob Bernoulli, Jacob Hermann, Johann Bernoulli, Varignon, or Hermann would serve my purpose, and indeed I plan to write a comprehensive study on the differential equations written after Leibniz's *Tentamen de motuum coelestium causis* (1689) and before Euler's *Mechanica* (1736). For our present purposes, one example will suffice. In this section, we shall consider the equation Johann Bernoulli formulated in 1710 to mathematize motion in a central force field (Bernoulli, 1710).[2] This equation and the techniques developed for its solution by Bernoulli are perceived as being "ours," and hence are appropriately included as "paradigmatic" – to use Thomas Kuhn's terminology – in the "normal science" taught in our university courses. However, when we examine this equation more carefully, we discover features that are "foreign" to our mathematical practice.

7.3.1 The equation and its domestication

Bernoulli considers a body accelerated by a central force. The problem is, of course, to determine the trajectory given the initial position and velocity.

[2] Bernoulli's equation is discussed in Speiser (1996) and in Nauenberg (2010). Especially this latter paper has been inspirational for me.

Bernoulli's equation (see Figure 7.1) is

$$dz = \frac{aacdx}{\sqrt{abx^4 - x^4 \int \phi\, dx - aaccxx}}. \tag{7.1}$$

This equation can be rewritten – or "domesticated," to use Lawrence Venuti's terminology (Venuti, 1995) – in a more familiar notation as:

$$\mathrm{d}\theta = \frac{l\mathrm{d}r}{\sqrt{2Er^4 - 2r^4 \int_r^\infty F\mathrm{d}r - l^2r^2}}, \tag{7.2}$$

after setting $x = r$, $z/a = \theta$, $\phi = 2F$, $ab = 2E$, $ac = l$, and with the assumption of unit mass: $m = 1$. Here we follow the conventional notation: F is the force's strength, E is the total energy, and l is the angular momentum. There will be more to say about this rewriting of Bernoulli's equation in modern notation. Here I wish to note that our constants are endowed with a very important physical meaning. Further, they stand for real numbers, once fundamental physical units have been defined.

Let us turn our attention to Bernoulli's text and continue to follow his reasoning in its original notation. How did he apply his equation to an inverse-square force? And how did he integrate the equation?

7.3.2 Reduction to quadrature

For an inverse-square force

$$\phi = \frac{a^2g}{x^2}, \tag{7.3}$$

where g is a constant.[3] Bernoulli obtained the following differential equation from (7.1):

$$dz = \frac{a^2cdx}{x\sqrt{abx^2 + a^2gx - a^2c^2}}, \tag{7.4}$$

which he could reduce to an easier form by substituting variables.

Bernoulli began (see Figure 7.2) by introducing a change of variables

[3] In a textbook in use today one might find a more straightforward choice of constants, for example: $a^2g = 2k$ for a force $F = -k/r^2$.

à bout. Afin de réduire cette valeur de dz à une formule differentielle ordinaire d'arc circulaire, ſoit $x = \frac{aa}{y}$: la ſubſtitution de cette valeur de x donnera $\frac{aacdx}{x\sqrt{abxx + aagx - aacc}}$ $= \frac{-acdy}{\sqrt{a^3b + aagy - ccyy}}$ (en ſuppoſant $y = \frac{aag}{2cc} - t$) $=$ $\frac{acdt}{\sqrt{a^3b + \frac{a^4gg}{4cc} - cctt}}$ (en ſuppoſant $cchh = a^3b + \frac{a^4gg}{4cc}$ pour abreger) $= \frac{adt}{\sqrt{hh - tt}} = dz$; & par conſéquent $\frac{dz}{a} = \frac{dt}{\sqrt{hh - tt}}$ $= \frac{1}{h} \times \frac{hdt}{\sqrt{hh - tt}}$, qui eſt une differentielle d'arc de cercle (dont le rayon eſt $= h$, & le ſinus $= t$) diviſé par ſon rayon.

Figure 7.2 Substitutions of variables. Source: Bernoulli (1710, p. 527).

via:

$$x = \frac{a^2}{y}, \tag{7.5}$$

followed by another one (note that t does not indicate the time):

$$y = \frac{a^2 g}{2c^2} - t. \tag{7.6}$$

The contemporary student of Newtonian mechanics will find these substitutions familiar! Then, "pour abréger," Bernoulli introduced a new constant h such that

$$c^2h^2 = a^3b + \frac{a^4g^2}{4c^2}, \tag{7.7}$$

which leads to the differential equation

$$\frac{dz}{a} = \frac{dt}{\sqrt{h^2 - t^2}}. \tag{7.8}$$

For Bernoulli, equation (7.8) was easy to solve. As we shall see in Section 7.5, his solution of (7.8) was attained as a geometrical construction, rather than as a formula. This marks a notable difference with respect to present-day mathematical practice. It was by these means that Bernoulli was able to show, of course, that a body acted upon by an inverse-square force necessarily moves along a conic section.

The substitutions of variables we have reviewed in this section (see

Figure 7.2) are essentially the method still used in textbooks on Newtonian mechanics (a somewhat ironic name for this discipline from the standpoint of the present account) to reduce the equation for central force motion in polar coordinates to a form that is easy to integrate. The fact that this Bernoullian technique has survived in calculus textbooks is telling of the "significance," in Eric Donald Hirsch's terms, that it possesses for us. But what about the "meaning" it had for its author?[4]

7.3.3 Solution

A twenty-first-century undergraduate student in mathematics can verify that integration of (7.8) leads to the equation for a family of conic sections written in the form:

$$x = \frac{c_1}{1 - c_2 \sin((z - z_0)/a)}, \qquad (7.9)$$

where c_1 and c_2 are constants.[5] This result has often been hailed as a turning point in the application of differential equations to the science of motion, since Bernoulli here proves that conic sections are necessary orbits in an inverse-square force field.

In their learned and informative study on the contributions to analytical mechanics made by the Bernoulli family and Hermann, David Speiser and Diarmuid Ó Mathúna (2019, p. 739) write:

> Johann Bernoulli responded [to Hermann] with a comprehensive analysis of the problem in its full generality. [...] [His] solution was expressed in terms of trigonometric functions of the true anomaly [$z/a = \theta$] – that is, in the form in use ever since. This response of Johann Bernoulli to Hermann may be considered to be the beginning of analytical mechanics.

This statement is an expression of how such anachronisms – indeed, a

[4] On "meaning" – i.e. what the text meant according to the author's intentions – versus "significance" – i.e. what the text came to mean for its readers in successive periods – see Hirsch (1967).

[5] We note that the constants c_1 and c_2 above depend on the initial conditions as follows: $c_1 = 2c^2/g$ and $c_2 = \sqrt{1 + 4bc^2/ag^2}$. Shifting to the "domesticating" modern notation ($ab = 2E$, $ac = l$, $ga^2 = 2k$): $c_1 = l^2/k$ and the eccentricity is $c_2 = \sqrt{1 + 2El^2/k^2}$, where E is the energy, l the angular momentum, and $F = -k/r^2$. Thus, taking into consideration the fact that $\sin(z/a) = \sin(\theta) = -\cos(\theta + \pi/2)$, the integral (7.9) is "domesticated" as:

$$1/r = (k/l^2)(1 + \sqrt{1 + 2El^2/k^2}\cos(\theta + \pi/2 - \theta_0)),$$

where $\theta_0 = z_0/a$. This is exactly what we will find in a modern textbook: see Goldstein *et al.* (2014, pp. 92–8).

venial sin when we consider the debt we owe to such wonderful scholars – lead to the received view on the "birth of analytical mechanics."

We should, however, note that – contra Speiser and Ó Mathúna – Bernoulli did not take this last step. We thus begin to appreciate the "otherness" of his mathematical practice. Bernoulli, like all his contemporaries, did not possess a calculus for trigonometric functions (Katz, 1987), but rather expressed (7.9) in geometric terms: he sought a solution in the form of a geometrical construction. This was one of his "tasks," to use Bos's terminology (Bos, 2004).[6] Bernoulli, closely followed by Varignon (1710), sought the solutions of their differential equations as constructions. And it is interesting to note that they rounded off their papers with a "demonstration of the construction," an expression reminiscent of the Euclidean canon, in which a *kataskeuē* (the construction) was to be followed by an *apodeixis* (the demonstration).

The fact that the geometrical constructions of the solutions of differential equations customary in this period have received so scant attention, as we have verified in Speiser and Ó Mathúna's commentary on Bernoulli (1710), is interesting. It is an effect of the habit to read past mathematics by concentrating on those aspects that are deemed interesting, profound, beautiful, and fruitful according to our standards. We focus on successes, innovations, and discoveries. We tend to be blind to the less attractive moments in mathematical practice, to the routine of "normal science." Consequently, as we have seen in the Introduction, historians of Descartes's *Géométrie* have devoted little attention to the Cartesian curve-tracing instruments. As we shall see in Section 7.5, Bernoulli's geometrical construction is not a good piece of mathematics: it is boring, difficult to follow, and we miss a sense that it has any purpose. Yet, this cumbersome mathematical practice was part of Bernoulli's canon, and for reasons that it is historically interesting to consider – most notably, the absence of a developed notation and calculus for trigonometric functions (Katz, 1987).[7]

In the following sections, I will discuss the role that geometrical

[6] I use the plural "tasks" here since, as I have argued in the Introduction, Bernoulli's paper might have been written under the assumption that more than one task should be pursued.

[7] The only extensive study I know of the role of constructions in the treatment of trigonometric, hyperbolic, and logarithmic functions in the Leibnizian calculus is Blåsjö (2017). From Blåsjö's viewpoint, it is quite evident why Bernoulli did not integrate his equation in terms of trigonometric functions. Leibniz and his immediate followers did not have a notion of function but rather a notion of plane curve, and they deployed a variety of construction tools in order to define and study transcendental curves.

interpretations (Section 7.4) and constructions (Section 7.5) still played in the handling of differential equations in the works of Leibniz and his immediate followers.

7.4 Equations and constants

A characteristic of the differential equations of Leibniz and of the first generation of his followers that distinguishes these equations from those in use nowadays concerns the use of constants. This may be illustrated by reconsidering Bernoulli's equation (7.1) for the solution of the inverse problem of central forces.

It must be observed that Bernoulli's equation has one constant more than the equation accepted today. Today, we write the equation for motion in a central force field for a unitary mass point ($m = 1$) by deploying two constants of great significance from the physical standpoint: energy E and angular momentum l. In Bernoulli's equation (7.1) we find three constants, a, b, and c, where we would use (equation (7.2)) only E and l. Why, then, does Bernoulli's equation have an extra constant? Bernoulli himself provides an answer. He writes:

> The equation [7.1] expresses the nature of the sought trajectory ABC, in which c is an arbitrary constant for making everything homogeneous.[8]

The constant c is introduced to guarantee the geometrical homogeneity of the terms occurring in the equation, so only symbols standing for geometrical magnitudes of the same dimension are added or subtracted.[9]

Let us analyze Bernoulli's use of an extra-constant by examining a simpler example (see Figure 7.3). In order to derive the differential equation (7.1) I have been discussing, Bernoulli expresses what we would call the conservation of mechanical energy:

$$\int \phi dx = ab - vv, \tag{7.10}$$

[8] "l'équation [7.1] exprime la nature de la trajectoire cherchée ABC, dans laquelle équation c est une constante arbitraire pour rendre le tout homogène" (Bernoulli, 1710, p. 526). See Figure 7.1.

[9] We have to pay attention here to the terminology used in Bernoulli's time, a terminology very much influenced by Viète. Today, a "homogeneous polynomial" is one in which all terms have the same degree, where the degree of a term is the sum of the exponents of the indeterminates in that term. Thus, from the contemporary standpoint, x^3 is a third-degree monomial, $3a^2x$ is a first-degree monomial, and the degree of ab^2 is zero. However, for Bernoulli all these terms are third degree and have a geometrical meaning: x^3 is a cube, $3a^2x$ and ab^2 are parallelepipeds, so the equation $x^3 + 3a^2x + ab^2 = 0$ is well formed.

ici $\frac{\varphi dx}{v} = -dv$, ou $\varphi dx = -v dv$, dont l'intégrale eſt $\int \varphi dx = ab - vv$: j'entens par *ab* une quantité conſtante quelconque, laquelle ſe peut ajoûter à volonté aux intégrales ſimples. Donc $v = \sqrt{ab - \int \varphi dx}$, qui eſt l'équation cherchée de la Courbe *DGH* des viteſſes.

Figure 7.3 Bernoulli's use of constants in writing the "equation of the curve of speeds," that is, the equation whose graph represents the variation of speed in function of radial distance x from the center of force. The equation is $v = \sqrt{ab - \int \phi dx}$. Source: Bernoulli (1710, p. 525).

where v is the speed, and a and b are two arbitrary constants of integration. The left-hand side of this equation expresses what we would call the work.

After setting $x = r$, $\phi = 2F$, and $ab = 2E$, we rewrite the equation (7.10) in contemporary language as:

$$\frac{v^2}{2} - \int_{r_1}^{r} \vec{F} \cdot d\vec{r} = E. \tag{7.11}$$

We use one constant: E. Bernoulli, instead, uses two: a and b. Our constant E is, of course, endowed with a very important physical meaning.[10]

Today, this equation appears in our textbooks and is taught to our students as expressing the conservation of mechanical energy. Bernoulli's equation (7.10), when translated as I have done above, is thus extremely familiar to contemporary mathematicians. But there is a detail that we should not overlook. Today, we would not write the constant of integration as the product of two constants, ab, but simply as one constant. Yet, for Bernoulli it is important to have a superfluous constant – "superfluous" from our point of view – because it is precisely this constant that guarantees the equation's geometric homogeneity.

Let me explain this last statement in some detail. Bernoulli conceives the intensity of force ϕ and speed v as geometrically represented by segments: they are the ordinates of graphs representing the variation of ϕ and v in function of the radial distance x from the center of force. Thus, we have to refer to two curves with abscissa x and ordinates ϕ and v. These curves visualize the dependence of force and speed from the radial

[10] Other notable differences that the historian should consider are: the different sign convention, the use of vector notation, the specification of the limits of integration. In this chapter, I ignore these differences. In a more complete treatment, they should be taken into due consideration.

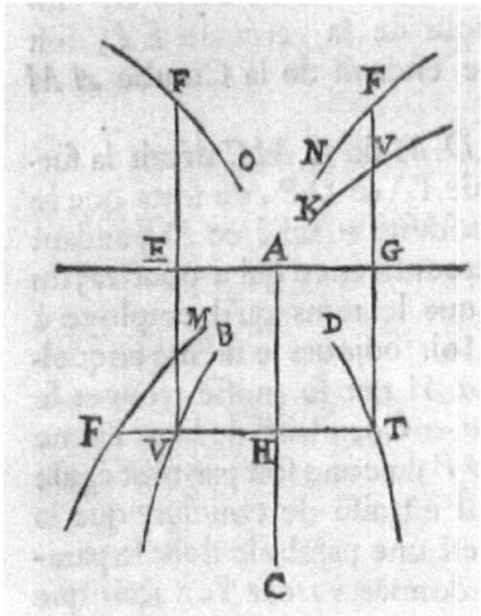

Figure 7.4 Varignon used these graphs to represent graphically the various combinations of functional dependency among the parameters of motion: the "curve of times" TD represents time as a function of distance; the "curves of speeds" VB and VK, the velocity as function of distance and time, respectively; and the "curves of force" FM, FN, and FO, the force as function of distance, time, and velocity, respectively. Quoted from Mahoney (1998, p. 747). Source: Varignon (1700, p. 22).

distance. We think in terms of numerical functions. Bernoulli thinks in terms of geometrical graphs. It is Pierre Varignon who systematized the use of graphs in analytical mechanics (see Figure 7.4).

The integral of $\phi\, dx$ thus stands for a surface: which is to say, the surface underlying the graph of ϕ over the x-axis. The left-hand side of (7.10), $\int \phi dx$, must therefore be made equal to a geometrically homogeneous magnitude. And this requirement is satisfied by the fact that the right-hand side is written as $ab - vv$, namely as the difference between two magnitudes that are two-dimensional. Geometrically, ab is a rectangle of sides a and b, and vv is a square of side v (bear in mind that v is a segment: the ordinate of the "curve of speeds"). All this makes little sense to us, since we read equation (7.11) numerically, namely as an equation between real-valued functions of a real variable, their derivatives and constants endowed with a physical meaning.

It is striking to note that, in the three pages of calculations that lead to the problem's solution, Bernoulli maintains the requirement of geometrical homogeneity with what I would call Euclidean rigour. For its author, the differential equations of analytical mechanics, rather than being endowed with a physical significance as it is for us, have a geometric one. The superfluous (from our standpoint) Bernoullian constant can take on an arbitrary numerical value – for the sake of simplicity, it can be made equal to 1. Its role appears wholly pleonastic, when the equations are read as dealing with numerical functions (say, functions from R to R). But the superfluous constant was important for Leibniz, Bernoulli, and the other first-generation Leibnizian mathematicians, precisely because (from their standpoint) the symbols occurring in equations were endowed with a geometric meaning, and it was therefore natural to manipulate them in such a way that equations express meaningful geometric relationships.

We can verify this by reconsidering equation (7.1)

$$dz = \frac{aacdx}{\sqrt{abx^4 - x^4 \int \phi\, dx - aaccxx}}.$$

The left-hand side is one-dimensional, since dz is an infinitesimal segment. The numerator of the right-hand side has dimension four, since it is the product of three finite segments aac multiplied by the infinitesimal segment dx. Under the square root we find a sum of six dimensional magnitudes. For example $x^4 \int \phi\, dx$ is the product of a four-dimensional magnitude, x^4, multiplied by $\int \phi\, dx$, which is a two-dimensional surface. The square root of a six-dimensional magnitude is three-dimensional. Thus, equation (7.1) can be read as a well-formed proportion between two ratios: the ratio between two infinitesimal magnitudes, dz/dx, is proportional to the ratio between two three-dimensional magnitudes (between two solids).

As John Roche (1998, pp. 86–116) has shown in his splendid study on the mathematics of measurement, mathematicians such as Johann and Jacob Bernoulli, Leibniz, and Varignon were stretching Euclidean proportion theory in order to apply it to complex functional relationships holding between continuous magnitudes. In classic Euclidean theory there are no ratios between non-homogeneous magnitudes. These mathematicians, however, following a tradition dating back to François Viète and Descartes, defined symbols as standing for segments, and defined the product, and division, between segments as leading to magnitudes of higher, and lower, dimension, so that, for example, the product of a and b is a two-dimensional surface ab, and the division of a solid by a surface is a segment. Further, this Viètian, so to speak, "algebra of segments" is well designed for the mathematization of dynamics since, once space and time are represented by segments, one can define velocity and acceleration as ratios (ds/dt, dv/dt) between infinitesimal segments. It was agreed that there is no ratio between space and time, or speed and time, but these ratios were accepted as well formed when understood as holding between infinitesimal segments. Indeed, in this way, the above ratios relate homogeneous magnitudes – segments standing for physical continuous magnitudes. For example, once a graph representing the variation of displacement s in function of time t is given, the ratio between infinitesimal segments ds/dt defines the ordinate v (actually the ratio between v and a unit segment) of another graph (the "curve of speeds") representing the variation of speed in function of time.

As Michael Friedman shows in his important monograph devoted to Immanuel Kant, this tradition was still alive at the end of the eighteenth century. Indeed, in certain areas of mathematical physics it survived even in the nineteenth century (Roche, 1998, p. 87). It is worth quoting Friedman's crystal-clear explanation:

> For us, a physical magnitude (such as mass) is a function from some set of physical objects (massive bodies) into the real number system, implemented by choosing some arbitrary unit object (a standard gram, for example) and then representing the ratios of this object to all other objects by real numbers. The output of our function is thus a dimensional real number encoding the system of units in question (grams). In the traditional theory, by contrast, there was not yet a single real number system. [. . .] Rather each type or kind of magnitude (lengths, areas, volumes, times, weights, and so on) was thought to form a system of its own, characterized by its own particular operation of *addition*. [. . .] Magnitudes are said to be *homogeneous*, then, when they are of the same dimension and can therefore be added together, and it is only magnitudes of the same dimension [. . .] that can meaningfully be said to have a *ratio* to one another. In this tradition, moreover, since what we would now express in terms of equations between magnitudes represented by real numbers is rather expressed in terms of proportionalities between ratios, there is no need for an arbitrary choice of unit (Friedman, 2013, pp. 53–4).

It is this traditional way of dealing with continuous magnitudes in terms of proportionalities between ratios of homogeneous magnitudes that structures the way of writing the equations in Bernoulli's paper. All his equations are subject to criteria of adequacy that make little sense for us, used to dealing with numerical functions. His different conception is also evident when we turn to the way in which Bernoulli sought a solution.

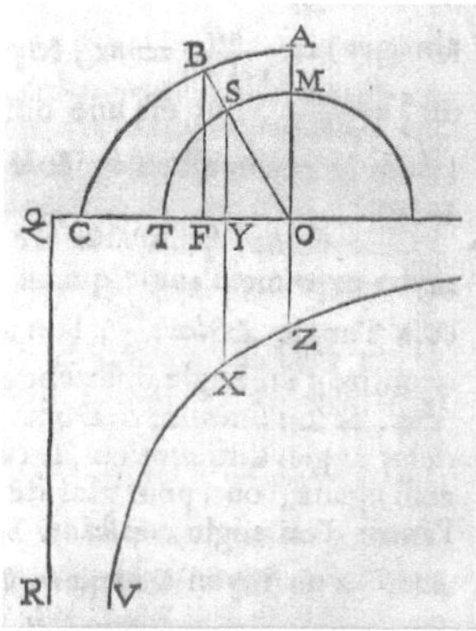

Figure 7.5 Construction of conic trajectories for an inverse-square force. Source: Bernoulli (1710, p. 528).

7.5 Solutions as constructions

Today, an ordinary differential equation is defined as an equation relating an unknown real-valued function of one independent real variable to its derivatives and some given functions of the same variable. To solve an ordinary differential equation means finding an algebraic expression of a family of functions, each of which is a particular solution of the equation. For Leibniz and his students, instead, this was not how things stood. Differential equations were equations relating continuous geometrical magnitudes and their differentials. Solutions, as we have already hinted at in Section 7.3.3, were sought as geometrical or mechanical constructions.

This also holds for Johann Bernoulli's paper on central forces. After reducing his equation (7.4), for an inverse-square force,

$$dz = \frac{a^2 c dx}{x\sqrt{abx^2 + a^2 gx - a^2 c^2}}$$

to the equation (7.8),

$$\frac{dz}{a} = \frac{dt}{\sqrt{h^2 - t^2}},$$

Bernoulli, rather than giving a solution in terms of trigonometric functions, provides a geometrical construction. This is interesting since an algebraic notation for the sine, cosine, tangent, etc., was available at the time. It is a calculus for trigonometric functions (and the notion of function itself) that was lacking.

Bernoulli defines the equation (7.8) as the "the differential of the arc of a circle whose radius is h and whose sine is t."[11] Bernoulli thinks geometrically and draws a circle MST with radius $OM = h$. He defines the angle $MOS = z/a$, so that $OY = t$ is the "sine" of MOS (see Figure 7.5).[12] This is a geometric representation of the relationship between variables z and t expressed by equation (7.8).

We should bear in mind that Bernoulli reduces equation (7.4) to equation (7.8) via two changes of variable: $x = a^2/y$ followed by $y = a^2 g/2c^2 - t$. In the figure, we see these variable substitutions implemented in a geometric diagram that delivers the sought trajectory ABC as a construction. The first substitution of variables is represented by the hyperbola VXZ. The second substitution is represented by a translation, as we read in the quotation below. Bernoulli (1710, p. 528) writes:

> Let the segment $OQ = aag/2cc$ be perpendicular at O on the segment AO; let VXZ be an equilateral hyperbola between the asymptotes QO and QR, where the rectangle of the coordinates $QY \times YX$ or $QO \times OZ = aa$; let an ordinate YX of the hyperbola be prolonged until it meets the circumference MST in S, and from that point S let OS be drawn, on which straight line (prolonged if necessary) let $OB = YX$ be taken. I say that point B will be one of those of the trajectory ABC.[13]

It is interesting to cite Bernoulli's words, since his language, for us somewhat arcane, is aimed at providing a geometrical construction. This way of writing was familiar to Bernoulli's contemporaries, while it is remote from what we are accustomed to when integrating a differential equation.

Let us dwell on the above quotation. Bernoulli uses a system of coordinates that is similar to our polar coordinates. Less anachronistically(!), the trajectory is given if we can associate to any angle MOS the radial

11 "qui $[dz/a]$ est une différentielle d'arc de cercle (dont le rayon est $= h$, & le sinus $= t$) divisé [sic] par son rayon" (Bernoulli, 1710, p. 527).

12 I put "sine" in inverted commas, because, according to our conventions, we would use this term for a unit circle. But in Bernoulli's times, given a circle of radius $OM = h$, such as the one shown in the figure, OY was defined as the sine of MOS. Today we would write $OY = OM \sin(MOS)$.

13 "Soit donc ici perpendiculairement en O sur AO la droite $OQ = aag/2cc$; & du centre Q entre les asymptotes QO, QR, une hyperbole équilatere VXZ, dont le rectangle (des coordonnées) QYX ou $QOZ = aa$ [$QY \times YX$ or $QO \times OZ = aa$]; soit prolongée une ordonnée quelconque YX de l'hyperbole jusqu'à ce qu'elle rencontre le cercle MST en S, par lequel point S soit menée OS, sur laquelle (prolongée si l'en est besoin) soit prise $OB = YX$. Je dis que le point B sera un de ceux de la trajectoire ABC."

distance OB, where B is the position of the body. Distance OB is constructed via the prescriptions quoted above. Indeed, the position B of the body is along the straight line passing through the given points O and S, and the radial distance is $OB = YX$ by construction.

To recapitulate: we can subdivide the construction into four steps.

Step 1 Bernoulli first represents the geometrical relationship between z and t expressed by equation (7.8).

Step 2 Then, Bernoulli constructs $y = QY$.

Step 3 Next, Bernoulli constructs $x = YX$.

Step 4 Segment YX is the sought distance OB. The last, fourth step consists in drawing the segment OB passing through the given point S.

The geometrical representation in Step 1 is obtained by drawing a circle of radius h, and by defining the angle MOS as equal to z/a. From this, it follows that $OY = \sin z/a = t$.

Step 2 is as follows. Segment QY is equal to $OQ - OY$, where, by construction, $OQ = a^2 g/2c^2$ and $OY = t$. As we know, $y = a^2 g/2c^2 - t$. We thus have a geometrical representation of the functional relationship between the variables t and y: $OY = t$ and $QY = y$.

Step 3 is as follows. YX is the ordinate of the hyperbola VXZ. By definition $YX = aa/QY$, and we know $x = a^2/y$. We thus have a geometrical representation of the functional relationship between the variables y and x: $QY = y$ and $YX = x$.

Step 4. The construction is now complete. Given any z/a (any angle MOS), we know how to construct $t = OY$. Given t, we know how to construct $y = QY$. Given y, we know how to construct $x = YX = OB$, the distance of the body from O.

We have not finished yet: we have to prove that the constructed trajectory ABC is a conic section! At this juncture, Bernoulli reverts to algebra. By using Cartesian coordinates $\xi = OF$ and $\chi = FB$, he verifies that the equation in ξ and χ is quadratic; in other words, that it is the equation of a conic section.

7.6 The slow transition from geometry to algebra

The period from Newton to Euler was one of transition. In dealing with the science of motion, with statics, optics, and other mixed mathematical sciences, mathematicians began to abandon the language of geometry, as found in Galileo's *Discorsi* (1638), in Huygens's *Horologium* (1673), and in Newton's *Principia* (1687). The Newtonian and the Leibnizian calculi began to be employed, differential and fluxional equations made their first appearance in the 1690s and became, especially in the Basel school, a major research area. This was a crucial phase in the birth of analytical mechanics, as Blay and Truesdell have pointed out. However, the passage from geometry to differential equations, as they are understood by us nowadays, did not occur in one step: it was a slow process that took about one century to be completed.

The magnitudes that were of interest for the early practitioners of the Leibnizian calculus are continuous magnitudes, such as times, angles, weights, distances, velocities, and accelerations. As late as the early eighteenth century, numbers were considered inadequate for representing these quantities. Indeed, continuous magnitudes could not be measured by numbers because of the problem with irrational ratios. The classic Euclidean theory of proportions, rather than the algebraic theory of equations, offered a rigorous tool for dealing with relationships between continuous magnitudes. The Euclidean theory was still highly valued: Galileo, Huygens, and, to some extent, Newton used proportion theory in their studies on mechanics. Their works were still exemplary well into the eighteenth century.

The theory of proportions, however, is subject to serious limitations. When the magnitudes are functionally related in a way that has some degree of complexity (e.g. when they are related by fractional exponents), one is almost compelled to use algebraic symbolism. Thus, the early practitioners of the differential and the fluxional calculus began writing equations, both differential and fluxional, rather than proportions, in order to deal with complex situations such as motion in resisting media. Most notably, when functional relationships were expressed by transcendental functions, as in the case of Bernoulli's study of central forces we have reviewed above, geometrical construction showed its limitations. A new algebraic symbolism for trigonometric, logarithmic, and hyperbolic functions began to emerge. This was a slow process that continued up to the middle of the eighteenth century (Katz, 1987).

At the turn of the eighteenth century, the symbols in the ordinary differential equations of Leibniz and the early Leibnizians, and in Newtonian fluxional equations as well, were understood as standing not for functions, but rather for finite and infinitesimal components of line-segments: i.e., the ordinates and abscissae of curves representing the functional relationships between variable continuous magnitudes.[14] Galileo, with his triangle representing the variation of speed in function of time in free fall had made these graphs famous. Mathematicians such as Evangelista Torricelli, Huygens, Isaac Barrow, Newton, Johann Bernoulli, and Varignon represented the functional dependence of continuous magnitudes, such as times, spaces, speeds, and forces, in terms of graphs: plane curves whereby, for example, the abscissa represents a distance and the ordinate the intensity of force, or the abscissa the time and the ordinate the speed. Varignon was particularly influential in promoting the systematic use of these graphs (see Figure 7.4).

As Roche (1998, p. 87) makes clear:

> [The representation of physical magnitudes by lines] allowed an algebra to be applied which treated physical quantities as though they were lines and subject to the geometrical algebra of Viète and Descartes. . . . Only in the late eighteenth century did functional algebraic equations, now interpreted numerically, begin to be regarded as acceptable statements of physical relationships in their own right.[15]

The fact that the symbols occurring in the equations stood for line segments prompted the mathematicians who were using them to pay due attention to geometrical homogeneity. Indeed, very much as in Viète's work, the multiplication of symbols was thought to lead to higher-dimensional magnitudes. That is to say, if a, b, and c stand for segments, then ab stands for a surface, and abc for a solid. Only magnitudes of the same dimension, i.e. homogeneous magnitudes, could be added or subtracted from one another. This requirement was considered a very important one, and, as we have seen in Section 7.4, was explicitly called into use.

The fact that Leibniz, Jacob Bernoulli, Johann Bernoulli, Hermann, Varignon, and even the young Euler in *Mechanica* (1736) used homogeneous equations is notable, since from Descartes's *Géométrie* it was clear how homogeneity could be dispensed with. In 1637, Descartes had explained how operations on segments could be defined as closed operations: so if x and y are two segments, xy is a segment too. In order to

[14] For the differential equations in Leibniz, Bernoulli, and Euler see Bos (1974). This paper also has a great deal to teach us about the requirement of homogeneity adopted by Leibniz and his acolytes.

[15] In writing this chapter, I owe a great debt to Roche's book.

express the product of x and y as a segment, he introduced a unit segment e, so that $e : x = y : xy$. Thus, from Descartes's viewpoint $xy + z = 0$, for example, is a perfectly meaningful equation in which two segments xy and z are summed. However, when applying algebra to geometry, Descartes himself wrote homogeneous equations, say, $xy + az = 0$. In the writings of Leibniz and the early practitioners of the calculus, especially in private correspondence and calculations, it is possible to find some occurrences of inhomogeneous differential equations.[16] The availability of a choice between two ways of writing differential equations makes it all the more interesting that Leibniz and his acolytes – after discussing at length what is, for us, a non-problem – in so many instances deliberately refrained from expressing inhomogeneous equations. One often finds them correcting an equation by apologetically multiplying a term by the necessary constant "in order to guarantee homogeneity." This was a deliberate choice, which Bos would call a "self-imposed criterion of quality control."

The advantages in constraining equations to geometrical homogeneity were the following.

- First, this constraint allowed a geometrical interpretation of the operations occurring in the equation to be confirmed (in the above example, one adds to surface xy a surface az, a perfectly meaningful geometrical construction).
- Second, solutions as constructions were much easier to find if, *ab initio*, all the operations are geometrically interpreted.
- Third, infinitesimal magnitudes could be compared to one another in a geometrically meaningful way.

This last point requires some explanation, since it is relevant to the handling of differential equations we have reviewed above.

In a manuscript treatise on the *mathesis universalis* Leibniz considers the formula

$$dx + d^2x + dxdx. \tag{7.12}$$

But what does (7.12) mean? If x represents a segment, dx and d^2x are two infinitesimal segments (the former a first-order infinitesimal, the latter a second-order infinitesimal). Yet, $dxdx$ represents the infinitesimal square

[16] I thank Viktor Blåsjö for drawing my attention to this very important point (private communication, November 19, 2017).

with side dx, and cannot be added to a segment. Instead, Leibniz claims, the practitioner of the calculus must observe a *lex homogeneorum*.[17] According to Leibniz's law, the correct expression is:

$$adx + ad^2x + dxdx, \tag{7.13}$$

an expression in which we sum three surfaces.

For Leibniz and for his immediate followers, differentials always had a geometric meaning. They did not understand dy/dx as the derivative d/dx of a function $y(x)$, but rather as a ratio between two infinitesimal magnitudes, dy and dx. Their differential equations were thus "equations relating differentials" in which, for the reasons discussed above, the constraint of geometrical homogeneity was important.

Slowly things began to change: in the 1720s and 1730s a more general, abstract notion of function began to emerge, especially in the Basel school. Homogeneity, it seems, was becoming a less compelling requirement. It is revealing to briefly consider the work of Euler. Homogeneous equations were still used in the *Mechanica*, where Euler sometimes used a constant, which he introduced in order to "guarantee" that the terms in the equation would be geometrically "homogeneous."[18] In this work, however, Euler consistently sought solutions of differential equations as algebraic expressions, not as constructions. Euler was eventually to discard the *lex homogeneorum*. In the second appendix to *Methodus Inveniendi*, where Euler derived the central force equations of motion for a particle from the principle of least action, using the differential variational equations that he produced in the treatise itself, he was not concerned about Vìetian-style matters of homogeneity.[19] Further, it was Euler who, in *Theoria motus corporum solidorum seu rigidorum*, systematically defined units for physical magnitudes (a set of fundamental ones, the meter, the second, etc.) and therefore for derived physical mag-

[17] See LH XXXV, I, 30, fol. 1–8 and 9–28 "Matheseos Universalis Pars Prima" (Niedersächsische Landesbibliothek) in Leibniz (2018, pp. 129–58, especially p. 146). I owe this reference to David Rabouin (private communication, December 15, 2017).

[18] See, for example, Vol. 1, Chapter 3, Prop. 25, §193 in Euler (1736). Once again, we must be cautious about using present-day terminology. The definitions of "homogeneous differential equation" today have nothing to do with the "homogeneity" in differential equations discussed by Leibniz and his acolytes, such as Varignon and the brothers Bernoulli. For the present-day definitions, see, for example, Simmons (1972, pp. 35–88, 74).

[19] Euler (1744, pp. 309–20). I thank Craig Fraser for this reference (private communication, November 5, 2017).

nitudes.[20] In this treatise, Euler came near to conceiving differential equations as relating a function and its derivatives.

Euler and other mathematicians of his generation, including d'Alembert, were not thinking about graphs representing the functional relationships between continuous magnitudes: they were not using curves in Varignon's style, but the algebraic concept of function.[21] Their differential equations were now understood in a more algebraic way and they consequently sought to integrate them in terms of classes of functions (as general integrals). The mid-eighteenth-century shift towards algebra was mostly due to the fact that the problems tackled became increasingly difficult. The concepts of a function of several variables and of partial derivative were introduced in order to mathematize continuum mechanics (the vibrating string, the mechanics of solid, fluid, and elastic bodies). Searching for solutions of differential equations in terms of geometrical or mechanical constructions was a hopeless task. Euler soon lost all interest in it. The habit of requiring differential equations to be geometrically homogeneous faded away too.

7.7 Some lessons

What historiographical lessons concerning anachronism can we draw from the case study we have reviewed in this chapter? We can gain some insights by following Bos's lead. Bos, as we have seen in the Introduction, claims that mathematicians in the past pursued different "self-imposed tasks" and different "criteria of adequacy or quality control" from the ones in use today (Bos, 2004).

Although the early Leibnizian calculus might at first sight appear modern to us, it was practiced by Leibniz and his early acolytes with a view to purposes (a notion of solution) anchored in the tradition of Viète, Descartes, Huygens, and Newton. One of the tasks of Leibnizian mathematicians was to provide solutions of ordinary differential equations via a construction, not via an equation. Their criterion of adequacy was the *lex homogeneorum*: sums and subtractions could only apply to homogeneous magnitudes. Avoiding a proleptic reading of the texts written by Leibniz and by his immediate followers at the turn of the eighteenth century allows us to appreciate that their authors were sharing tasks and criteria of adequacy different from ours.

20 Euler (1765, I, p. 71).
21 I thank Christian Gilain for noting this (private communication, December 10, 2017).

The differential equations of Leibniz and his acolytes look modern: they are deceptively similar to the equations written in our textbooks. It is easy to massage them so as to make the similarity to our own mathematical language even more apparent. We have performed this seemingly innocuous domesticating translation with Bernoulli's equation for central force motion. When a mathematical theory, algorithm, or solution enjoys a particularly fortunate reception and comes to occupy a prominent position in the canon, paradigm, or framework taught today, our reading of the texts in which these theories, algorithms, and methods were originally expressed risks projecting our own modernity, our own "tasks" and our own "criteria of adequacy" onto them.

We ran precisely this risk in the present chapter when, at the outset of Section 7.3, we translated Bernoulli's equation for central force motion into our own language, using physical constants for energy, mass, and angular momentum, although Bernoulli was referring to geometrical magnitudes. This possibility, the ease with which this translation takes place, makes us forget that what is "lost in translation" – maybe a tiny detail concerning the occurrence of an extra-constant – might be the one detail revealing that otherness of the text which, as historians, we should regard as important. Indeed, one might object that the difference we have highlighted between Bernoulli's equation and our own is trifling and that we should not worry about it. The point is that this difference was considered crucially important by the historical actors, who discussed the *lex homogeneorum* at length. We should take seriously what was important for them.

Yet, we should not speak too loudly against domesticating translations "à la Whiteside," as I have dubbed them in the Introduction. The fact that these translations are possible is worth noting. It is an effect of the extraordinary stability of mathematics: a feature of mathematical knowledge that as historians we have to capture as a notable fact concerning the development of mathematics. Indeed, if it is important to recognize the otherness of the past, it is equally of import to highlight the continuity and analogies between past and present. Without continuity with the present, the past would be totally opaque: we could not even begin to understand what Bernoulli was up to. This is not to deny, however, that we learn about his equation also by noting the differences with its modern counterpart, as I have claimed above.

Domesticating translations have a number of other advantages. They allow historians to write in a way that is meaningful to mathematicians, who enjoy entertaining a conversation with past authors bringing their ideas to life, in such a way that these ideas can become part of present-day mathematical research and teaching. Domesticating translations shed light on textual reception. Once domesticated, Bernoulli's equation looks very much the form of it fixed at the end of the eighteenth century. By domesticating it, we learn what happened to it in the decades following its publication in the *Mémoires* for 1710. Further, we can apply our present theory of differential equations and use present-day knowledge in order to evaluate Bernoulli's equation. For example, we now know that the equation for central force motion is analytically integrable only for a few central forces. Once we use this knowledge, not available to Bernoulli, we understand why elliptic integrals were developed in the eighteenth century, for they were needed in order to extend the class of central forces for which the equation is integrable.[22]

Last but not least, one might view the case study considered in this chapter from the standpoint of the eminent Russian mathematician Vladimir Arnold, rather than that of Blay and Truesdell. Arnold's campaign in favor of a qualitative approach to the theory of dynamical systems quite explicitly structured the way he understood the development of rational mechanics from Newton to Henri Poincaré. His admiration for the geometrical style in both the *Principia* (1687) and in Huygens's *Horologium* (1673) clearly reflected the research approach he endorsed as a practicing mathematician (Arnold, 1990). It would be interesting to compare Arnold's views on the development of geometry and dynamics with Truesdell's authoritative efforts aimed at "rediscovering the rational mechanics of the age of reason" (Truesdell, 1960b). As in Arnold's case, Truesdell's historical studies were largely influenced by his own research in rational thermodynamics, the work for which he was best known as a scientist. For Truesdell, Newton's *Principia* belonged to a preliminary stage in the mathematical science of motion. This stage served as the background that eventually led to the advent of rational mechanics as

[22] For $F = kr^n$, elementary integrations in terms of circular or hyperbolic functions are possible only for $n = 1$, $n = -2$, and $n = -3$. There are other integer exponents for which an exact solution is possible in terms of elliptic functions ($n = 5, 3, 0, -4, -5, -7$): these were not known when Bernoulli wrote his paper (Bernoulli, 1710). Some fractional exponents can be shown to lead to elliptic functions, and other exponents can be expressed in terms of the hypergeometric function. In general, one has to make recourse to numerical integration (Goldstein *et al.*, 2014, p. 89). Knowledge of this result – not accessible to Bernoulli – can help us to appreciate the development of Bernoulli's mathematical theory of central force motion.

epitomized in the work of Leonhard Euler. In the case at hand, Truesdell considered Bernoulli's calculus approach to central force motion to be a major breakthrough, rather than an unimportant translation into symbols of Newton's geometrical reasoning, as Arnold would claim. As I have written in the Introduction, the past can be evaluated from different present viewpoints. Mathematicians, not surprisingly, look at the past through the lens of their own research interests, which reflect a diversity of aims and methods. As a historian, I am interested in how they go about reconstructing the past. The views of mathematicians turned historians are often prejudiced by their research interests, but I do not share the skepticism, sometimes even derision, so commonly felt by professional historians when they accuse mathematicians who turn to history of producing work that is hopelessly naïve because it is often based on anachronistic translations and evaluations.

Domesticating translations and anachronistic evaluations are indeed indispensable tools in our understanding of past mathematical texts. As a first interpretative move, we have to make ourselves familiar with these texts, domesticating them into our own language. But then – for sure – we have to foreignize the translation, in order to capture the "otherness" of the tasks and criteria of our predecessors. Through this hermeneutic work, we add new layers of translation: we rewrite the text, striving to teach the foreign idioms spoken in Bernoulli's "other present" to our readers. Translations and rewritings have occurred throughout the history of mathematics. As historians, we have to be aware that we are located at the end of a long succession of translations and transformations. This succession is at the same time a distorting lens through which we look at the past, a tool for investigation, and an object of historical study.

Acknowledgements Some of the themes in this chapter have been discussed with David Rowe in writing Guicciardini (2015) and with Dana Jalobeanu and David Marshall Miller in writing Guicciardini (2022). I thank them, as well as the participants to the workshops held in Bergamo (October 5–7, 2017) and at Iowa State University (July 26–28, 2018), for all the help they have generously given to me. I owe a debt of gratitude to George E. Smith (Tufts) for a challenging and enlightening discussion on the significance of the inverse problem of central forces in the works of Newton and Johann Bernoulli. George gave me a lesson in scholarship and intellectual integrity I will never forget. I thank Sergio Knipe for his careful revision of my English.

References

Arnold, Vladimir I. (1990). *Huygens and Barrow, Newton and Hooke: Pioneers in Mathematical Analysis and Catastrophe Theory from Evolvents to Quasicrystals*. Translated from the Russian by Eric J.F. Primrose. Basel, Boston, Berlin: Birkhäuser.

Bernoulli, Johann (1710). Extrait de la réponse de M. Bernoulli à M. Herman, datée de Basle le 7 Octobre 1710. *Mémoires de l'Académie des Sciences*, 521–33.

Blåsjö, Viktor (2017). *Transcendental Curves in the Leibnizian Calculus*. Amsterdam: Academic Press.

Blay, Michel (1992). *La Naissance de la Mécanique Analytique: La Science du Mouvement au Tournant des XVIIe et XVIIIe Siècles*. Paris: Presses Universitaires de France.

Bos, Henk J.M. (1974). Differentials, higher-order differentials and the derivative in the Leibnizian calculus. *Archive for History of Exact Sciences* **14**, 1–90.

Bos, Henk J.M. (2004). Philosophical challenges from history of mathematics. In: *New Trends in the History and Philosophy of Mathematics*, Tinne Hoff Kjeldsen, Stig Arthur Pedersen, and Lise Mariane Sonne-Hansen (eds). Odense: University Press of Southern Denmark, 51–66.

Costabel, Pierre (1967). Newton's and Leibniz's dynamics. *The Texas Quarterly* **10** (3), 119–26.

Euler, Leonhard (1736). *Mechanica, sive motu scientia analytice exposita*. St. Petersburg: Ex typographia academiae scientiarum.

Euler, Leonhard (1744). *Methodus inveniendi lineas curvas maximi minimive proprietate gaudentes*. Lausanne and Geneva: Bousquet.

Euler, Leonhard (1765). *Theoria motus corporum solidorum seu rigidorum*. Rostock and Greifswald: litteris et impensis A.F. Röse.

Friedman, Michael (2013). *Kant's Construction of Nature: a Reading of the Metaphysical Foundations of Natural Science*. Cambridge: Cambridge University Press.

Goldstein, Herbert, Charles Poole and John Safko (2014). *Classical Mechanics*, third edition. Upper Saddle River, NJ: Pearson/Addison Wesley.

Guicciardini, Niccolò (2015). Proofs and contexts: The debate between Bernoulli and Newton on the mathematics of central force motion. In: *A Delicate Balance, Global Perspectives on Innovation and Tradition in the History of Mathematics: A Festschrift in Honor of Joseph W. Dauben*, David E. Rowe and Wann-Sheng Horng (eds). Basel, Boston, Berlin: Birkhäuser, 67–102.

Guicciardini, Niccolò (2022). Mathematical innovation and tradition: the Cartesian common and the Leibnizian new analyses. In: *The Cambridge History of Philosophy of the Scientific Revolution*, Dana Jalobeanu and David Marshall Miller (eds). Cambridge: Cambridge University Press.

Hirsch, Eric Donald (1967). *Validity in Interpretation*, New Haven: Yale University Press.

Katz, Victor J. (1987). The calculus of the trigonometric functions. *Historia Mathematica* **14**, 311–24.

Leibniz, Gottfried Wilhelm (1689). Tentamen de motuum coelestium causis. *Acta eruditorum* **8**, 82–96.

Leibniz, Gottfried Wilhelm (2018). *Mathesis Universalis: Écrits sur la Mathématique Universelle*. David Rabouin (ed). Paris: Vrin.

L'Hôpital, Guillaume F.A. de (1696). *Analyse des Infiniment Petits pour l'Intelligence des Lignes Courbes*. Paris: Imprimerie Royale.

Mahoney, Michael Sean (1998). The mathematical realm of nature. In: *The Cambridge History of Seventeenth-Century Philosophy*, Daniel Garber and Michael Ayers (eds). Cambridge: Cambridge University Press, 702–55.

Nauenberg, Michael (2010). The early application of the calculus to the inverse square force problem. *Archive for History of Exact Sciences* **64**, 269–300.

Roche, John J. (1998). *The Mathematics of Measurement: A Critical History*. London: The Athlone Press.

Rossi, Paolo (1999). *Un Altro Presente: Saggi sulla Storia della Filosofia*. Bologna: Il Mulino.

Simmons, George F. (1972). *Differential Equations: With Applications and Historical Notes*. New York: McGraw-Hill.

Speiser, David (1996). The Kepler problem from Newton to Johann Bernoulli. *Archive for History of Exact Sciences* **50**, 103–16.

Speiser, David and Diarmuid Ó Mathúna (2019). Newton's influence on the mathematicians and physicists of the Bernoulli family and on Jacob Hermann. In: *The Reception of Isaac Newton in Europe*, Scott Mandelbrote and Helmut Pulte (eds). London: Bloomsbury Academic, 735–47.

Truesdell, Clifford A. (1954). Rational fluid mechanics, 1687–1765. In L. Euler, *Opera Omnia*. Lausanne, 2/12: pp. i–cxxv.

Truesdell, Clifford A. (1960a). The rational mechanics of flexible or elastic bodies 1638–1788. In L. Euler, *Opera Omnia*. Zürich, 2/11, §2.

Truesdell, Clifford A. (1960b). A program toward rediscovering the rational mechanics of the Age of Reason. *Archive for History of Exact Sciences* **1**, 3–36.

Varignon, Pierre (1700). Manière générale de déterminer les forces, les vitesses, les espaces et les tems [sic], une seule de ces quatre choses étant donnée dans toutes sortes de mouvemens rectilignes variés à discrétion. *Mémoires de l'Académie des Sciences*, 22–27.

Varignon, Pierre (1710). Des forces centrales inverses. *Mémoires de l'Académie des Sciences*, 533–44.

Venuti, Lawrence (1995). *The Translator's Invisibility: A History of Translation*. London: Routledge.

8

Euler and analysis: case studies and historiographical perspectives

Craig Fraser
University of Toronto

Andrew Schroter
Branksome Hall, Toronto

But [the historian's] particular business lies, not with this bare and general similarity, but with the detailed dissimilarity of past and present. He is concerned with the past as past, and with each moment of the past in so far as it is unlike any other moment (Oakeshott, 1933, p. 106).

Abstract: Two parts of analysis to which Leonhard Euler contributed in the 1740s and 1750s are the calculus of variations and the theory of infinite series. Certain concepts from these subjects occupy a fundamental place in modern analysis, but do not appear in the work of either Euler or his contemporaries. In the case of variational calculus there is the concept of the invariance of the variational equations; in the case of infinite series there is the concept of summability. However, some modern mathematicians have suggested that early forms of these concepts are *implicitly* present in Euler's writings. We examine Euler's work in calculus of variations and infinite series and reflect on this work in relation to modern theories.

8.1 Introduction

The present study explores the notion of anachronism in the history of mathematics in relation to some mathematical work of Leonhard Euler. The focus is on the 1740s and early 1750s, during Euler's Berlin period, when he was approaching the height of his mathematical powers and productivity. We consider aspects of two subjects that he investigated in an original and ground-breaking way: calculus of variations and the

[a] From *Anachronisms in the History of Mathematics: Essays on the Historical Interpretation of Mathematical Texts*, edited by Niccolò Guicciardini

theory of infinite series. Certain concepts occupy a fundamental place in the modern subject, but do not appear in the work of either Euler or his contemporaries. In the case of variational calculus there is the concept of the invariance of the variational equations; in the case of infinite series there is the concept of summability. While both concepts are a product of research since the later part of the nineteenth century, modern historical commentators have discerned the presence of intuitions or embryonic ideas of invariance and summability in Euler's writings. Our goal will be to look more closely at what Euler did and to evaluate the historical claims being made for his mathematical prescience.

8.2 Euler and the invariance of the variational equations

8.2.1 Euler's variational equation

In 1744 Euler published a major book on what would later be called the calculus of variations, his *Methodus Inveniendi Lineas Curvas Maximi Minimive Proprietate Gaudentes* (hereafter referred to as *Methodus Inveniendi*).[1] In Chapter 2 of this work we are given the general problem of finding the curve that maximizes or minimizes a definite integral $\int_a^b Zdx$, where Z is a function of x, y, and $p = \frac{dy}{dx}$. The curve is given in an orthogonal Cartesian system, depicted in Figure 8.1. Suppose that $dZ = Mdx + Ndy + Pdp$. Through a process of reasoning that involved disturbing a single ordinate of the curve, Euler was able to show that the optimizing curve satisfies the differential equation $N - \frac{dP}{dx} = 0$. In modern notation this equation is written as $\frac{\partial Z}{\partial y} - \frac{d}{dx}\frac{\partial Z}{\partial y'} = 0$, and is known as the Euler equation or the Euler–Lagrange equation for the variational problem.

In §33 of Chapter 2 Euler considers the problem of the shortest distance between two points in the plane. In an orthogonal coordinate system the differential element of path length is $\sqrt{(dx^2 + dy^2)} = \sqrt{(1 + pp)}dx$.[2] Hence $Z = \sqrt{(1 + pp)}$ and the length of a curve joining the two points is $\int_a^b \sqrt{(1 + pp)}dx$. We have $dZ = \frac{p}{\sqrt{(1+pp)}}dp$ and the equation $N - \frac{dP}{dx} = 0$

[1] For the development of Euler's researches leading up to this work see Fraser (1994).

[2] The derivative p is given by Euler in the form $pdx = dy$. An oddity of his notation is that he writes dx^2 and dy^2, but always writes pp rather than p^2. Thus in §20 of Chapter 1 we have $\sqrt{(dx^2 + dy^2)} = \sqrt{(1 + pp)}dx$.

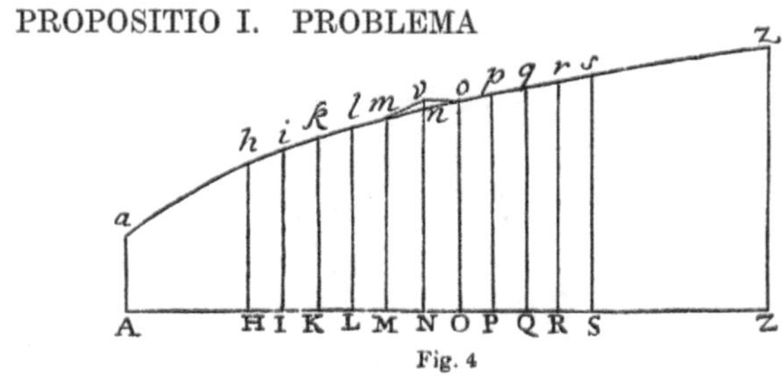

Figure 8.1 Euler (1744, Tabula I, Fig. 4).

reduces to $\frac{p}{\sqrt{(1+pp)}}$ = constant, or p = constant, which is the equation of the line $y = a + nx$, where a and n are constants.

In the fourth chapter of the *Methodus Inveniendi* Euler begins with a proposition, the intent of which is to assert a very wide interpretation for the methods he had introduced in the earlier chapters. He observes that if we have any equation between two variables x and y, we can always consider these variables as the orthogonal coordinates of a curve defined by the equation. Hence if we are given a function Z of x, y, and p (where $p = \frac{dy}{dx}$) we can apply the earlier methods to find the particular equation between y and x (the function y of x) that maximizes or minimizes the integral $\int_{a^0}^{a} Zdx$. In a corollary Euler elaborates on the significance of this finding:

> Therefore the method previously presented may be applied widely to find the equations between the coordinates of a curve, so that the expression $\int Zdx$ is a maximum or minimum. Indeed, it extends to any two variables, whether they belong to any given curve, or are only conceived of in analytical abstraction[3] (Euler, 1744, p. 130).

Euler is observing that logically his methods do not depend on any particular geometric coordinate system but are part of pure analysis.

In examples that follow in the chapter, Euler (1744, pp. 134–144) illustrates this conclusion by deriving the variational equation for examples involving curves given in non-orthogonal coordinate systems. The first example concerns the problem of the shortest distance between two

3 "Methodus ergo ante tradita multo latius patet, quam ad aequationes inter coordinatas curvarum inveniendas, ut quaepiam expressio $\int Zdx$ fiat maximum minimumve. Extenditur scilicet ad binas quascunque variabiles, sive eae ad curvam aliquam pertineant quomodocunque, sive sola analytica abstractione versentur."

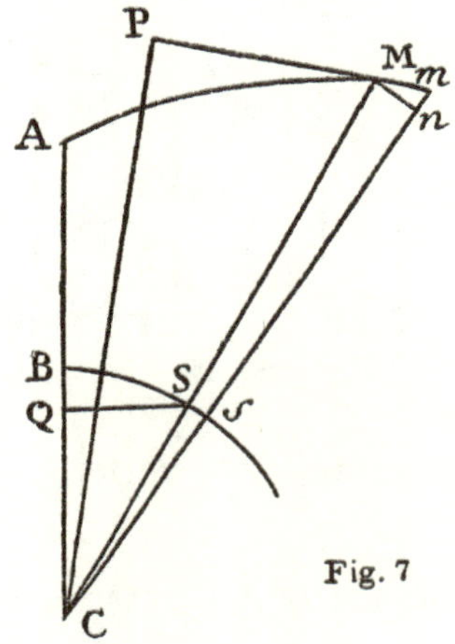

Figure 8.2 Euler (1744, Tabula I, Fig. 7).

points, where the problem is formulated in polar coordinates. In Figure 8.2, it is necessary to find the curve joining the points A and M of least length. Euler sets the angle ACM equal to x, and the radius CM equal to y. In the triangle nmM we have $Mn = ydx$ and $mn = dy$, so the infinitesimal pathlength Mm is $Mm = dx\sqrt{(yy+pp)}$. Thus, the length along the curve from A to M is

$$\int dx\sqrt{(yy+pp)},$$

where the integral is evaluated from $x = 0$ to the angle x corresponding to the point M. With $Z = \sqrt{(yy+pp)}$ and $dZ = Mdx + Ndy + Pdp$ we have

$$M = 0, N = \frac{y}{\sqrt{(yy+pp)}}, P = \frac{p}{\sqrt{(yy+pp)}}.$$

Because Z does not contain x we see immediately that a first integral of the equation $N - \frac{dP}{dx} = 0$ is $Z + C = Pp$, where C is a constant.[4] Given the expressions for Z and P above this equation may be simplified

[4] Euler is using the result that if $Z = Z(y, y')$ does not contain x then the equation $\frac{\partial Z}{\partial y} - \frac{d}{dx}\frac{\partial Z}{\partial y'} = 0$ is integrable. We have

$$\frac{dZ}{dx} = \frac{\partial Z}{\partial y}y' + \frac{\partial Z}{\partial y'}y'' + \frac{d(\frac{\partial Z}{\partial y'})}{dx}y' + \frac{\partial Z}{\partial y'}y'' = \frac{d(\frac{\partial Z}{\partial y'}y')}{dx}.$$

Hence $Z + C = \frac{\partial Z}{\partial y'}y'$, where C is a constant.

to $\frac{yy}{\sqrt{(yy+pp)}} = const = b$. The triangles Mnm and CPM are similar and we have the proportional equality $Mm : Mn = MC : CP$. Letting $Mm = dx\sqrt{(yy+pp)}, Mn = ydx, CM = y$ we obtain $CP = \frac{yy}{\sqrt{(yy+pp)}}$, which is a constant. Hence the perpendicular CP from C to the tangent to M is a constant, and the curve AM must be a straight line.[5] Since two constants are available in the integration of $Ndx - dP = 0$ the straight line AM is a solution to the problem.[6]

8.2.2 Invariance in calculus of variations and analytical dynamics

The examples just presented are at the center of some modern claims about Euler's intuitive familiarity with the concept of invariance. *Invariance* (or covariance) has different meanings in different areas of mathematics – algebraic forms, projective geometry, differential geometry, topology and functional analysis, to name a few. As far as variational equations are concerned, the relevant historical domain of research involved work in analytical mechanics and the calculus of variations in the second half of the nineteenth century. The subject of invariance came to the fore in the researches of Carl Jacobi in analytical dynamics. Jacobi began with the canonical equations of motion and considered transformations of the variables that would preserve the canonical form of these equations. He was able to show that such transformations can be given in terms of what became known as a generating function. The methods and ideas he pioneered were taken up further by researchers in celestial mechanics, most importantly by Henri Poincaré, the Swedish astronomer Ludwig Charlier, and the English mathematician Edward Whittaker. A key step was to show that a solution is effected by taking the generating function to be a solution of the Hamilton–Jacobi partial differential equation.

In the early years of the twentieth century German physicists working in quantum physics realized that canonical transformations and the associated Hamilton–Jacobi theory provided exactly the mathematical tools needed to investigate the physical systems that were of interest to them. The physicist Arnold Sommerfeld (1923, pp. 555–6) wrote: "Up to a

[5] In modern terminology CM and CP are the pedal coordinates of the point M on the curve AM. This terminology was not used by Euler.

[6] Today we would integrate $\frac{yy}{\sqrt{(yy+pp)}}$ = constant and obtain a solution of the form $y = \frac{A}{\cos x + B}$ for the polar-coordinate equation of the straight line joining A to M.

few years ago it was possible to consider that the methods of mechanics of Hamilton and Jacobi could be dispensed with for physics and to regard it as serving only the requirements of the calculus of astronomic perturbations and the interests of mathematics." As a result of the rapid development of quantum theory the situation had changed dramatically. Sommerfeld continued: "... it seems [today] almost as if Hamilton's method was expressly created for treating the most important problems of physical mechanics."

In twentieth-century literature on mechanics, canonical transformations occupy a prominent place. However, the subject in the early years of the century was a peripheral one within the calculus of variations. It did not appear at all in Oskar Bolza's comprehensive *Vorlesungen über Variationsrechnung* of 1909, although this book did deal in detail with some parts of Hamilton–Jacobi theory. In the years following the publication of this book, the situation began to change, as mathematicians turned their attention to those transformations that were of such interest to celestial mechanicians and quantum physicists. One such figure was the Munich mathematician Constantin Carathéodory, who wrote a chapter on the calculus of variations for Phillip Frank and Richard v. Mises's 1925 *Die Differential- und Integralgleichungen der Mechanik und Physik*. Consider a dynamical problem described by canonical equations of motion for a given coordinate system. A transformation from this set of coordinates to a new set of coordinates is canonical if the equations of motion in the new coordinate system are also canonical. A key result proved by Carathéodory was that the canonical equations of motion are preserved under a specified class of transformations, given in terms of a suitable generating function. The invariance of the canonical equations under a canonical transformation is the fundamental key to the utility of Hamilton–Jacobi methods in describing dynamical systems.[7]

In 1935 Carathéodory published his *Variationsrechnung und Partielle Differentialgleichungen Erster Ordnung*, which contained his distinctive blend of partial differential equations, tensor analysis and the calculus of variations. The author displayed an incomplete grasp of the history, apparently unaware of Hamilton's dynamical memoirs of the 1830s that stimulated Jacobi's famous 1837 paper. He seemed to believe that Hamil-

[7] There are other forms of invariance that are of interest with respect to canonical transformations, such as the Poincaré integral invariants and Lagrange and Poisson brackets as canonical invariants (see Goldstein, 1950, pp. 247–58). The invariance that is germane to our discussion is the property of preserving the form of Hamilton's equations under transformation.

ton's contributions were solely to optics, and there is nothing to indicate any substantial familiarity with Jacobi's original papers.[8] The overall aim of the book was to investigate in a systematic way connections between partial differential equations and variational analysis. In part one on partial differential equations he developed the theory of canonical transformations, and presented a rather abstract and difficult-to-follow proof of the invariance of the Lagrangian equations of dynamics under a transformation (more precisely, he presented a general theorem from which this result is said to follow).[9]

Carathéodory was a cosmopolitan figure known for his facility with languages and broad appreciation of culture. From the 1930s until the end of his life he engaged in the study of the eighteenth-century history of the calculus of variations leading up to Euler's 1744 treatise as well as some of Euler's later memoirs on the subject. In his obituary of Carathéodory, Oskar Perron (1952, p. 42) noted, "he understood masterfully how to extract from the deficient methods of each period fruitful approaches for the exact treatment of the problems raised."[10] Carathéodory himself provided an elegant statement of the promise of history in an address he delivered in 1936 to a meeting of the Mathematical Association of America at Harvard University:

> It may happen that the work of the most celebrated men may be overlooked. If their ideas are too far in advance of their time, and if the general public is not prepared to accept them, these ideas may sleep for centuries on the shelves of our libraries. Occasionally, as we have tried to do to-day, some of them may be awakened to life. But I can imagine that the greater part of them is still sleeping and is awaiting the arrival of the prince charming who will take them home (Carathéodory, 1937, p. 233).

Carathéodory published two scholarly articles on the early history of the calculus of variations, but his main contribution was an introduction he wrote to Euler's *Methodus Inveniendi*, an undertaking carried out

8 Carathéodory incorrectly gives the year of Jacobi's paper as 1836. He could not have been very familiar with the paper, since Jacobi's opening sentence consists of an acknowledgment of Hamilton's papers on mechanics of 1834 and 1835. Carathéodory's neglect of history here stands in contrast to his scholarly investigations at this time of the early-eighteenth-century history of the calculus of variations.

9 In a review of the 1965 English translation of part one of Carathéodory's book, Richard Courant (1967) writes "The original [1935] book is a masterpiece of mathematical writing." An historical appraisal of its contents and influence remains to be written.

10 "verstand er es auch meisterhaft, aus den unzulänglichen Methoden jener Zeit den fruchtbaren Kern herauszuschälen und wertvolle Ansätze zu finden zu einer exakten Behandlung der aufgeworfenen Probleme."

in what must have been difficult circumstances and connected to the planning for the re-publication of Euler's book in series one of his *Opera omnia*. Carathéodory completed this introduction by 1946, but volume 24 of the *Opera* with this introduction would not appear until 1952, two years after his death in February of 1950.

Carathéodory's account of Euler's book tended to have a modernizing quality, at times attributing to Euler twentieth-century methods and ideas that were understood to be implicit or at least foreshadowed in his analysis. This tendency is apparent in his account of Example 7 to Proposition III from Chapter 2 of Euler's 1744 book. The problem consists of maximizing or minimizing the integral $\int_a^b (x^2+y^2)^n \sqrt{(1+p^2)}dx$, where $p = \frac{dy}{dx}$.[11] Euler calculates N and P and considers the equation $N - \frac{dP}{dx} = 0$ for three cases: $n = 1$, $n = \frac{1}{2}$, and $n = \frac{3}{2}$. He shows in each case that the differential equation can be integrated by quadratures. In his solution Euler introduces complex numbers in order to integrate rational expressions in a way that had become fairly standard by that time. According to Carathéodory (1952, p. xxxvii), this problem today would be reduced by means of a conformal mapping to the problem of the shortest distance in the plane. While Carathéodory conceded that Euler did not notice this fact, he found it "astonishing" that in his introduction of complex numbers in the solution Euler followed a method that "in principle agrees with the one that we would use today."[12]

Carathéodory provided a classification of the different variational

[11] This problem for $n = 1$ was analyzed in parametric form by Tonelli (1923, pp. 430–435), who referred to Euler. Tonelli presented it as a minimization problem for moment of inertia. Carathéodory (1935, pp. 307–309) formulated the general problem in the complex plane where the real and complex parts are given parametrically. Euler presents the variational integral in the form $\int (xx+yy)^n dx\sqrt{(1+pp)}$.

[12] "... der in Prinzip mit demjenigen, den man heute benutzen würde, übereinstimmen." The following is a reconstruction of the train of thought that may have led Carathéodory to this improbable conclusion. In Section IV of his introduction Carathéodory began by describing a paper which Euler wrote near the end of his life and which was published in Euler (1810). Carathéodory refers to it only as E 731. In fact it was written some 35 years after the publication of the *Methodus Inveniendi*. The problem under consideration is a generalization of Example 7 from Chapter 2 of the 1744 book. Euler showed how the use of polar coordinates simplifies the integration. In the introduction Carathéodory then continued with a discussion of Example 7. Unmentioned by Carathéodory was the fact that he had analyzed in some detail a version of Example 7 in his 1935 book, where he took a complex number in polar form and transformed the real and imaginary parts to end up with an expression (in parametric form) formally similar to the one in Example 7. He was apparently impressed by the fact that in the 1810 paper Euler had used polar coordinates, that he himself used the polar representation of complex numbers in his 1935 book, and that Euler had in 1744 in his solution to Example 7 in the *Methodus Inveniendi* used complex numbers to integrate rational expressions. These considerations seem to have led Carathéodory to the above conclusion.

problems that Euler addressed. The problem of the shortest distance between two points in a polar-coordinate system, discussed above, was grouped with other examples from Chapter 4 of the *Methodus Inveniendi* under the title "Covariant transformation of variational problems." He wrote that these examples "may be viewed as evidence for the covariance of Euler's equations for arbitrary coordinate transformations."[13] He concluded with the observation: "Thus, in Euler's book, we find the first indications of a theory that has only been systematically developed in our time."[14]

In his *History of the Calculus of Variations* Herman Goldstine (1980, p. 84) was inspired to include a whole section of the chapter on Euler under the heading "Invariance Questions." Here are presented the examples from the first part of Chapter 4 of the *Methodus Inveniendi*. Goldstine acknowledged a suggestion he had received from André Weil that "Euler's interest in this topic probably stemmed from Leibniz's inquiries into the behavior of $\int ydx$ under coordinate transformations." Goldstine stated, echoing Carathéodory, "It is truly in keeping with Euler's genius that he should have worked at ideas that were only to be satisfactorily and completely discussed in modern times."

8.2.3 Some critical reflections

Euler first formulated and proved fundamental theorems about variational integrals with a general integrand function $Z(x, y, y')$. The further elaboration of this theory took the form of the derivation and solution of the Euler variational equation for a range of problems. A variational problem is posed and gives rise to an integrand function $Z(x, y, y')$. It is necessary to find $y = y(x)$ such that $\int_a^b Zdx$ = an extremum. If we formulate the problem analytically in other variables u and v in a different coordinate system then the variational problem gives rise to an integrand function $W(u, v, v')$ and it is necessary to find $v = v(u)$ such that $\int_a^b Wdu$ = an extremum. One always begins with the problem, followed by an analytical description, followed by the Euler variational equation, followed by a solution.

13 "können als Proben für die Kovarianz der Eulereschen Gleichungen bei beliebigen Koordinaten-Transformationinen bewertet werden."

14 "Somit finden wir im Eulerschen Buche die ersten Ansätze zu einer Theorie, die erste in unseren Tagen systematisch entwickelt worden ist."

One could begin with the condition $\int_a^b Zdx$ = an extremum, independent of any motivating problem and ask if a transformation of variables from x, y to u, v leads to a differential equation that may also be solved. This is apparently what Carathéodory and Goldstine had in mind when they credited Euler with having the intuition or some inkling of modern ideas of invariance. The problem of the shortest distance between two points was formulated in a standard Cartesian coordinate system and reduced to a differential equation by the variational process. The same result was then obtained using a polar-coordinate system, and again reduced to a differential equation and solved. While Euler did not employ a transformation from Cartesian to polar variables, the ideas of transformation and invariance were implicit in his analysis. The difficulty with this interpretation is that it projects onto the original analysis a way of thinking about the subject that is not present either as a potential idea or as an unrealized intuition.

As a point of comparison consider canonical transformations in Hamilton–Jacobi theory. For any dynamical system one can take a set of coordinate variables and investigate the system from first principles using them. One would end up with a Hamiltonian and canonical equations for these variables. No transformations are involved. Why then are transformations useful? The answer is that one can find canonical transformations using a generating function, and in the new coordinates so obtained the equations will be canonical and may be easier to integrate. Indeed, by taking the generating function to be a solution of the Hamilton–Jacobi partial differential equation for the problem, one is able to transform the original coordinates to ones that are constant, and the problem is solved. The invariance of the canonical equations under transformations provides a coherent and effective tool for integrating the differential equations that describe the dynamical system.

The situation with the Euler equation is rather different. In textbooks on the calculus of variations, the Euler equation is typically obtained for a given problem using a suitable selection of variables. There seems to be no advantage in beginning with a set of variables and transforming to new ones to obtain a transformed Euler equation. One could simply formulate the problem directly in terms of the new variables.

It is certainly of mathematical interest to investigate the invariance of the Euler equation for a given set of variables, independent of how the variational integral was obtained or any geometric or physical significance one attaches to the variables. No one in the eighteenth century did this, and few modern textbooks do it. In 1969 the Austrian–American mathematician Hans Sagan published a textbook in which he gave a detailed account of the invariance of the Euler equations.[15] Having developed the basic variational theory in the usual way as part of real analysis, Sagan (1969, p. 108) makes the remarkable and incorrect assertion that the derivation of the Euler equation "was essentially based on the fact that the coordinates were cartesian coordinates." Nevertheless, he does provide a detailed and useful account of the issues – not altogether simple – arising in any transformation and states conditions under which the families of comparison arcs in the two systems are comparable. Sagan (1969, pp. 108–115) shows that under "fairly liberal conditions on the integrand, the extremal and the transformation itself" invariance will hold.

8.2.4 Euler and the foundations of analysis

In Leibniz's original paper of 1684 on the calculus he considered the problem of finding the path followed by a light ray in going from two points *A* and *B* across an optical interface. The time of transit is connected by an equation to a spatial coordinate variable. The relationship between the time and the spatial variable can then be expressed by a curve, and one is able to apply the differential algorithm that he developed for curves. In the early years of the next century mathematicians such as Pierre Varignon used a comparison orthogonal Cartesian graph in investigating curves given in polar coordinates. There was a pervasive use of geometrical diagrams and representations in investigating what today would be called functional relationships between variables. (See Fraser (2003) for more details.)

The *Methodus Inveniendi* was an important step in Euler's program to separate analysis from geometry, and here lies the significance of the first part of Chapter 4 of that work. An equation between two variables is the basic object of study, and could be conceived of and investigated independently of any particular geometrical interpretation or coordinate representation. Although the function concept is not explicitly intro-

[15] Hans Sagan received his PhD in 1950 in calculus of variations from the University of Vienna and had a career in the United States at North Carolina State University. He was a commentator for the collected works of Johan Radon, Hans Hahn, and Karl Menger. See "History of the Math Dept at NCSU," at `http://www4.ncsu.edu/~njrose/Special/Bios/Sagan.html`.

duced in 1744, it is certainly implicit in Euler's investigation, and would be the next step in the elaboration of his program of analysis.

8.3 Euler and divergent series

8.3.1 Convergence and rigor

In the years following the publication of the *Methodus Inveniendi*, Euler pursued a range of subjects in analysis. A prominent area of investigation was infinite series, mainly but not exclusively power series. Detailed accounts were presented in his *Introductio in Analysin Infinitorum* of 1748 and in the multi-volume tomes on calculus that he published in the 1750s. During this period he began to investigate in a more serious way series that do not converge arithmetically but nonetheless have interesting properties. Euler's most productive effort in this direction was a paper on divergent series that he submitted first to the Berlin Academy in 1746, then to the St. Petersburg Academy in 1753, and that was finally published in 1760 in the memoirs of the St. Petersburg Academy for 1754–1755.[16]

Among modern mathematicians there is often a sense that Euler's use of divergent series was naïve and cavalier regarding convergence. Typical is C.N. Moore's (1932, p. 64) remark, "It is apparent that the procedure of Leibnitz and Euler in the case of the simple series $1 - 1 + 1 - 1 + \cdots$ is entirely out of harmony with present day notions of rigor in analysis." However, to one recent commentator Euler's reputation has been unjustly "tarnished" and should be redeemed "as a result of recent developments" in the theory (Kowalenko, 2011, p. 370). According to this view, the route to Euler's vindication is provided by the modern theory of *summability*. Summability, developed in the 1890s by Ernesto Cesàro and others, established a rigorous foundation for divergent series. This leads to the question: Was Euler a summability theorist ahead of his time?

8.3.2 Infinite series in the eighteenth century[17]

Early-eighteenth-century mathematicians had intuitive notions of convergence and divergence. Due to the prevailing view that mathematics was about *quantity*, one was free to obtain the "development" of any

[16] The chronology for the publication of this paper is given by Faber (1935, p. lxxv).
[17] In this section we follow Ferraro (2008) and Kline (1972).

quantity in the form of a power series (assumed to converge for at least some values of the variable) but could leave aside the question of convergence until applying the series to a geometric problem. This principle of "infinite extension" thus allowed for the formal manipulation of both finite and infinite series to occur prior to questions of convergence. The mid-century shift in calculus from its geometric foundation to a form of algebraic analysis gave a further boost to mathematicians' confidence in formulas. Under this philosophy, the general applicability of any method derived from the generality of its *object*. Since formulas were objectively given as part of algebra, their generality of usage was assured, even if this gave rise to divergent series. Formalism thus became untethered from geometry while remaining subtly connected with intuitive notions of quantity. As the most prolific practitioner of this new analysis, Euler's willingness to pursue formalism's implications for infinite series brought this tension to the foreground.

Consider the series $1 - 1 + 1 - 1 + \cdots$, which had been studied by several mathematicians before Euler. Clearly it did not become infinite. Grouping the terms $(1-1)+(1-1)+\cdots$ seemed to make the sum zero, but when writing $1-(1-1)-(1-1)-\cdots$ it appeared the sum was 1. In 1703, Guido Grandi argued that the series representation

$$\frac{1}{1+x} = 1 - x + x^2 - x^3 + \cdots$$

yields the answer $\frac{1}{2} = 1 - 1 + 1 - 1 + \cdots$ upon letting $x = 1$. Leibniz concurred, but he made his case probabilistically: since the result was either 0 or 1 depending on the number of terms summed, one should take the answer to be the average. When Euler turned to this series, he began with the expansion

$$\frac{1}{1-x} = 1 + x + x^2 + x^3 + \cdots$$

which he saw as a valid formal development of the "quantity" $\frac{1}{1-x}$. Upon letting $x = -1$, it yields $\frac{1}{2} = 1 - 1 + 1 - 1 + 1 - \cdots$, in agreement with Grandi and Leibniz. But Euler went on to let $x = 2$, which gave him $-1 = 1 + 2 + 4 + 8 + \cdots$. This put him in new territory, as the series is no longer bounded but clearly divergent in the infinite sense. In a similar vein, the substitution of $x = -1$ into the development

$$\frac{1}{1+x} = 1 - x + x^2 - x^3 + \cdots$$

led Euler to write $\infty = 1+2+3+4+\cdots$. Upon comparing these latter two results, Euler concluded that -1 must be greater than ∞, hence ∞ must serve as a kind of boundary between the positive and negative numbers. These results were not without criticism even in his own time. Ferraro (2008, p. 216) highlights the following comment Nicolaus Bernoulli made to Euler in 1743, "I cannot persuade myself that you think that a divergent series . . . provides the exact value of a quantity which is expanded into the series." Bernoulli's caution may ring true to modern mathematicians because of the hindsight afforded by the nineteenth century's turn from formalism to rigor (Fuss, 1843, pp. 701–702).[18]

8.3.3 Cauchy's new definitions

The decline of formalism stemmed mainly from its limitations as a means of generating useful results. Moreover, as methods began to change, an awareness of formalism's apparent difficulties and even contradictions lent momentum to efforts to rein it in. Euler had been confident that the "out-there" objectivity of algebra secured the generality of his formal techniques, but Cauchy demanded that generality be found within mathematical methods themselves. In his *Cours d'analyse* of 1821, Cauchy rejected formalism in favor of a fully quantitative analysis. Rather than the formulas themselves, the "quantities" became the individual values of the expressions when the variable took on a certain value. Hence the statement $f(x) = g(x)$ was not a formal relationship holding for indeterminate x but was a quantitative statement holding only for specific values of x. The statement $\frac{1}{(1-x)} = 1 + x + x^2 + x^3 + \cdots$ was true only for the values for which the series converged. Otherwise (say, when $x = 2$) it was meaningless. Cauchy specified that given the sum of the first n terms of a series

$$s_n = \sum_{i=0}^{n-1} u_n,$$

[18] Commenting on the work of eighteenth-century mathematicians but referring specifically to Euler, Kline (1983, p. 307) writes, "Their efforts to justify their work, which we can now appraise with the advantage of hindsight, often border on the incredible." Wonderment at the reasonings of the mathematical masters of the past occurs not infrequently in modern historical commentaries. Consider the following comments of James Pierpont (1928, p. 32) on Lagrange's expansion of functions by Taylor series: "When a modern reader looks over reasoning like this and bears in mind that Lagrange was one of the greatest mathematicians of all time, he is amazed. The great gulf that separates mathematical reasoning of to-day from that of date 1813 is brought home very clearly to him."

if the partial sums s_n approach a limit s as n increases, then the series converges with s as its sum. If the limit does not exist, the series diverges and there is no sum. The question of convergence now came first, and the manipulations the formalists had assumed valid for both finite and infinite series now depended on whether the series converged. Niels Abel concurred with the new ideas and in a letter of 1828 to Bernt Holmboe declared, "Divergent series are in general something very fatal, and it is a shame to base on them any demonstration whatsoever. By using them, one may draw any conclusion one wishes and it is these series that have produced so many misfortunes and given birth to so many paradoxes" (Abel, 1881, pp. 256–7).[19]

It should be noted that mathematicians in England and Germany continued to study divergent series well into the middle decades of the nineteenth century. Eberhard Knobloch (2015, p. 501) has suggested that the theoretical predilections of English formalists such as George Peacock and Augustus De Morgan were aligned with Euler's conception of divergent series. Certainly, the English formal school was more favorably disposed to divergent series than were the French rigorists. (For a survey of these developments see Burkhardt (1910). Compare also Fraser (2003, pp. 325–7).) However, it is fair to say that the mainstream of analysis with its emphasis on rigor led as the nineteenth century progressed to the marginalization of divergent series as an area of mathematical investigation. The notion that rigor rescued mathematics from disarray has become a rather common view and might be considered a part of the *heritage* of the modern mathematician, leading to the discomfort over Euler's status. If one views the concepts "series" and "sum" as cumulatively improving entities gradually unveiled for us over hundreds of years, then one might well say that Cauchy's refinements do eclipse Euler's results.[20] However, Cauchy did not instill rigor by *introducing* convergence, as if Euler had failed to consider it. Cauchy changed the notion of what an infinite series is – no longer an algebraic object subject

19 "Les séries divergentes sont en général quelque chose de bien fatal, et c'est une honte qu'on ose y fonder aucune démonstration. On peut démontrer tout ce qu'on veut en les employant, et ce sont elles qui ont fait tant de malheurs et qui ont enfanté tant de paradoxes." These sentences are translated by Kline (1972, p. 973) as follows: "The divergent series are the invention of the devil, and it is a shame to base on them any demonstration whatsoever. By using them, one may draw any conclusion he pleases and that is why these series have produced so many fallacies and so many paradoxes."

20 This line of thought may even be natural to, say, a calculus teacher who needs to justify why convergence matters.

to formal manipulation, but a relationship between definite numerical series subject to convergence first. Euler and Cauchy had distinct notions of "series," so it is dubious to charge Euler with mere negligence.[21] One might instead pivot to the idea that Euler's reputation was unfairly stained by "victors" who rewrote mathematics at his expense. Consider Hardy's (1949, p. 17) remark: "Mathematics after Euler moved slowly but steadily towards the orthodoxy ultimately imposed on it by Cauchy, Abel and their successors, . . . after Cauchy, the opposition seemed definitely to have won." The belief that Euler's successors treated him poorly might lead the historian today to assess his theory of series in a more sympathetic way.

8.3.4 Summability theory

The predominant mathematical trend in the nineteenth century was to support Cauchy's thinking about infinite series. The sum of a convergent series *existed* and the sum of a divergent series did not *exist*. To be sure there were English and German exceptions, and their researches provided impetus to continue to think about divergent series. Also notable was evidence for the utility of divergent series, such as in the asymptotic approximation of certain functions. Research in complex analysis (then called the theory of analytic functions) suggested a possible need. Consider an analytic function represented by a power series on an open disk but not for values on the disk's boundary: a new concept of "sum" would allow for the assignment of a value for the function on the boundary.

Alongside these new research concerns, shifts occurred in the foundational methods of mathematics. Mathematicians began to see their theories not as descriptions of reality but as syntactic structures whose theorems were logical derivations from axioms and definitions. Support for this conception came from the non-Euclidean geometries, which showed that the postulates of Euclid were not necessary to produce a consistent geometry. Kline (1972, p. 1097) writes: "The mathematicians slowly began to appreciate that mathematics is man-made and that Cauchy's definition of convergence could no longer be regarded

[21] Fraser (1989, pp. 323–4) notes, "The significant change in the theory of infinite series, however, was not so much that classical analysis brought rigour to the subject by paying attention to convergence, but that an arbitrary series whose individual terms were specified at will now became, subject to convergence over some domain, implicitly an object of mathematical study. The understanding of what an infinite series was had undergone a substantial transformation."

as a higher necessity imposed by some superhuman power." The first suggestive attempts to produce a new theory of series came from Ferdinand Georg Frobenius (1880) and Otto Hölder (1882). In 1890, Ernesto Cesàro provided the first modern concept of the sum of a divergent series, known as the theory of summability.[22]

According to the theory one is free, in principle, to select any desired procedure P for the summation of a series. This choice is merely conventional and has no intrinsic connection to the series. There is no sense of finding the "true" sum or any anchoring to an external reality. Hence the sum of a series "exists" (or not) only relative to the procedure P; if it does exist, it is called the P-sum. In practice, P should give fruitful consequences. For example, it is desirable that the P-sum agree with the usual sum for a convergent series – this is called *regularity*. Consider, then, the summation of the previously considered series $1 - 1 + 1 - 1 + \cdots$ according to the following method. Given the series $s_n = u_0 + u_1 + \cdots + u_n$, if

$$\lim_{n\to\infty} \frac{s_0 + s_1 + \cdots + s_n}{n+1}$$

exists and is equal to s, then we call s the (C, 1) sum of the series (C stands for Cesàro). Now considering our series $s_n = \sum_{i=0}^{n}(-1)^i$ we see that the "sum of the partial sums" is

$$s_0 + s_1 + \cdots + s_n = \begin{cases} 1 + 0 + 1 + \cdots + 1 & = \frac{n+2}{2} \quad \text{for even } n \\ 1 + 0 + 1 + \cdots + 0 & = \frac{n+1}{2} \quad \text{for odd } n \end{cases}.$$

But in either case we get

$$\lim_{n\to\infty} \frac{s_0 + s_1 + \cdots + s_n}{n+1} = \frac{1}{2}$$

so the (C, 1)-sum of the series is $\frac{1}{2}$. This agrees, of course, with Euler's answer.

For a second example (Hardy, 1949, pp. 7–8), consider another definition as follows. If the power series $\sum a_n x^n$ converges in the usual sense for small x and defines a function $f(x)$ (subject to some additional conditions) and $f(1) = s$, then we call s the $\mathfrak{E}$-sum of the series $\sum a_n$ (the symbol $\mathfrak{E}$ is for Euler). But then for

$$f(x) = \frac{1}{1-2x} = 1 + 2x + 4x^2 + 8x^3 + \cdots$$

[22] This exposition follows Ferraro (1999).

we have $s = f(1) = -1$, so the $\mathfrak{E}$-sum of the series $1 + 2 + 4 + 8 + \cdots$ is -1, which again agrees with Euler's answer.

Despite these advances, Abel's distrust of divergent series seems to have remained fairly widely held. In the preface to Hardy's *Divergent Series* (1949, p. vii) J.E. Littlewood remarked: "In the early years of the century the subject, while in no way mystical or unrigorous, *was* regarded as sensational, and about the present title, now colourless, there hung an aroma of paradox and audacity." However, by the time of Hardy's volume, summability was well established, making divergent series mundane and leading to a reassessment of past results. In fact, Hardy (1949, p. 47) states that Leibniz, without actually specifying a definition, nevertheless employed precisely the (C, 1) procedure, and that the E-procedure is based on Euler's principles. His remarks are those of a mathematician who tends to see distant predecessors, in Fried's (2018) parlance, as "mathematical colleagues." Thus Hardy (1949, p. 15) says, "It is a mistake to think of Euler as a 'loose' mathematician, though his language may sometimes seem loose to modern ears; [it] somehow suggests a point of view far in advance of the general ideas of his time ... language which might almost have been used by Cesàro or Borel."[23]

Hardy's perception would be echoed in historians that followed him. In the article on Euler in the *Dictionary of Scientific Biography* A.P. Yushkevich (2008, p. 473) writes:

> But he also was a creator of new and important notions and methods, the principal value of which was in some cases properly understood only a century or more after his death. Even in areas where he, along with his contemporaries, did not feel at home, his judgment came, as a rule, from profound intuition into the subject under study. His findings were intrinsically *capable of being grounded in the rigorous mode of demonstration* that became obligatory in the nineteenth and twentieth centuries [emphasis added].

Euler biographer Ronald Calinger (2016, p. 93) observes that "In some cases it would take a century for scientists to grasp the proper use of his procedures." On the subject of infinite series Kline (1972, p. 453) lauds Euler's formal manipulations.

It is certainly true that Euler displayed a remarkable mathematical inventiveness and employed a range of methods in his study of divergent series. Kline (1972, p. 1110) concludes that Euler's results awaited

[23] Hardy's remarks here are also quoted by Varadarajan (2006, p. 130), who adds, "It is therefore clear that Euler had an understanding of the issues involving divergent series that was very much ahead of his time."

only a rigorous confirmation, which vindicated them. He states directly: "With hindsight we can see that the notion of summability was really what the eighteenth- and early nineteenth-century men were advancing." These comments seem to ascribe to Euler an unaccountable prescience that saw the "right" foundational theory on some imagined future horizon. However, as we shall now try to show, the approach of a modern mathematician differs substantially from that of an eighteenth-century formalist.

8.3.5 Different kinds of definition

The contrast in understanding between an eighteenth-century mathematician such as Euler and a modern researcher appears most strikingly in the notion of *definition*. On this point, Hardy concedes the methodological gulf between Euler and himself. Regarding eighteenth-century mathematicians he says:

> They had not the habit of definition; it was not natural to them to say, in so many words, 'by X we *mean* Y'... Mathematicians before Cauchy asked not 'How shall we *define* $1 - 1 + 1 - \cdots$?' but 'What *is* $1 - 1 + 1 - \cdots$?', ... This habit of mind led them into unnecessary perplexities (Hardy, 1949, p. 6).

Similarly, Konrad Knopp (1928, p. 457) puts his finger on the issue: "In our exposition, *the symbol for infinite sequences was created* and then worked with; it was not so originally, these sequences *were there*, and the question was, what could be done with them" (emphasis in original). For mathematicians of the modern era, a symbol has meaning only when one assigns it such. Definitions are acts of the will, which construct *a priori* the very objects of study and thus cannot be "true" or "false." One defines basic terms (or primitives) implicitly via the axioms, while other terms are defined explicitly simply as symbolic abbreviations. A theory consists of symbols arranged in a syntactic structure, whose elements can take on any interpretation one wishes; hence, for example, one is free to define a summation procedure P and see what happens.

This practice was foreign to Euler. Eighteenth-century mathematicians did make definitions, yet they were of the "Euclidean" sort, where the appropriate response to their question "What is X?" was a *descriptive* definition that gave an accurate account of X's true nature. Objects of mathematical study were taken to have a real existence in the world. The purpose of a definition was to specify and clarify an object to enable its investigation.[24] Thus while a P-sum is self-consciously "man-made," an Eulerian sum is an attempted observation of nature. Euler's thinking was therefore more in line with Cauchy's than with that of a modern summabilist. Both Euler and Cauchy wielded the "Euclidean" notion of definition as elucidation of "the unique and necessary 'truth' that already existed in nature" (Ferraro, 2008, p. 222). For Euler, objective truth was a necessary criterion for a proper definition.

8.3.6 Euler's definitions

Euler's formalism derived from the objective reality of the rules of algebra. Consider his treatment of the statement $d(\log x) = \frac{dx}{x}$. Leibniz held that this is meaningful only for positive real x, but Euler disagreed:

> For, as this calculus concerns variable quantities, that is, quantities considered in general, if it were not generally true that $d \cdot lx = \frac{dx}{x}$, whatever value we give to x, either positive, negative, or even imaginary, we would never be able to make use of this rule, the truth of the differential calculus being founded on the generality of the rules it contains (Euler, 1751, p. 143, translation by Fraser, 1989, p. 331).[25]

For Euler, the applicability of the calculus stemmed from the general character of its formulas and rules, which were true and given as part of the subject of mathematics. Yet with analysis detached from its original anchoring in geometry, the status of its *objects* of study was not entirely clear. Euler's results on divergent series had strained the connection between formalism and "quantity" and required clarification. But Euler did not make an arbitrary definition subject only to logical consistency; this would have been unthinkable. Instead, he sought a concept of "sum" that would avoid disputes and provide a basis for further research; it must *not* be arbitrary. Euler dealt with the matter in his *Institutiones calculi differentialis* (1755) and *De seriebus divergentibus* (1760). His finding that -1 exceeded ∞ *was* sensibly quantitative, he said, on the grounds that infinity, in analogy with zero, was a transition from positive to negative. This could be defended by the "law of continuity and geometry." But

[24] Naturally Hardy does not consider such an act to be a "definition," as the meaning was different by his time.

[25] "Car comme ce calcul roule sur des quantités variables, c. à d. sur des quantités considérées en général, s'il n'etoit pas vrai généralement, qu'il fût $d \cdot lx = \frac{dx}{x}$, quelque quantité qu'on donne à x, soit positive ou négative, ou même imaginaire, on ne pourrait jamais se server de cette régle, la verité du calcul differential étant fondée sur la généralité des regles, qu'il renferme."

upon seeing that

$$\frac{1}{(1-x)^2} = 1 + 2x + 3x^2 + 4x^3 + \cdots$$

yields $1 = 1+4+12+32+\cdots$ for $x = 2$ but gives $\infty = 1+2+3+4+\cdots$ for $x = 1$, it now seemed that even 1 must be greater than ∞, which was a clear difficulty. Hence Euler (1755) admitted that a basically quantitative understanding of divergent series could not be sustained. In his *Institutiones calculi* he extended the notion in a more formal direction:

> Let us say, therefore, that the *sum* of any infinite series is the finite expression, by the expansion of which the series is generated. In this sense, the sum of the infinite series $1 - x + x^2 - x^3 + \cdots$ will be $\frac{1}{(1+x)}$, because the series arises from the expansion of the fraction, whatever number is put in place of x.

This statement agrees with the above comment about $d(\log x)$. He continues:

> If this is agreed, the new definition of the word sum coincides with the ordinary meaning when a series converges; and since divergent series have no sum, in the proper sense of the word, no inconvenience can arise from this new terminology. Finally, by means of this definition, we can preserve the utility of divergent series and defend their use from all objections (Euler, 1755, pp. 78–9, translation by Bromwich, 1908, p. 266).[26]

Barbeau and Leah (1976, p. 142) interpret these passages to mean that Euler "distinguishes between convergent and divergent series along modern lines ... Thus his assignment of a sum to a divergent series is a matter of conscious decision, made on pragmatic grounds and defensible by the consistency of mathematical analysis." They note that while Euler often is perceived as misunderstanding infinite series, his ideas were vindicated by Hardy. It is certainly true that Euler was creative and versatile in his investigation of such series. Yet there is a clear difference between Euler's thinking and Hardy's. Euler was not seeking an artificial

[26] "Dicamus ergo seriei cuiusque infinitae summam esse expressionen finitam, ex cuius evolutione illa series nascatur. Hocque sensu seriei infinitae $1 + x + x^2 + x^3 +$ &c. summa revera erit $= \frac{1}{1-x}$, quia illa series ex huius fractionis evolutione oritur; quicunque numerus loco x substituatur. Hoc pacto, si series suerit convergens, ista nova vocis summae definitio, cum consueta congruet; & quia divergentes nullas habent summas proprie sic dictas, hinc nullum incommodum ex nova hac appellatione orietur. Denique ope huis definitionis utilitatem serierum divergentium tueri, atque ab omnibus iniuriis vindicare poterminus." Difficulties with this definition in the context of some of Euler's posthumously published work on trigonometric series are discussed by Faber (1935, p. lxiv). See also Knobloch (2015, p. 501).

way to retain the use of divergent series but was trying to capture what a "sum" really was, to clarify the nature of the objects of his study. For Euler the sum of a series was a function if and only if the series resulted from the formal development of that function by the principle of infinite extension. He defended his definition:

> If therefore we change the accepted notion of sum to such a degree that we say the sum of any series is a finite expression out of whose development that series is formed, all difficulties vanish of their own accord. For first that expression from whose expansion a convergent series arises displays the sum, this word being taken in its ordinary sense; and if the series is divergent, the search cannot be thought absurd if we hunt for that finite expression which expanded produces the series according to the rules of analysis[27] (Euler, 1760, p. 212, translation by Barbeau and Leah, 1976, p. 148).

Again, a cursory reading might see Euler's comments as making an arbitrary definition. But there is a distinction: Euler's definition is grounded in the "rules of analysis". As he saw it, the true meaning of "sum" had to accommodate divergent series. He believed that the use of divergent series could never lead to an error (Ferraro, 2008, p. 225).

Euler was trying to study the objects that nature had thrust upon him, by making and testing hypotheses. An example of this heuristic is his consideration of the series

$$s = x - (1!)x^2 + (2!)x^3 - (3!)x^4 + \cdots$$

which diverges for all x except 0. This series formally satisfies the differential equation $ds + \frac{sdx}{x^2} = \frac{dx}{x}$; however, the equation can be integrated to obtain the solution

$$s = e^{\frac{1}{x}} \int_x^0 \frac{e^{\frac{-1}{t}}}{t} dt.$$

Euler maintained that the infinite series must be the expansion of this solution. Letting x equal 1 we obtain a value for the divergent hypergeometric series of alternating factorials:

$$1 - 1! + 2! - 3! + \cdots = e \int_0^1 \frac{e^{\frac{-1}{x}}}{t} dt,$$

27 "Si igitur receptam summae notionem ita tantum immutemus, ut dicamus cuisque seriei summam esse expressionem finitam, ex cuius evolutione illa ipsa series nascatur, omnes difficultates, quae ab utraque parte sunt commotae, sponte evanescent. Primo enim ea expressio, ex cuius evolution nascitur series convergens, eius simul summam, voce hac vulgari sensu accepta, exhibit, neque, si series fuerit divergens, quaestio amplius absurda reputari poterit, si eam indagemus expressionem finitam, quae secundum regulas analyticas evoluta illam ipsam seriem producat."

with the integral on the right being approximately 0.59637.[28] This estimate for the sum of the series was confirmed by other methods Euler introduced to find its value (the "Euler summability method" and partial fractions; see Barbeau and Leah, 1976, pp. 149–53). Thus, he saw that his definition of sum was supported by results from different approaches, confirming the objectivity and correctness of his proposal to define the sum in this way.[29]

Despite this success, Euler's attempt to capture the true nature of divergent series encountered difficulties. In 1797 J.F. Callet pointed out that given

$$\frac{1+x}{1+x+x^2} = \frac{1-x^2}{1-x^3} = 1 - x^2 + x^3 - x^5 + x^6 + \cdots,$$

when $x = 1$ we have

$$\frac{2}{3} = 1 - 1 + 1 - 1 + \cdots,$$

which does not agree with Euler's previously determined value $\frac{1}{2}$. In some of his posthumously published writings Euler himself drew attention to "paradoxes" that arise in the study of infinite series (see the footnote 18 on page 236). These kinds of problems would eventually lead to Cauchy's attempt at a novel definition for the "true" sum of a series. While Cauchy was initiating a new epoch in the history of calculus, a belief in objective truth was something he continued to share with the older researchers. By contrast the summabilists, whose work traced the fault lines of nineteenth-century upheavals in the foundations of mathematics, introduced definitions that had no concern with truth beyond logical coherence.

28 Interestingly, although the power series diverges, nonetheless its successive partial sums give rather good approximations of the integral for a given x. For example, if one lets $x = 0.1$ and takes the successive partial sums of s one obtains a good approximation to the integral from the seventh sum to the fourteenth sum; the sums remain close until the twentieth sum after which the factorials begin to dominate and the expansion diverges. Euler did not comment on the asymptotic character of the expansion although it is not unlikely that he was aware of it. Considerations related to asymptoticity arise in the memoir Euler (1750); see Faber (1935, pp. cxi–cxii) and Barbeau and Leah (1976, p. 150).

29 While philosophical analysis is beyond the scope of this chapter, it could be maintained that the contents of Euler (1760) are characterizable in terms of the "quasi-empiricism" identified by philosopher Hilary Putnam (1975). Another article that touches on Euler's work on divergent series in relation to the philosophy of mathematics and mathematics education is Schroter (2018).

8.4 Conclusion

There is more than one way to view the mathematics of the past. Ivor Grattan-Guinness (2004) identifies a disjunction between *heritage* (our tracking of a particular concept's journey along the "royal road" from the past to the present) and *history* (our attempt to explain why a certain mathematical development happened). The "heritage" approach evaluates past mathematics in light of recent theories, looking for similarities that reveal the gradual unveiling of a mathematical concept. Conversely, "history" instinctively looks for differences and discontinuities. Michael Fried (2018) extends Grattan-Guinness's idea by positing a spectrum of viewpoints. He describes *mathematicians*, who see their work as entirely coextensive to that of earlier mathematicians; *mathematician–historians*, trained mathematicians to whom modern mathematics provides the privileged perspective with which to assess the past; and *historians of mathematics*, to whom "the past is a problem" that stands in contrast with the present. Fried comments, "Faced with a mathematical text, historians of mathematics try not to coordinate the text with the mathematics of the present, but to set it out from the present; they try to make it not more familiar but rather more strange, more foreign . . . bring out its identity." Fried notes that these sentiments are echoed in political philosopher Michael Oakeshott's concern for the past as past, and with each moment of the past in so far as it is unlike any other moment.

Therefore, one's assumptions and mode of thought become paramount. When studying past mathematics, should we take it to be an historically evolving subject or understand it as the unfolding of a timeless whole? The mathematics teacher or theorist may view it in the latter way, at least psychologically; mathematics (all of it) is what mathematicians do, so it must possess common styles of thought, procedures, inferences, and rules for mathematical advance. By contrast, history looks for differences. A historian can remain agnostic on philosophy of mathematics while an historically minded mathematician generally may not feel so inclined. Each mode of thought has distinct objectives.

Historically speaking, then, the perception of Euler as a visionary may obscure the actual character of his work and its foundational import. The significance of the examples in Chapter 4 of the *Methodus Inveniendi* derives from their place in Euler's evolving program to separate analysis from geometry, not in any glimpse of some future notion of invariance.

The significance of Euler's formal approach to divergent series is not in the way it foreshadowed modern theories of summability, but rather in the latitude it provided him to obtain actual numerical values for divergent series. The modern conception of mathematical invariance or summability, as well as the associated philosophical commitments, are very different from Euler's own beliefs and outlook.

Claims that Euler grasped invariance, or was a summabilist, thus are anachronistic. On one hand, anachronistic approaches find productive uses in the classroom. After all, we are able to recognize from our point of view that both Euler and Hardy were engaged in studying "divergent series." Hence we may draw on "heritage" for didactic purposes – say, to teach about convergence. Indeed, there is some appeal to tracing the "history of a concept" – a directional journey from a "period of indecision," to the clear present. On the other hand, for one who takes our modern concepts and methods to be correct, it is easy to slip from a view that Euler *ought* to have used them to a claim that he *did* somehow use them. It is then that anachronism reaches the end of its utility: a more historical lens is required to help us see the "past as past" and understand Euler's achievements in their own context.

References

Abel, Niels Henrik (1881). *Oeuvres complètes de Niels Henrik Abel, Nouvelle Edition Vol. 2*. Christiania: Grøndahl & Søn.

Barbeau, Eduard J. and P.J. Leah (1976). Euler's 1760 paper on divergent series. *Historia Mathematica* **3**, 141–60.

Bolza, Oskar (1909). *Vorlesungen über Variationsrechnung*. Leipzig and Berlin: B.G. Teubner.

Boyer, Carl B. (1939). *The Concepts of the Calculus: A Critical and Historical Discussion of the Derivative and the Integral*. New York: Columbia University Press.

Bromwich, Thomas J. (1908). *An Introduction to the Theory of Infinite Series*. London: Macmillan.

Burkhardt, Heinrich (1910). Über den Gebrauch divergenter Reihen in der Zeit von 1750–1860. *Mathematische Annalen* **70**, 169–205.

Calinger, Ronald S. (2016). *Leonhard Euler: Mathematical Genius in the Enlightenment*. Princeton, NJ: Princeton University Press.

Carathéodory, Constantin (1925). Variationsrechnung. In *Die Differential- und Integralgleichungen der Mechanik und Physik*, Philipp Frank and Richard von Mises (eds). Braunschweig: Friedrich Vieweg & Sohn. Second edition (1930), pages 227–79.

Carathéodory, Constantin (1935). *Variationsrechung und Partielle Differentialgleichungen erster Ordnung*. Leipzig and Berlin: B.G. Teubner.

Carathéodory, Constantin (1937). The beginning of research in the calculus of variations. *Osiris* **3**, 224–40.

Carathéodory, Constantin (1952). Einführung in Eulers Arbeiten über Variationsrechnung. In *Leonhardi Euleri Opera Omnia* Ser. 1 V. 24, viii–lxii. Bern: Orell Füssli.

Carathéodory, Constantin (1965). *Calculus of Variations and Partial Differential Equations of the First Order. Part I: Partial Differential Equations of The First Order*. Translation of the first part of Carathéodory (1935) by Robert B. Dean and Julius J. Brandstatter. San Francisco, London, and Amsterdam: Holden Day Inc.

Courant, Richard (1967). Review of Carathéodory (1965). *Mathematical Reviews*, MR0192372.

Euler, Leonhard (1744). *Methodus Inveniendi Lineas Curvas Maximi Minimive Proprietate Gaudentes sive Solutio Problematis Isoperimetrici Latissimo Sensu Accepti*. Lausanne and Geneva: Marcum–Michaelem Bousquet & socios. Reprinted as *Leonhardi Euleri Opera Omnia Ser. 1 V. 24* (1952), Bern: Orell Füssli. There is a German translation of Chapters 1, 2, 5, and 6 by Paul Stäckel in *Ostwalds Klassiker der exakten Wissenschaften* Nr. 46, Abhandlungen über Variations-Rechnung. Erster Theil Joh. Bernoulli (1696), Jac. Bernoulli (1697), Leonhard Euler (1744), Leipzig: Wilhelm Engelmann (1894). There is an English translation by Ian Bruce at `http://www.17centurymaths.com/contents/Euler'smaxmin.htm`.

Euler, Leonhard (1750). Consideratio progressionis cuiusdam ad circuli quadraturam inveniendam idoneae. *Commentarii Academiae Scientiarum Petropolitanae* **11** (1739), 116–127. In *Leonhardi Euleri Opera Omnia* Ser. 1 V. 14 (1925), Leipzig and Berlin: B.G. Teubner, pages 350–363.

Euler, Leonhard (1751). De la controverse entre Mrs. Leibnitz & Bernoulli sur les logarithmes des nombres negatifs et imaginaires. *Memoires de l'Académie des Sciences de Berlin* 5 (1749), 139–171. In *Leonhardi Euleri Opera Omnia* Ser. 1 V. 17 (1914), Leipzig and Berlin: B.G. Teubner, pages 195–232.

Euler, Leonhard (1755). *Institutiones calculi differentialis cum eius usu in analysi finitorum ac doctrina serierum*. St. Petersburg: Imperial Academy of Sciences. Reprinted as *Leonhardi Euleri Opera Omnia Ser. 1 V. 10* (1911), Leipzig and Berlin: B.G. Teubner.

Euler, Leonhard (1760). De seriebus divergentibus. *Novi Commentarii Academiae Scientiarum Petropolitanae* **5**, 205–237. In *Leonhardi Euleri Opera Omnia Ser. 1 V. 14* (1925), Leipzig and Berlin: B.G. Teubner, pages 585–617.

Euler, Leonhard (1810). Solutio problematis ob singularia calculi artificia memorabilis. *Mémoires de l'académie de St.-Pétersbourg* **2** (1807/8), 3–9. In

Leonhardi Euleri Opera Omnia Ser. 1 V. 25 (1952), Bern: Orell Füssli, pages 280–285.

Faber, Georg (1935). Übersicht über de Bände 14, 15, 16, 16* der ersten Serie, *Leonhardi Euleri Opera Omnia Ser. 1 V. 16,2* Carl Boehm (ed.). Basel: Orell Füssli, pages vii–cxii.

Ferraro, Giovanni (1999). The first modern definition of the sum of a divergent series: An aspect of the rise of 20th century mathematics. *Archive for History of Exact Sciences* **54**, 101–35.

Ferraro, Giovanni (2008). *The Rise and Development of the Theory of Series up to the Early 1820s*. New York: Springer.

Fraser, Craig (1989). The calculus as algebraic analysis: some observations on mathematical analysis in the 18th century. *Archive for History of Exact Sciences* **39**, 317–335.

Fraser, Craig (1994). The origins of Euler's variational calculus. *Archive for History of Exact Sciences* **47**, 103–141.

Fraser, Craig (2003). Mathematics. In: *The Cambridge History of Science. Volume 4: Eighteenth-century Science*, Roy Porter (ed). Cambridge: Cambridge University Press, pages 305–27.

Fried, Michael (2018). Ways of relating to the mathematics of the past. *Journal of Humanistic Mathematics* **8** (1), 3–23.

Fuss, Paul Heinrich (1843). *Correspondance mathématique et physique de quelque célèbres géomètres du XVIIIème siècle*, St. Pétersbourg: Académie Impériale des Sciences.

Goldstein, Herbert (1950). *Classical Mechanics*. Reading, MA: Addison-Wesley.

Goldstine, Herman H. (1980). *A History of the Calculus of Variations from the 17th through the 19th Century*. New York: Springer-Verlag.

Grattan-Guinness, Ivor (2004). The mathematics of the past: distinguishing its history from our heritage. *Historia Mathematica* **31**, 163–185.

Hardy, Godfrey H. (1949). *Divergent Series*. Oxford: The Clarendon Press.

Kline, Morris (1972). *Mathematical Thought from Ancient to Modern Times*. New York: Oxford University Press.

Kline, Morris (1983). Euler and infinite series. *Mathematical Magazine* **56**, 307–14.

Knobloch, Eberhard (2015). Euler transgressant les limites l'infini et la musqiue. In: *Leonhard Euler Mathématicien, Physicien et Théoricien de la Musique*, Xavier Hascher and Athanase Papadopoulos (eds). Paris: CNRS Editions, pages 491–505.

Knopp, Konrad (1928). *Theory and Application of Infinite Series*. London and Glasgow: Blackie & Son Limited.

Kowalenko, Victor (2011). Euler and divergent series. *European Journal of Pure and Applied Mathematics* **4**, 370–423.

Moore, Charles N. (1932). Summability of series. *The American Mathematical Monthly* **39**, 62–71.

Oakeshott, Michael (1933). *Experience and its Modes*. Cambridge: Cambridge University Press.

Perron, Oskar (1952). Constantin Carathéodory. *Jahresbericht der Deutschen Mathematiker-Vereinigung* **55**, 39–51.

Pierpont, James (1928). Mathematical rigor, past and present. *Bulletin of the American Mathematical Society* **34**, 23–53.

Putnam, Hilary (1975). What is mathematical truth? Chapter 4 of *Mathematics, Matter and Method*. London and New York: Cambridge University Press.

Sagan, Hans (1969). *An Introduction to the Calculus of Variations*. New York: McGraw Hill.

Schroter, Andrew (2018). In defence of Platonism in the mathematics classroom. *Philosophy of Mathematics Education Journal* **33** (online).

Sommerfeld, Arnold (1923). *Atomic Structure and Spectral Lines*. London: Methuen & Co. Ltd. (Second edition 1928).

Tonelli, Leonida (1923). *Fondamenti del calcolo delle variazioni. Volume secondo*. Bologna: Nicola Zanichelli Editore.

Varadarajan, Veeravalli Seshadri (2006). *Euler through Time: A New Look at Old Themes*. Providence, RI: American Mathematical Society.

Yushkevich, Adolf P. (Youschkevitch) (2008). Euler, Leonhard. *Complete Dictionary of Scientific Biography, vol. 4*, pages 467–484. New York: Charles Scribner's Sons. Gale Virtual Reference Library. Originally published in 1974.

9

Measuring past geometers: a history of non-metric projective anachronism

Jemma Lorenat
Pitzer College

In this work I have sought to make the geometry of situation an independent science, which does not require measurement (von Staudt, 1847, p. iii).

Abstract: In 1847 Karl Georg Christian von Staudt framed his *Geometrie der Lage* as completing the newly emerging distinction between geometry of situation and geometry of measurement.

By the end of the century, historians of mathematics had accepted von Staudt's diagnosis of his predecessors and solution in separating the two areas of study: geometry of situation and metric geometry. In the case of projective geometry, these changes within mathematics resulted in a distorted view of history. A recurring historical narrative depicts Jean-Victor Poncelet, Michel Chasles, Jakob Steiner, and other early-nineteenth-century geometers as striving and failing to create a non-metric projective geometry. According to this historiographical view, only in mid-century with von Staudt would projective geometry be liberated from its ties to measurement. This claim for geometers before von Staudt is what I will call *the non-metric projective anachronism.* This chapter will consider how and why pure geometers of the early nineteenth century came to be seen as opposed to measurement. I will argue that the non-metric projective anachronism occurred as a two-part process. First, the geometry of Poncelet, Steiner, Chasles, and von Staudt (among others) began to be classified as projective geometry. Following this unification into projective geometry, historical accounts began to interpret past "projective geometers" as all pursuing a non-metric pro-

[a] From *Anachronisms in the History of Mathematics: Essays on the Historical Interpretation of Mathematical Texts*, edited by Niccolò Guicciardini

jective geometry. This is the second, and more historically problematic, part of the non-metric projective anachronism. While the non-metric projective anachronism can be found across many kinds of mathematical literature (textbooks, introductions to research articles, lectures on history, etc.), Felix Klein features as one very vocal supporter of a unified geometry grounded in pure projective geometry. A focus on Klein will capture features of late-nineteenth-century mathematics that made the non-metric projective anachronism so appealing.

9.1 Introduction

By the beginning of the twentieth century, projective geometry was entirely non-metric. In their axiomatic treatment, Oswald Veblen and John Wesley Young distinguished projective geometry from ordinary geometry *because* the former had no place for metric properties.

> It is evident that no properties that involve essentially the notion of measurement can have any place in projective geometry as such; hence the term *projective*, to distinguish it from the ordinary geometry, which is almost exclusively concerned with properties involving the idea of measurement (Veblen and Young, 1910, p. 14).

Pure projective geometry was the "natural" foundation from which metric geometries could later be derived. This hierarchy, the authors claimed, had both pedagogical and scientific benefits (pp. iii–iv).

A naïve glance at the beginning of the nineteenth century might suggest that the lack of measurement had long been a feature of projective geometry. In 1817 Jean-Victor Poncelet distinguished between geometry of situation and ordinary geometry.

> This problem, thus stated in a general manner, is essentially composed of two distinct parts; the one belongs to *geometry of situation*, and the other depends simply on *ordinary geometry* (Poncelet, 1817, p. 146).[1]

However, as will be shown, in 1822 Poncelet presented his studies on the projective properties of figures as encompassing both "metric" and "descriptive" properties.

Certainly the definitions of mathematical subjects change over time. As evidence of the change in geometry over the course of the nineteenth

[1] Ce problème, énoncé ainsi d'une manière générale, se compose essentiellement de deux parties distinctes; l'une appartenant à la *géométrie de situation*, et l'autre dépendant simplement de la *géométrie ordinaire*.

century, compare entries in the *Encyclopaedia Britannica*. In the ninth edition (1877), Olaus Henrici described the foundations of geometry with respect to axioms based in experience of space.

> The axioms are obtained from inspection of space and of solids in space – hence from experience (Henrici, 1877, p. 376).

By the beginning of the next century, the unsigned entry on Geometry offered several alternative means of generating axioms.

> The geometrical axioms are merely conventions; on the one hand, the system may be based upon inductions from experience, in which case the deduced geometry may be regarded as a branch of physical science; or, on the other hand, the system maybe formed by purely logical methods, in which case the geometry is a phase of pure mathematics (Anonymous, 1911, p. 675).

In the case of projective geometry, changes within mathematics resulted in a distorted view of history. A recurring historical narrative depicted Poncelet, Michel Chasles, Jakob Steiner, and other early-nineteenth-century geometers as striving and failing to create a non-metric projective geometry. According to this historiographical view, only in mid-century with von Staudt would projective geometry be liberated from its ties to measurement. This claim for geometers before von Staudt is what I will call *the non-metric projective anachronism*.

This chapter will consider how and why some geometers of the early nineteenth century came to be seen as opposed to measurement. I will argue that the non-metric projective anachronism occurred as a two-part process. First, the geometry of Poncelet, Steiner, Chasles, and von Staudt (among others) began to be classified as projective geometry. A non-metric projective geometry accorded with late-nineteenth-century foundational developments. In particular, a non-metric projective geometry supported the emergence of non-Euclidean geometries as sub-geometries based on a common projective foundation. Following this unification into projective geometry, historical accounts began to interpret past "projective geometers" as all pursuing a non-metric projective geometry. This is the second, and more historically problematic, part of the non-metric projective anachronism.

The non-metric projective anachronism can be seen most prominently in general histories of mathematics. Section 9.2 will briefly establish the existence and persistence of this anachronism through the historiography of geometry during the twentieth century. Then, to see the emergence of the non-metric projective anachronism, one must first have a sense of how geometers before von Staudt approached the study of projective properties of figures. As will be shown in Section 9.3, this research included both metric and non-metric properties.

Section 9.4 will examine von Staudt's presentation of the intended goals and audience for *Geometrie der Lage*. In contrast to the historical record, von Staudt introduced his non-metric *Geometrie der Lage* as the culmination of failed past attempts to segregate measurement and position. By reshaping the aims of previous geometers' research, von Staudt created support for *Geometrie der Lage* as the proper introduction to the study of geometry.

These preliminaries will serve as the background to the history of projective geometry – a discipline that was first coined in 1859. Only in the 1870s did Alfred Clebsch and Felix Klein provide an extensive description of "projective geometry" following von Staudt's *Geometrie der Lage*. This new projective geometry in turn engendered a historiography that traced the discipline back to the early nineteenth century. However, as will be considered in Section 9.5, the Clebsch/Klein conception of a unified non-metric projective geometry competed with other contemporary notions of projective geometries that aligned with the research of Poncelet.

While the non-metric projective anachronism can be found across many kinds of mathematical literature (textbooks, introductions to research articles, lectures on history, etc.), Klein features as one very vocal supporter of a unified geometry grounded in pure projective geometry. Moreover, he has been credited with first distinguishing non-metric geometry and first illuminating von Staudt's contributions (Enriques, 1907, p. 63 and Hartshorne, 2008, p. 299). Even though Klein was not the first in either of these pronouncements, the attributions highlight his outsized influence in the historiography of nineteenth-century geometry. Following Klein will illuminate developments and propagation of the non-metric projective anachronism. Section 9.6 will document the strong connection between Klein's mathematical research on deriving non-Euclidean geometry from projective geometry and his later historical publications on nineteenth-century mathematical developments.[2] Finally, Section 9.7 will consider the wider implications of Klein's historical vision as reflected in the *Encyklopädie der mathematischen Wis-*

[2] On Klein's use of von Staudt in his research and the Erlangen Program, see Gray (2015).

senschaften, in many ways a product of his editorial direction. A focus on Klein will capture features of late-nineteenth-century mathematics that made the non-metric projective anachronism so appealing.

9.2 The non-metric projective anachronism in the historiography of geometry

Notably, the non-metric projective anachronism is not a feature of more local historical analyses. Indeed, studies that focus on the history of early-nineteenth-century geometers confirm the importance of metric considerations in this time period.[3] Nevertheless, the anachronism continues to appear in general histories of mathematics, particularly those written in English during the twentieth century. A few excerpts will serve to illustrate common points in the narrative. In *A History of Geometrical Methods*, Julian Lowell Coolidge (1940) frames von Staudt as perceiving "essential weaknesses in the synthetic projective geometry studied by his predecessors." In particular, Steiner improperly relied on distance.

> The basis of the geometry of Steiner was the projective relation between the fundamental one-dimensional forms, but it was not properly defined. How can we develop this relation in a purely descriptive fashion, independent of distance? (ibid.)

Criticisms that negatively compare earlier geometers to von Staudt form a pattern in this literature. In more extreme versions, the aims of von Staudt are transposed to Poncelet, Steiner, or Chasles. For instance, Eric Temple Bell (1940, p. 39) asserts that Poncelet "banned" metric properties concerning distance and angles. In a similar manner, Roger Cooke (2011, p. 367) attributes metric-free geometry to Steiner in *The History of Mathematics: a Brief Course*:

> He [Steiner] sought to restore the ancient Greek "synthetic" approach to geometry, the one we have called metric-free, that is, independent of numbers and the concept of length.

3 While there are many excellent early-twentieth-century accounts (some of which will be discussed in what follows), the following references focus on more recent historical studies. The most comprehensive recent history of projective geometry is Bioesmat-Martagnon (2011). Of the above mentioned geometers, Poncelet is by far the best documented, see Chemla (1998), Nabonnand (2011b,a, 2015, 2016), Friedelmeyer (2011, 2016), Gray (2005b), Belhoste (1998), Billoux and Devilliers (1998), Gouzévitch and Gouzévitch (1998). On Chasles, see Chemla (2016), Nabonnand (2011b, 2016). On Steiner, see Blasjö (2009), Lorenat (2016).

By contrast to this specificity, in *A History of Mathematics: An Introduction* Victor Katz (2009, p. 855) claims metric independence for past projective geometry in general:

> Interestingly, although projective geometry aimed to study properties of figures not dependent on such concepts as that of length, the very basis of the definition of a cross ratio was in fact the length of a line segment. It was Christian von Staudt (1798–1867) who was able to correct this problem in 1847 by outlining an axiomatic system for projective geometry, based on the notion of a projective mapping as one that preserved harmonic tetrads.

Further, since this quote immediately follows a discussion of Chasles and the cross ratio, Katz thus seems to suggest that Chasles and von Staudt had the same aim. In these accounts, projective geometry before von Staudt appears as improper, impure, defective, and incorrect because of metrical considerations. The implication, sometimes made explicit, is that von Staudt's predecessors had tried to develop a non-metric projective geometry.

From the standpoint of mathematics, the axioms of projective geometry are independent from metric axioms. However, this view has shaped history to discount how historical actors valued measurement. Further, interpreting historical descriptions of geometrical methods as equivalent to later non-metric methods obscures how the practices and values of these methods have changed over time. Digging deeper into the origins and persistence of the non-metric projective anachronism uncovers expectations of mathematics that developed toward the end of the nineteenth century and do not represent the expectations of the early-nineteenth-century geometers here described.

9.3 The distinction between metric and the non-metric among early-nineteenth-century geometers

Tracing the development of the non-metric projective anachronism must begin by understanding the positions of earlier geometers with respect to the role of measurement. These positions varied between geometers and across their research publications. Since historical treatments cite Poncelet as emphasizing the difference between metric and descriptive properties, it is important to see how this distinction informed his research in the projective properties of figures.[4] Further, we will consider how measurement featured in Poncelet's delineation of pure geometry as distinct from analytic geometry. Finally, a survey of what kinds of research was labeled as *Geometrie der Lage* or *géométrie de situation* in the first half of the nineteenth century will serve to situate the unique nature of von Staudt's 1847 text.

Poncelet (1817) had already explicitly distinguished "ordinary geometry" and "geometry of situation". This categorization emerged in a solution to the problem of circumscribing an m-vertexed polygon to a conic with vertices lying on m given lines.[5] The first part of the solution reduced to determining how many different types of polygons could be formed with vertices on m given points. Poncelet solved this problem through what he called *géométrie de situation*. From considering the cases of three or four given points, Poncelet concluded that one would arrive at $3 \cdot 4 \cdot 5 \cdots (m-1)$ possible arrangements of m given points. Thus, Poncelet saw geometry of situation as one part of his research, which might be described today as combinatorial.[6]

Just as Poncelet did not exclusively engage in geometry of situation, he also did not limit his study of projective properties to non-metric properties. In his 1822 *Traité des propriétés projectives des figures* he described his study of the projective properties of figures as encompassing multiple recent theories and combining descriptive and metric relations.

> Since the *Geometry of the ruler*, or *Theory of concurrent points*, on the one hand, is only concerned with descriptive or positional properties of systems of indefinite straight lines, and the *Geometry of transversals*, on the other hand, only treats metric relations relative to figures likewise composed of systems of indefinite straight lines, cut in any manner by lines or curves called, for this reason, transversals; one understands *a priori* that they are included implicitly in the definition we have given of relations and *projective figures* [...] (Poncelet, 1822, pp. 76–7).[7]

[4] Many authors have noted Poncelet's attention to the difference between metric and descriptive projective properties, for instance Schoenflies (1909) will be discussed later in this regard.

[5] For a more detailed exposition of the following, see Friedelmeyer (2011).

[6] Likewise, the full title of Steiner's monograph situates *Geometrie der Lage* as only one aspect of his research: *Systematische Entwicklung der Abhängigkeit geometrischer Gestalten von einander, mit Berücksichtigung der Arbeiten alter und neuer Geometrie über Porismen, Projections-Methoden, Geometrie der Lage, Transversalen, Dualität, und Reciprocität, etc.*

[7] La *Géométrie de la règle*, ou *Théorie des points de concours*, ne s'occupant, d'une part, que des propriétés descriptives ou de situation des systèmes de lignes droites indéfinies, et la *Géométrie des transversales* n'ayant pour objet, d'une autre, que les relations métriques relatives aux figures composées également de systèmes de lignes droites indéfinies, coupées d'une manière

Poncelet consistently presented his work as engaged with both metric and descriptive properties. In fact, the very first example of a projective property that he provided in his *Traité* concerned how a ratio of distances cut by a transversal remained constant under projection (ibid., p. xxx).

Poncelet particularly emphasized his interest in metric properties when contrasting the results of his theory of polar reciprocity with Joseph-Diez Gergonne's theory of duality (Poncelet, 1827a,b, 1828). Whereas Gergonne had claimed that duality only concerned geometry of situation, Poncelet showed how the theory also applied to "the metric relations of distances, angles and trigonometric lines" (Poncelet, 1827a, p. 127).

Pure geometry, for Poncelet, excluded computations that "permit momentarily to lose view of the figure with which one is occupied" (Poncelet, 1817). Measurement could remain a feature of pure geometry as long as it was defined in relation to the figure. Thus, Poncelet's decision in 1864 to publish the underlying algebraic computations – *Applications d'analyse et de géométrie qui ont servi de principal fondement au Traité des propriétés projectives des figures* – does not damage his claims to be practicing pure geometry, as these computations could be avoided by assuming the principle of continuity (Poncelet, 1864). In defining pure geometry in this manner, Poncelet was less concerned with the foundations than with the experience of proving theorems and solving problems.

Though Chasles and Steiner did not align with all of Poncelet's principles of geometry – for instance, both rejected Poncelet's principle of continuity – they also employed measurement in their research. For instance, their respective definitions of *rapport anharmonique* and *Doppelverhältnis* relied on lengths of lines.[8] Consider how Steiner defined *Doppelverhältnis* in 1832. With reference to a figure showing a pencil

quelconque par des droites ou courbes appelés, pour cette raison, transversales; on conçoit, à priori, qu'elles se trouvent toutes deux comprises implicitement dans la définition que nous avons donnée des relations et des *figures projectives* [...]

[8] In his *Aperçu historique sur l'origine et le développement des méthodes*, Chasles claimed that the metric relations of homographic figures are "a consequence" of their descriptive relations (Chasles, 1837, p. 700). However, later in the same text he contrasted the importance and utility of metric relations over descriptive relations and suggested an inverse relationship:

"In general, the metric relations of figures are still more important and more useful to know than their purely descriptive relations, because they are susceptible to a great number of applications, and moreover, they almost always suffice to reach an understanding of descriptive relations" (ibid., p. 775).

(En général, les relations métriques des figures sont encore plus importantes et plus utiles à connaître que leurs relations purement descriptives, parce qu'elles sont susceptibles d'un plus of rays and corresponding points on a straight line, Steiner (1832, p. 14) introduced a sine relationship.

This expression can be kept here, without preventing the manner of consideration (the method) from being synthetic, because it indicates only a certain relation, which can be represented by two straight lines and determines the angle concerned.[9]

For Steiner, the use of measurement and the sine ratio – grounded in the straight lines and angles of a given figure – was part of "synthetic" geometry.

In fact, of all the early-nineteenth-century geometers cited as predecessors to von Staudt, it was Gergonne who most enthusiastically embraced a metric-free geometry.[10] In the seventeenth volume of his *Annales des Mathématiques Pures et Appliquées*, Gergonne (1827, p. 383) introduced a new category with which to classify articles in his journal – geometry of situation – which he defined in the table of contents:

One understands here, under the title of Geometry of situation, all the parts of geometry that depend neither on ratios of angles nor on ratios of length, and of which the geometry of the ruler is only a small part.[11]

Some of the research on projective properties of conic sections by Chasles and Steiner was classified in the *Annales* as geometry of situation (Poncelet did not publish in the *Annales* after 1826). However, a substantial portion of the contents falling under geometry of situation concern the relationship between vertices, edges, and faces of a polyhedra. Gergonne's geometry of situation shares properties with von Staudt's *Geometrie der Lage*, but goes well beyond what would be considered as projective geometry by the end of the century.

In the early nineteenth century, the study of projective properties of figures and geometry of situation were overlapping, but distinct, areas of research. Moreover, metric relations had significant applications and

grand nombre d'applications, et que d'ailleurs elles suffisent presque toujours pour arriver à la connaissance des relations descriptives.)

This sentiment is reiterated in Chasles (1852).

9 Diese Bezeichnung kann hier beibehalten werden, ohne dass dadurch die Art der Betrachtung (die Methode) aufhört synthetisch zu sein, weil durch dieselbe nur ein gewisses, durch zwei Gerade darstellbares, den gedachten Winkel bestimmendes Verhältniss angedeutet wird.

10 On Gergonne, see Gérini (2010), Otero (1997).

11 On comprend ici, sous le titre de Géométrie de situation, toute cette partie de la géométrie qui ne dépend ni des rapports d'angles ni des rapports de longueur et dont la géométrie de la règle n'est qu'une faible partie.

remained an active avenue of geometrical research. Further, the persistence of measurement in geometry provided an important link to older traditions. Philippe Nabonnand (2015) has shown how Poncelet intended his so-called "modern pure geometry" as a continuation of the pure geometry of the eighteenth and seventeenth centuries. Similarly, Steiner advertised that his systematic development of geometry encompassed the problems of ancient geometry (for instance, see Steiner, 1826). Early-nineteenth-century geometers drew inspiration and research questions, including problems concerning measurement, from "géométrie ancien" or "alter Geometrie." By contrast, von Staudt appeared to sever all relations to past geometers.[12]

9.4 *Geometrie der Lage* as an introduction to the geometry of modern times

Von Staudt introduced his 1847 *Geometrie der Lage* with a claim for "modern times" in which the geometry of situation has been distinguished from geometry of measurement.[13]

> In modern times, the Geometry of Situation has rightly been distinguished from metric geometry, and yet theorems, in which no quantity is mentioned, nevertheless are often proved by the consideration of ratios. I have sought in this text to make Geometry of Situation an independent science, which requires no measurement[14] (von Staudt, 1847, p. iii).

Since von Staudt only vaguely situated his work with respect to predecessors and provided no references, it is not clear whether he saw himself as contributing to the same geometry as, for instance, Poncelet and Steiner. He at once praised the attempts of recent geometers as "rightly" distinguishing metric and non-metric geometries and critiqued their failure in successfully separating the two geometries. Certainly, his vocabulary choices reflect a familiarity with the work of Steiner and Poncelet, and

[12] The role of citations in connecting contemporary projective geometry to ordinary geometry of the past is emphasized by Luigi Cremona (1873) in the introduction to *Elementi di Geometria Projettiva*.

[13] Historical and technical expositions of von Staudt's *Geometrie der Lage* and *Beiträge zur Geometrie der Lage* can be found in Nabonnand (2016), Bioesmat-Martagnon (2011), Voelke (2008), Reich (2005), Scott (1900), Kötter (1901).

[14] Man hat in den neuern Zeiten wohl mit Recht die Geometrie der Lage von der Geometrie des Masses unterschieden, indessen gleichwohl Sätze, in welchen von keiner Grösse die Rede ist, gewöhnlich durch Betrachtung von Verhältnissen bewiesen. Ich habe in dieser Schrift versucht, die Geometrie der Lage zu einer selbstständigen Wissenschaft zu machen, welche des Messens nicht bedarf.

his other publications do explicitly mention these authors. In the context of his *Geometrie der Lage*, the absence of citations may have served to reinforce von Staudt's claims for novelty.

Indeed, von Staudt's interest in a purely non-metric geometry was new. He justified establishing such an independent science as pedagogically beneficial, explaining that learning geometry of situation before introducing measurement would lead to a correct understanding of the whole of geometry.

> Perhaps this work will motivate some teachers to preface their teaching of metric geometry with the essential part of Geometry of Situation, so that their pupils will first gain an overview of the science, without which correctly understanding the relationship of individual theorems to the whole is impossible[15] (ibid.).

For von Staudt, a lack of measurement (and a lack of drawn figures) also served to develop student intuition. He continued by explaining the teaching value of beginning with geometry of situation before introducing measurement, in contrast to "most textbooks." Thus, von Staudt presented *Geometrie der Lage* as the correct way to begin learning geometry. Theoretically, the text required no prior geometric knowledge and thus was foundational in the sense of being introductory. Indeed, by the 1850s von Staudt's approach had entered the textbook literature. Yet despite his claims to write for teachers of geometry, other textbook authors noted that beginners had great difficulty in comprehending von Staudt (for instance, see Zech (1857) and Witzschel (1858)).

One especially successful adaptation of von Staudt is Theodor Reye's *Die Geometrie der Lage*, which first appeared as a textbook in 1866. Reye favored von Staudt's approach over "all other writers upon Modern Geometry" and presented von Staudt's treatment as the best suited to the educational goal of geometric visualization.

> That is to say, he excludes all calculations whether more or less complicated which make no demands upon the power of representation, and to whose comprehension there is required instead a certain mechanical skill having little to do with geometry in itself[16] (Reye, 1866, p. vii).

[15] Vielleicht wird diese Schrift einige Lehrer bestimmen, ihrem Unterrichte in den Geometrie des Masses des Wesentliche aus der Geometrie der Lage voranzuschicken, damit ihre Schüler gleich Anfangs denjenigen Ueberblick über die Wissenshaft gewinnnen, ohne welchen das rechte Verständen der einzelnen Sätze und ihrer Beziehung zum Ganzen nicht wohl möglich ist.

[16] Er schliesst nämlich, alle mehr oder minder complicirten Rechnungen aus, welche die Vorstellungskraft nicht beanspruchen, zu deren Verständiss vielmehr eine gewisse mechanische Fertigkeit erforderlich ist, die mit der Geometrie an sich wenig zu schaffen hat.

This passage neatly aligns with von Staudt's claims for teaching geometry of situation before ordinary geometry. Yet, Reye explicitly distinguished geometry of situation from the modern geometry practiced by von Staudt's predecessors, and consequently avoided judging others by von Staudt's vision of non-metric geometry. Reye's *Die Geometrie der Lage* was well received and quickly followed by several editions published into the early twentieth century as well as translations into Italian, French, and English. Though by this time research mathematicians used "projective geometry" to denote the works of von Staudt and Reye (as will be discussed in the subsequent section), Reye never adopted this terminology in describing his or von Staudt's contributions.

Like the teachers and pupils of geometry for whom von Staudt professed to write, Klein admitted that he initially could not digest von Staudt's style. He only began to take an interest in von Staudt through the work of his friend, Otto Stolz (see Klein, 1926a, p. 133). In his research in analytic geometry, Stolz described von Staudt's research as a "purely geometrical method" and the foundation for "modern synthetic geometry," but also important as a foundation to coordinate systems.

> v. Staudt's research, which has provided the long-missing foundation of modern synthetic geometry, has not yet been duly appreciated within it. May the present paper, illuminating the same from another side, serve to emphasize anew its fundamental importance [. . .][17] (Stolz, 1871, p. 417).

Though Stolz himself only included this historical nod in his prefatory remarks, his positive assessment of von Staudt as fundamental to conceptions of modern and pure geometry carried into Klein's mathematical and historical writings. So while students struggled with *Geometrie der Lage* as a foundation from which to learn geometry, research mathematicians promoted *Geometrie der Lage* as a foundation on which to build modern geometry.[18] The next section will consider how Klein promoted non-metric projective geometry as the basis of all geometry.

[17] v. Staudts Forschungen, welche der neueren synthetischen Geometrie die lange vermisste Grundlage geschaffen, sind trotzdem von derselben noch nicht gebührend gewürdigt worden. Möge der vorliegende Aufsatz, der dieselben von einer anderen Seite beleuchtet, dazu dienen, ihre fundamentale Wichtigkeit neuerdings zu betonen [. . .]

[18] The engineer Karl Culmann initially assigned von Staudt's *Geometrie der Lage* as a prerequisite for his course on graphical statics. As evidence of the difficulty students encountered with this text, he admitted in the second edition (Culmann, 1875) that some found von Staudt "too abstract" and instead recommended Reye's text.

9.5 The beginning of "projective geometry"

The first known use of the expression "projective geometry" only dates back to Olry Terquem (1859). In his journal *Nouvelles Annales de Mathématiques*, Terquem provided a list of eight geometries "distinguished from each other by logical differences" and designated with "names of French geometers who have published special works." Notably, though Chasles is also on the list, Poncelet was the only geometer associated with projective geometry. Further, the list does not include geometry of situation.

In the early 1870s, projective geometry appeared in the obituary of Plücker by Alfred Clebsch (1872) and Klein's *Vergleichende Betrachtungen über neuere geometrische Forschungen* (the Erlangen Program paper) (Klein, 1872). The novelty of "projective geometry" is apparent in Clebsch's *Vorlesungen über Geometrie* from the winter semester of 1871/1872:

> The whole geometry then divides into two parts: the one (simpler) is independent of the parallel axioms and comprises the so-called projective geometry; the other is based on the parallel axiom, and therefore adds something new, and is accordingly to be described as more complicated, even if, as a result of the usual course of geometry, which is provisionally connected with the familiar concepts of distance and angle, it first appears to be simpler[19](Lindemann, 1891, p. 433).

In this classification, geometry was either projective geometry or a Euclidean geometry that included measurement. However, Arthur Cayley (1859) had already shown that "Metrical geometry is [. . .] a part of descriptive geometry and descriptive geometry is all geometry." Klein extended these results even further, showing that projective geometry also contained non-Euclidean geometry.

This new finding was advertised in Klein's Erlangen Program, where he promoted projective geometry as the basis of a recently possible approach to unify previously disparate methods.

> Among the advances of the last fifty years in the field of geometry, the development of *projective geometry* occupies the first place. Although it seemed at

[19] Die ganze Geometrie zerfällt hiernach in zwei Theile: der eine (einfachere) ist unabhängig vom Parallelenaxiome und umfasst die sogenannte projectivische Geometrie; der andere stützt sich auf das Parallelenaxiom, nimmt also etwas Neues hinzu, ist demgemäss als der complicirtere zu bezeichnen, wenn derselbe auch in Folge des in der Geometrie üblichen Lehrganges, der mit Recht an die naheliegenden Begriffe von Entfernung und Winkel vorläufig anknüpft, zunächst als der einfachere erscheint.

first as if the so-called metrical relations were not accessible to its treatment, as they do not remain unchanged by projection, we have learned recently to consider them from the projective point of view, so that now the projective method embraces the whole of geometry[20] (Klein, 1872, p. 3).

Klein's fifty-year timeline would situate the opposition between projective and metric in the 1820s, when, as shown above, metric properties could also be projective properties.

Around the same time as Klein's Erlangen Program, Luigi Cremona (1873) carefully addressed the problem of naming a different kind of projective geometry in his *Elementi di Geometria Projettiva.* As evidenced by his title, Cremona chose to call his subject projective geometry and began his text by explaining why.

Various names have been given to this set of geometric doctrines of which we are about to develop the fundamental principles here. I prefer not to adopt that of geometria superiore (*Géométrie supérieure, höhere Geometrie*), because that which at one time seemed higher, may to-day have become very elementary; nor that of geometria moderna (*neuere Geometrie, modern Geometry*), which in like manner expresses a purely relative idea; and is moreover open to the objection that although the methods may be regarded as modern, yet the matter is to a great extent old. Nor does the title geometria di posizione (*Geometrie der Lage*) as used by STAUDT seem to me a suitable one, since it excludes the consideration of the metrical properties of figures. I have instead chosen the name of geometria projettiva, as expressing the true nature of the methods, which are based essentially on central projection or perspective;[21] (Cremona, 1873, p. viii).

So for Cremona, like Terquem, projective geometry was the geometry of Poncelet, which, unlike von Staudt's *Geometrie der Lage*, included

[20] Unter den Leistungen der letzten fünfzig Jahre auf dem Gebiete der Geometrie nimmt die Ausbildung der projectivischen Geometrie die erste Stelle ein. Wenn es anfänglich schien, als sollten die sogenannten metrischen Beziehungen ihrer Behandlung nicht zugänglich sein, da sie beim Projiciren nicht ungeändert bleiben, so hat man in neuerer Zeit gelernt, auch sie vom projectivischen Standpuncte aufzufassen, so dass nun die projectivische Methode die gesammte Geometrie umspannt.

[21] Diversi nomi erano stati dati a quell'insieme di dottrine geometriche di cui qui si pongono i primi fondamenti. Non mi piacque accogliere quello di geometria superiore (*Géométrie supérieure, höhere Geometrie*), perché in sostanza ciò che una volta poté parere elevato, ora è divenuto elementarissimo; né quello di geometria moderna (*neuere Geometrie, modern Geometry*), che esprime del pari un concetto puramente relativo; e d'altronde la materia è in gran parte vecchia, sebbene i metodi si possano considerare come recenti. Anche il titolo di geometria di posizione (*Geometrie der Lage*) nel senso di STAUDT non mi parve meglio conveniente, per ciò che esso esclude la considerazione delle proprietà metriche delle figure. Ho invece preferito quello di geometria projettiva, col quale vocabolo si enuncia la vera natura de' metodi, che essenzialmente si fondano sulla projezione centrale o prospettiva;

the study of metrical properties of figures. Thus Cremona directly contradicted Klein's nomenclature.

An even more disjointed classification of geometries appeared in Paul Terrier's French translation of Antonio Favaro's *Lezioni di statica grafica* (Favaro, 1877, 1879). Terrier provided a lengthy historical introduction in which Poncelet, Steiner, and von Staudt are all considered as practicing pure modern geometry. However, Terrier only classified Poncelet and Steiner as projective geometers, with Poncelet as the creator of projective geometry and Steiner as his "most faithful continuator" (ibid., pp. xiii–xv). By contrast, Terrier used the original author's own terms in designating Chasles's research as higher geometry and von Staudt's as geometry of situation. While Terrier drew on all these geometers, he called his lessons simply *Géométrie de position.*

Unlike Terquem, Cremona, or Terrier, Klein aimed to unite disparate geometries. Klein's lifelong efforts in this regard are well known in the historical literature, particularly with respect to the Erlangen Program. For instance, David Rowe (1992, p. 47) describes Klein's view of geometry in the early 1870s:

> He hoped thereby to restore some order to a field that had become notoriously disjointed, with practitioners who sometimes behaved as if geometry had descended to a veritable Tower of Babel condition.

This description of a Tower of Babel might align with any of the above lists of possible geometries. Further, as Jeremy Gray (2005a, p. 546) has noted, Klein was among the first to take this broad view of a single geometry:

> He claimed no novelty for the way he treated specific topics; what was original was the unified viewpoint he offered and its suggestions for the direction of future work.

As the nineteenth century progressed, Klein's original vision of a nonmetric projective geometry seemed to succeed.[22]

The choice of the expression "projective geometry" in this capacity may reflect a concern for avoiding confusions in translation between practitioners and publications. In claiming that projective geometry was all of geometry, Klein was following Cayley. But Cayley (1859) had written (in English) that "descriptive geometry is all geometry". On the

22 For an example, consider the history of the fundamental theorem of projective geometry (Voelke, 2008).

continent, as Cremona had noted, descriptive geometry had a particular meaning as the orthogonal projective methods codified by Gaspard Monge in the late-eighteenth century. Similarly, in French *géométrie de position* might imply a connection to Lazare Carnot's *Géométrie de position* (Carnot, 1803).[23] As a relatively new expression, projective geometry thus had less baggage and could well serve to bring together various modern geometries.

Klein's particular conception of the non-metric goals of projective geometry reflects his research on non-Euclidean geometry. In Klein (1871, 1873), he expressed interest in establishing a non-metric projective geometry to serve as a foundation for non-Euclidean geometry in order to make that subject more "intuitive" (Klein, 1871, p. 573). Thus non-Euclidean geometry provided a new motivation to separate projective geometry from metric geometry. Further, von Staudt's contributions in "pure" or "ordinary" (gewöhnlich) projective geometry connected to Klein's broader programmatic goals in uniting mathematics and axiomatic geometry. With respect to the latter, Klein (1873, p. 115) pointed to the advantage of reducing the number of axioms of projective geometry, providing another motivation for why projective geometry *should* be free from measurement. Klein would later use his conception of projective geometry as a standard with which to assess past geometers and their work.

While specialized historical studies can focus on the specific practices of individual geometers, this is not possible within the limited space available in general histories of mathematics. In this latter medium, the history of early-nineteenth-century geometries often became the history of projective geometry. Yet, the development sketched above shows that the consolidation was *not* actually inevitable. Klein's contemporaries, less interested in a unified geometry, presented alternative categorizations.

One consequence of Klein's understanding of projective geometry was that Steiner, Poncelet, and Chasles – now categorized as projective geometers – could be seen as subscribing to a non-metric geometry. Though Klein may not have been the first to draw these conclusions in his

[23] The potential ambiguity between these similar names is noted by David Eugene Smith (1906, p. 46) in his *History of Modern Mathematics*:
"But although Carnot's work was important and many details are now commonplace, neither the name of the theory nor the method employed have endured. The present Geometry of Position (Geometrie der Lage) has little in common with Carnot's *Géométrie de Position*."

historical writing, discussed below, his presentation of this anachronistic judgment is representative of contemporary and later iterations in the historiographical literature.

9.6 Klein's historiography of projective geometry

The most comprehensive source for Klein's historical exposition is his *Vorlesungen über die Entwicklung der Mathematik im 19. Jahrhundert*, which was published in 1926 based on lectures given in 1919 (Klein, 1926a). Though this work only appeared posthumously, edited by Richard Courant and Otto Neugebauer, Klein maintained an interest in learning and disseminating history throughout his career (such as in Klein, 1893, 1926b).

The title of Klein's lectures are important in the emphasis on development limited to the nineteenth century. With this approach, early-nineteenth-century geometers represent a beginning. As in the 1870s, Klein here portrayed projective geometry as gathering together many past sub-geometries. Yet, he also drew a distinction between those that contributed to projective geometry and those that were "true" projective geometers. This judgment is even leveled against Klein's much-admired doctoral advisor, Julius Plücker, who had "clung to the concrete" in the style of eighteenth-century geometers with his "detailed investigations" of the asymptotic behavior of curves at infinity (Klein, 1926a, p. 126).

Similarly, Klein concluded that Steiner was not "the one-side and systematic projectivist" (ibid., p. 131). First, Steiner's *Systematische Entwickelung* fell short of his systematic goal in lacking "a possibility of general formulation." Further, Steiner, along with Poncelet and Möbius, defined the Doppelverhältnis "auf metrische Weise" and so left the relationship between metric and projective geometry "unexplained" (*unaufgeklärt*). Klein diagnosed this as a "problem" that would only be solved later in the century by von Staudt, who appeared as "the first to liberate synthetic geometry from these and other imperfections"[24] (ibid., p. 130). For Klein, von Staudt was the quintessential synthetic and projective geometer, and in these roles served as a standard with whom to contrast his contemporaries and predecessors who worked toward, but ultimately fell short of, projective geometry. According to Klein, the

[24] Von diesen und anderen Unvollkommenheiten ist die synthetische Geometrie erst durch von Staudt befreit worden.

former worked toward, but ultimately fell short of, projective geometry. Interests in other areas of geometry, like asymptotes for Plücker or metric considerations for Steiner, appear as backward-looking.

In the subsequent chapter, Klein offered an even stronger claim about the failed efforts of early projective geometry.

> The first and most important point, toward which the whole development of the preceding decades had been straining [. . .] was to establish a projective geometry independent of all metrical considerations. As we have seen, the projective geometry of Poncelet and of Steiner contained a fatal inconsistency if its goal was completely to put aside metric geometry or, as indeed soon happened, to make it but a special part of projective geometry. The most important concept of projective geometry, the cross-ratio, and with it the general projective coordinate system, still rested on a metric-based definition[25] (ibid., p. 133).

Not only Steiner, but "the whole development of the preceding decades" moved inevitably, but stumblingly, toward non-metric projective geometry. Beyond critiquing the use of the metric within projective geometry, Klein put forward a historical assertion about the goals of past geometers.

As for Chasles, Klein mostly dismissed him as derivative of Möbius and Steiner and supplanted by von Staudt.

> Perhaps Chasles already had possessed these ideas for a long time; in any event, he came forward with them only after they had taken shape in Germany a long time earlier, and the only merit that remains to him, in this respect, is to have imported them to France and, through the far-reaching influence of French instruction, to have spread them to England, Italy, Scandinavia, etc.[26] (ibid., p. 141).

The national bias in Klein's assessment reveals further potential impetus for celebrating von Staudt's German contributions. With even more geographic specificity, Klein glorified Erlangen (where he held von Staudt's

[25] Der erste wichtigste Punkt, auf den die ganze Entwicklung der letzten Jahrzehnte hindrängte [. . .] ist die von metrischen Betrachtungen unabhängige Begründung der projektiven Geometrie. Wie wir sahen, enthielt nämlich die projektive Geometrie Poncelets und Steiner eine verhängnisvolle Inkonsequenz, wenn als Ziel ins Auge gefasst war, die metrische Geometrie ganz beiseite zu schieben oder gar, wie es nun bald darauf gelang, sie als einen besonderen Teil der projektiven Geometrie einzuordnen. Der wichtigste Begriff der projektive Geometrie, das Doppelverhältnis, und mit ihm das allgemeine projektive Koordinatensystem ruhte noch auf einer aus der Metrik stammenden Definition.

[26] Vielleicht hat Chasles diese Ideen bereits lange in sich getragen; hervorgetreten ist er jedenfalls erst damit, nachdem sie in Deutschland längst Gestalt gewonnen hatten, und es bleibt ihm in dieser Hinsicht nur das Verdienst, die Übertragung nach Frankreich und, an der Hand des weitreichenden französischen Unterrichts, die Weiterverbreitung nach England, Italien, Skandinavien usw. geleistet zu haben.

former chair between 1872 and 1875) as a location perfectly suited to mathematical fruition.

In the quiet and simplicity of the former states of Erlanger, yet untouched by the great life, Staudt found the peace and seclusion that was necessary to work out his ideas undisturbed[27] (ibid., p. 32).

Klein implied that a more bustling city, like Poncelet's Paris or Steiner's Berlin, might thus impede scientific progress.[28]

In his mathematical writings of the 1870s, Klein had suggested the axiomatic value of von Staudt's work and by the early twentieth century research on the foundations of geometry had become increasingly important. In his history, Klein connected von Staudt to the recent research of David Hilbert (such as in Hilbert (1899)).

One establishes projective geometry without speaking of metric relations, according to von Staudt's ideas. To express myself in the way that has become usual following Hilbert, one posits the axioms of ordering, connection, and continuity, and erects a geometry upon them[29] (ibid., pp. 150–151).

Klein could then position von Staudt as one of the forefathers of the mathematics of Göttingen by reading *Geometrie der Lage* as unifying nineteenth-century geometry and by promoting an axiomatic approach. Yet the non-metric projective anachronism was not limited to Klein, Göttingen, or even Germany. Klein's research and his conception of history

[27] In der Stille und Einfachheit der damaligen Erlanger Zustände, die vom grossen Leben nicht berührt wurden, fand Staudt die Ruhe und Abgeschlossenheit, die zum ungestörten Ausspinnen der eigenen Gedankenwelt nötig ist.

[28] This regional tension is perhaps best exemplified by Hankel (1875), who also (not coincidentally) had a position in Erlangen.
"The French certainly accomplishes no less in the exact sciences than the German, but he uses aids where he finds them; he does not sacrifice the intuitiveness for systematicity, and does not exchange ease for purity of method. There in the quiet Erlangen, v. Staudt could develop his own scientific system, only once in a while he developed one or two listeners at his desk; but there in Paris in lively exchanges with colleagues and numerous listeners the formulation of the system would have been impossible."
(Der Franzose leistet sicherlich in den exacten Wissenschaften nicht weniger als der Deutsche, aber er nimmt die Hilfsmittel, wo er sie findet; er opfert nicht die Anschaulichkeit der Systematik auf, und gibt nicht die Leichtigkeit für die Reinheit der Methode preis. Da in dem stillen Erlangen konnte wohl v. Staudt für sich und in sich abgeschlossen sein wissenschaftliches System entwickeln, das er nur hie und da einmal ein oder zwei Zuhörern an seinem Schreibtische entwickelte; dort aber in Paris in lebendigem Verkehr mit Collegen und zahlreichen Zuhörern wäre die Ausarbeitung des Systemes unmöglich gewesen.)

[29] Man begründet, ohne von metrischen Beziehungen zu sprechen, in Anlehnung an die von Staudt gegebenen Gedanken, die projektive Geometrie. Um mich in der durch Hilbert üblich gewordenen Weise auszudrücken: man stellt die Axiome der Anordnung, Verknüpfung und Stetigkeit auf und errichtet auf ihnen eine Geometrie.

spread widely (Rowe, 2001). One path on which to follow the propagation of the non-metric projective anachronism is in the *Encyklopädie der mathematischen Wissenschaften*, directed in large part by Klein.[30]

9.7 The historiography of projective geometry in the *Encyklopädie der mathematischen Wissenschaften*

Though each article of the *Encyklopädie* offers a different author's perspective, the overall enterprise projects a unified mathematics as the culmination of past contributions. Towards these goals, the editors – an evolving group led by Felix Klein and Franz Meyer – sent contributors "general principles" to follow (von Dyck, 1904). Several of these principles concern the historical contents. For instance, in principle 6, authors were recommended to begin their historical accounts at the start of the nineteenth century, rather than seeking earlier origins. This temporal focus thus links early nineteenth-century mathematicians to their successors as opposed to their predecessors.[31]

The guidelines also included an admonition with respect to certain judgment claims. While descriptions like "epoch-making", "ingenious", "great", "classical" were discouraged as meaningless, the authors were encouraged to show progress through explaining newer results, greater rigor, or more complete systematicity.

The *Encyklopädie* contains over thirty articles on geometry, the first of which are dated from 1902. These first appeared as independent articles and were then gathered into the third volume, which began publication in 1923.

In the foreword to this volume, the geometry editors Meyer and H. Mohrmann described the contents as "a fully satisfying overview of all geometry"[32] (Meyer and Mohrmann, 1923, p. xi). Historical references juxtaposed the present unity to perceived divisions from the previous century. For instance, the editors critiqued unnamed representatives of "so-called pure Geometrie der Lage" who uphold the discipline as "the ideal of an 'autochthonous' science" and disdain "mixed methods" (ibid., p. viii). Instead, they asserted that achieving this metric-independence is only initially important. First, construct projective geometry with-

[30] For more on the history of the *Encyklopädie* see Tobies (1994).

[31] On the problems raised by creating a strict divide between the eighteenth and nineteenth centuries, see Gilain and Guilbaud (2015).

[32] [. . .] einem vollbefriedigenden Überblick über die gesamte Geometrie.

out "metrische Hilfsmittel" and afterward one can follow an analytic treatment as desired. According to the editors, only after Klein's 1872 group-theoretic Program could geometry be held together despite its "many-sidedness" (Vielseitigkeit).

This image of a uniform projective geometry progressing toward, but also grounded by, geometry of situation appears in numerous contributions to the Geometry volumes. The inevitability of this progression is explained metaphorically as "the flower of descriptive geometry" becoming "the ripe fruit of geometry of situation" in Erwin Papperitz's "Darstellende Geometrie" (Papperitz, 1909, p. 523). Von Staudt marked a turn from the mathematics of the first half of the nineteenth century, serving as an inflection point. This intermediate position can be seen in the article "Elementargeometrie und elementare nichteuklidische Geometrie in synthetischer Behandlung," where Max Zacharias describes the defining feature of past projective geometry as "the unity of its structure" and celebrates von Staudt as the first to develop a projective geometry independent from the metric relations of elementary geometry (Zacharias, 1913, pp. 905, 899).

Beyond these short mentions, two articles included more detailed histories of early-nineteenth-century geometry: Gino Fano's historical article on the opposition between synthetic and analytic geometry during the nineteenth century (Fano, 1907), and Arthur Moritz Schoenflies's article on projective geometry (Schoenflies, 1909).

In "Gegensatz von synthetischer und analytischer Geometrie in seiner historischen Entwicklung im 19. Jahrhundert" Fano defined synthetic geometry as that which "considers figures in themselves" and designated Poncelet's *Traité* as the first text in modern synthetic geometry (Fano, 1907, p. 231):

> Poncelet's "Traité" marks the transition from old to new synthetic geometry, where for the first time certain modern concepts and viewpoints occur in a certain form and find rational application.[33]

Further, in a footnote, Fano recognized that for Poncelet projective properties were properties "of situation" along with special "metric" properties. Fano offered no judgment as to Poncelet's research on the metric properties that remain invariant under projection. However, he inter-

33 Poncelet's ≪Traité≫ markiert den Übergang von der älteren zur neueren synthetischen Geometrie, indem hier zum ersten Male gewisse moderne Begriffe und Gesichtspunkte in bestimmter Form auftreten und rationelle Anwendung finden.

preted Steiner's geometry as grounded in metric concepts. Because of this Fano criticized Steiner as lacking complete generality (ibid., p. 237):

> But it was in the nature of things that certain evils could not be avoided and a full generality in the figures produced could not always be achieved. The concept of cross-ratio, which originates from the consideration of the lengths and angles, that is, based on metric concepts, was still fundamental in Steiner's theory of projective relations;[34]

Fano then compared Steiner with von Staudt, whom he praised as simpler and more rational.

Fano's article contrasted the opposition of the past with the unity of the present, but geometry could not be unified as part of projective geometry until projective geometry could be made independent from measurement. This narrative of unity is absent from Schoenflies's article, which examined contemporary practices of projective geometry and their historical origins and is organized into short historical expositions of certain principles, problems, or results. Because Schoenflies is more interested in a historical survey than a progressive history, he only treats "the possibility" of projective geometry independent of the metric towards the end of his paper (Schoenflies, 1909, p. 465). He thus avoids the non-metric projective anachronism.

In contrast to Fano, for Schoenflies, von Staudt's generality was not the only form of generality. He praised Steiner's study of geometrical structures as intuitive and "the most generalizable" (ibid., p. 415). Later in the text he compared a French and German school of thought with respect to generality (ibid., p. 465):

> Another interesting question, concerning the methods of projective geometry, is whether it is permissible, and to what extent it can be permitted, to *particularize* the figures or problems, without diminishing the generality of the results obtained in any way. For a long time, the French school has thoughtlessly situated itself directly on the basis of the principle of continuity, considering in particular the behavior of curves and surfaces towards g_∞ or ϵ_∞. In the same period, the teachers of the German school thought to the contrary that such a specialization should be considered as an affront on the generality of research.[35]

[34] Es lag aber in der Natur der Sache, dass gewisse Übelstände nicht vermieden und eine volle Allgemeinheit in den erzeugten Figuren nicht immer erreicht werden konnte. Der Begriff des Doppelverhältnisses, welcher von der Betrachtung der Längen und Winkel, d.h. von metrischen Begriffen ausging, war in der Steiner'schen Theorie der projektiven Beziehungen immer noch grundlegend;

[35] Von methodischem Interesse ist auch die Frage, ob und inwieweit es gestattet ist, die Gebilde und Probleme zu *spezialisieren*, ohne dass die für sie abgeleiteten Resultate die allgemeine

He noted that in the French school, geometers focus on particular results that could be generalized through projection, grounded in the principle of continuity. Schoenflies thus seemed to equate the French school with Poncelet, as the principle of continuity was not well received in France. By contrast, Schoenflies noted, for the German school this approach was considered too specialized until Klein showed the analogy to using particularly well-chosen coordinates in analytic geometry. So Schoenflies presented the value of generality as something that might take multiple forms and with different standards over time and space.[36]

These national differences are also apparent in a comparison between the original German and the French adaptations. In the French *Encylopédie*, edited by Jules Molk, more significant French amendments were set off with asterisks.[37] Within the asterisks emerges a somewhat different conception of the historical role of measurement, particularly in emphasizing the contributions of French geometers.

For instance, in an asterisked footnote, Sauveur Carrus (adapting Fano) distinguished Chasles and von Staudt with respect to Chasles's adherence to metric considerations (Carrus and Fano, 1915, p. 206):

> But while M. Chasles can be reproached [. . .] for the constant use of the anharmonic ratio, which gives to the "Géométrie supérieure" an almost exclusively metric character, K.G.Chr. von Staudt more closely approaches J.V. Poncelet by endeavoring to establish a geometry freed from any metric relations and based solely on relationships of situation.[38]

This summary is perhaps an even more anachronistic portrait of Poncelet than the non-metric projective anachronism. Carrus critiqued the metric character of Chasles, but his praise of von Staudt is couched as being closer to Poncelet rather than as an independent liberator.

The *Encyklopädie* mostly conveyed a vision of a unified geometry

Geltung verlieren. Die französische Schule hat sich lange Zeit unbedenklich direkt auf den Boden des Prinzips der Kontinuität gestellt und für die Erörterung von Kurven und Flächen inbesonderheit ihr Verhalten gegenüber g_∞ bzw. ϵ_∞ in Betracht gezogen. Hingegen haben die führenden deutschen Geometer gemeint, jede Spezialisierung als einen Verstoss gegen die Allgemeinheit der Untersuchung verschmähen zu sollen.

36 Karine Chemla (1998, 2016) has analyzed the different concepts of generality among French geometers in the early nineteenth century.

37 A more detailed analysis of the differences between the French and German editions can be found in Gispert (1999).

38 Mais tandis qu'on peut reprocher à M. Chasles, [. . .] l'emploi constant du rapport anharmoniques, ce qui donne à la ≪Géométrie supérieure≫ un caractère presque exclusivement métrique, K.G.Chr. von Staudt se rapproche plus nettement de J.V. Poncelet en s'attachant à constituer une géométrie affranchie de toute relation métrique et basée uniquement sur des rapports de situation.

based in non-metric projective geometry. In this aspect, the project can be viewed as a successful instantiation of the views put forth by Klein in 1872 and reflected in his historical writings. The difference between authors lay in their subtle historical assessments. In their attention to Chasles, the French authors consequently presented more examples of metric geometry, particularly Chasles's use of anharmonic ratios (Doppelverhältnis/rapport anharmonique), though (as Carrus added) it was not a "purely geometric notion" (Carrus and Fano, 1915, p. 32). Even in the German edition, Klein's historical vision did not have absolute authority on this narrative. As Schoenflies's article demonstrates, detailed attention to the history of past geometers showed multiple understandings of the subject matter, some of which included measurement. By contrast, Fano's inquiry into the division between analysis and synthesis in the early-nineteenth century, illuminated the unity of present geometers. Whether intentional or not, this portrayal supported Klein's understanding of mathematics and its history.

9.8 Conclusion

Mathematicians employed projective geometry to reinforce two strong mathematical trends of the late-nineteenth and early-twentieth centuries: the unification of seemingly disparate parts of mathematics and growing interest in foundations. The early-twentieth century also witnessed a flowering of historical studies, often written by practicing mathematicians whose research interests shaped their historical analyses.[39]

Concurrently, Klein played an important role in the development of research mathematics in the United States (Parshall and Rowe, 1994). Thomas Holgate was a new faculty member in the mathematics department of Northwestern University when Klein gave his *Evanston Lectures* in 1893. In 1898, Holgate published an English translation of Reye's *Lectures on the Geometry of Position* where he distinguished the "development of modern pure geometry" as rooted in the difference between metric and positional properties (Reye, 1898, translator's preface).[40] Similarly, in 1894, George Bruce Halsted's review of Klein's *Evanston Lectures* celebrated how Klein had elevated von Staudt.

[39] On the growth of the history of mathematics in the late-nineteenth and early-twentieth centuries see Dauben and Scriba (2002) and Gispert (2001).

[40] In Thomas McCormack's review of Holgate's translation he makes an even stronger claim: "The real distinction between the old and the new geometry [...] can be summed up in the

As geometer supreme of the moderns he gives not Steiner but von Staudt, my own ideal, still the man of the future, whose pure system has never yet been given in English, but must now be, since no other will serve as foundation for projective metrics and projective non-Euclidean geometry (Halsted, 1894).

Klein's version of the history of mathematics substantially informed his American contemporaries' histories of mathematics in which we still find the non-metric projective anachronism today.

A linear story of projective geometry from Poncelet, through von Staudt, and up to contemporary research in axiomatic systems provided a simplified view of the present as the successful culmination of past failed efforts. Further, historically oriented texts written in the late nineteenth century still serve as references to general histories of mathematics. So even if the unification of Euclidean and non-Euclidean geometries as encompassed by projective geometry is no longer such a lively research area, it continues to shape present conceptions of the history of mathematics.

References

Anonymous (1911). Geometry. In *Encyclopaedia Britannica: a dictionary of arts, sciences, and general literature. Volume 11. Franciscans to Gibson.* Cambridge: Cambridge University Press, 675–735.

Belhoste, Bruno (1998). De l'École Polytechnique à Saratoff, les premiers travaux géométriques de Poncelet. *Bulletin de la Sabix* **19**, 9–29.

Bell, Eric Temple (1940). *The Development of Mathematics.* New York: McGraw Hill.

Billoux, Claudine and Devilliers, Jean-Paul (1998). Jean-Victor Poncelet: Esquisse d'un portrait. *Bulletin de la Sabix* **19**, 1–5.

Bioesmat-Martagnon, Lise (ed.) (2011). *Eléments d'une Biographie de l'Espace Projectif.* Nancy: Presses universitaires de Nancy.

Blasjö, V. (2009). Jakob Steiner's "Systematische Entwickelung": the culmination of classical geometry. *The Mathematical Intelligencer* **31** (1), 21–29.

Carnot, Lazare (1803). *Géométrie de Position.* Paris: Duprat.

Carrus, Sauveur and Fano, Gino (1915). Exposé parallèle du développement de la géométrie synthétique et de la géométrie analytique pendant le 19ième siècle. In *Encyclopédie des Sciences Mathématiques Pures et Appliquées*, volume 3. Paris: Gauthier-Villars, 185–259.

fact that while the former was almost entirely metrical, the latter is almost purely positional" (McCormack, 1898, p. 465).

Cayley, Arthur (1859). A sixth memoir upon quantics. *Philosophical Transactions of the Royal Society of London* **149**, 61–90.

Chasles, M. (1837). *Aperçu Historique sur l'Origine et le Développement des Méthodes en Géométrie particulièrement de celles qui se rapportent à la Géométrie Moderne*. Paris: Gauthier-Villars.

Chasles, Michel (1852). *Traité de Géométrie Supérieure*. Paris: Bachelier.

Chemla, Karine (1998). Lazare Carnot et la généralité en géométrie. Variations sur la théorème dit Menelaus. *Revue d'Histoire des Mathématiques* **4**, 163–190.

Chemla, K. (2016). The value of generality in Michel Chasles's "Historiography of Geometry". In Karine Chemla, Renaud Chorlay and David Rabouin (eds). *The Oxford Handbook of Generality in Mathematics and the Sciences*. Oxford: Oxford University Press, 47–89.

Clebsch, Alfred (1872). *Zum Gedächtnis an Julius Plücker*. Göttingen: Dieterichschen Buchhandlung.

Cooke, Roger (2011). *The History of Mathematics: a Brief Course*. Second edition. Hoboken: Wiley-Interscience.

Coolidge, Julien Lowell (1940). *A History of Geometrical Methods*. Oxford: Oxford University Press.

Cremona, Luigi (1873). *Elementi di Geometria Projettiva*. Roma: G.B. Paravia e comp.

Culmann, Carl (1875). *Die Graphische Statik*. Second edition. Zürich: Meyer and Zeller.

Dauben, Joseph W. and Scriba, Christoph J. (eds.) (2002). *Writing the History of Mathematics: its Historical Development*. Basel: Birkhäuser.

Enriques, Federigo (1907). Prinzipien der geometrie. In *Encyklopädie der mathematischen Wissenschaften mit Einschluss ihrer Anwendungen*, volume III. Leipzig: Teubner, 1–129.

Fano, Gino (1907). Gegensatz von synthetischer und analytischer geometrie in seiner historischen entwicklung im 19. jahrhundert. In *Encyklopädie der mathematischen Wissenschaften*, volume III. Leipzig: Teubner, 221–288.

Favaro, Antonio (1877). *Lezioni di Statica Grafica*. Padova: F. Sacchetto.

Favaro, Antonio (1879). *Leçons de Statique Graphique*. Paris: Gauthier-Villars.

Friedelmeyer, Jean-Pierre (2011). L'impulsion originelle de Poncelet dans l'invention de la géométrie. In Bioesmat-Martagnon, Lise (ed.) *Eléments d'une Biographie de l'Espace Projectif*. Nancy: Presses universitaires de Nancy, 55–158.

Friedelmeyer, J.-P. (2016). Quelques jalons pour l'histoire des transformations au 19e siècle, les contributions de Poncelet et Möbius. In Bioesmat-Martagnon, Lise (ed.) *Eléments d'une Biographie de l'Espace Projectif*. Nancy: Presses universitaires de Nancy, 15–68.

Gergonne, Joseph-Diez (1827). Table des matières contenues dans le XVII.e volume des Annales. *Annales de Mathématiques Pures et Appliquées* **XVII**, 381–386.

Gérini, Christian (2010). Pour un bicentenaire: polémiques et émulation dans les Annales de Mathématiques Pures et Appliquées (1810–1832) de Gergonne, premier grand journal de l'histoire des mathématiques. *IREM Caen Revue*, 1–16.

Gilain, Christian and Guilbaud, Alexandre (eds.) (2015). *Sciences Mathématiques 1750–1850: Continuités et Ruptures*. Paris: CNRS Éditions.

Gispert, Hélène (1999). Les débuts de l'histoire des mathématiques sur les scènes internationales et le cas de l'entreprise encyclopédique de Felix Klein et Jules Molk. *Historia Mathematica* **26**, 344–360.

Gispert, Hélène (2001). The German and French Editions of the Klein–Molk Encyclopedia: contrasted images. In Bottazzini, Umberto and Dahan-Dalmédico, Amy (eds). *Changing Images in Mathematics: from the French Revolution to the New Millennium*. London: Routledge, 93–112.

Gouzévitch, Irina and Gouzévitch, Dmitri (1998). La guerre, la captivité et les mathématiques. *Bulletin de la Sabix* **19**, 30–68.

Grattan-Guinness, Ivor (ed.) (2005). *Landmark Writings in Western Mathematics*. Amsterdam, Oxford: Elsevier.

Gray, Jeremy (2005a). Felix Klein's Erlangen program 'Comparative Considerations of Recent Geometrical Researches' (1872). In Grattan-Guinness, Ivor (ed.) *Landmark Writings in Western Mathematics*. Amsterdam, Oxford: Elsevier, 544–552.

Gray, Jeremy (2005b). Jean Victor Poncelet, 'Traité des Propriétés Projectives des Figures'. In Grattan-Guinness, Ivor (ed.) *Landmark Writings in Western Mathematics*. Amsterdam, Oxford: Elsevier, 366–376.

Gray, Jeremy (2015). Klein and the Erlangen programme. In Lizhen Ji and Papadopoulos, Athanase (eds.) *Sophus Lie and Felix Klein: The Erlangen Program and Its Impact in Mathematics and Physics*. Strasbourg: European Mathematical Society, 59–76.

Halsted, George Bruce (1894). Klein's Evanston lectures. *Annals of Mathematics* **8**, 100–102.

Hankel, Hermann (1875). *Die Elemente der Projectivischen Geometrie in Synthetischer Behandlung*. Leipzig: Teubner.

Hartshorne, Robin (2008). Publication history of von Staudt's 'Geometrie der Lage'. *Archive for the History of the Exact Sciences* **62**, 297–299.

Henrici, Olaus (1877). Geometry. In Baynes, Thomas Spencer and Smith, William Robertson (eds.) *Encyclopaedia Britannica: a dictionary of arts, sciences, and general literature. Volume 10*. Edinburgh: A. and C. Black, 376–420.

Hilbert, David (1899). *Grundlagen der Geometrie*. Leipzig: Teubner.

Katz, Victor (2009). *A History of Mathematics: an Introduction*. Third edition. Boston: Addison-Wesley.

Klein, Felix (1871). Ueber die sogenannte nicht-Euklidische geometrie. *Mathematische Annalen* **4**, 573–625.

Klein, Felix (1872). *Vergleichende Betrachtungen über Neuere Geometrische Forschungen. Programm zum Eintritt in die philosophische Facultät und den Senat der k. Friedrich-Alexanders-Universität zu Erlangen.* Erlangen: Andreas Deichert.

Klein, Felix (1873). Ueber die sogenannte nicht-Euklidische geometrie (zweiter aufsatz). *Mathematische Annalen* **6**, 112–145.

Klein, Felix (1893). *The Evanston Colloquium: Lectures on Mathematics*. New York: American Mathematical Society.

Klein, Felix (1926a). *Vorlesungen über die Entwicklung der Mathematik im 19. Jahrhundert*. Berlin: Springer.

Klein, Felix (1926b). *Vorlesungen über Höhere Geometrie*. Berlin: Springer.

Kötter, Ernst (1901). Die Entwickelung der synthetischen geometrie von Monge bis auf Staudt (1847). *Jahresbericht der Deustchen Mathematiker-Vereinigung* **5**(2), 1–484.

Lindemann, Ferdinand (ed.) (1891). *Vorlesungen über Geometrie, unter besonderer Benutzung der Vorträge von Alfred Clebsch*. Leipzig: Teubner.

Lorenat, Jemma (2016). Synthetic and analytic geometries in the publications of Jakob Steiner and Julius Plücker (1827–1829). *Archive for the History of Exact Sciences* **70**, 413–462.

McCormack, Thomas (1898). Review of 'Lectures on the Geometry of Position'. By Theodor Reye, Professor of Mathematics in the University of Strassburg. Translated and edited by Thomas F. Holgate. *The Monist* **9**, 465–466.

Meyer, Franz and Mohrmann, Hans (1923). Vorrede zum dritten Bande. In *Encyklopädie der mathematischen Wissenschaften mit Einschluss ihrer Anwendungen*, volume III. Leipzig: Teubner, v–xi.

Nabonnand, Philippe (2011a). Deux droites coplanaires sont sécantes. In Bioesmat-Martagnon, Lise (ed.) *Eléments d'une Biographie de l'Espace Projectif*. Nancy: Presses universitaires de Nancy, 159–195.

Nabonnand, Philippe (2011b). L'argument de la généralité chez Carnot, Poncelet et Chasles. In *Justifier en Mathématiques*. Paris: Éditions de la Maison des sciences de l'homme, 17–47.

Nabonnand, P. (2015). L'étude des propriétés projectives des figures par Poncelet: une modernité explicitement ancrée dans la tradition. In Gilian, Christian A. G. and Guilbaud, Alexandre (eds.) (2011). Sciences Mathématiques 1750–1850: Continuités et Ruptures. Paris: CNRS Éditions, 381–402.

Nabonnand, Philippe (2016). Utiliser des éléments imaginaires en géométrie: Carnot, Poncelet, von Staudt et Chasles. In Bioesmat-Martagnon, Lise (ed.) *Eléments d'une Biographie de l'Espace Projectif*. Nancy: Presses universitaires de Nancy, 69–106.

Otero, Mario H. (1997). Joseph-Diez Gergonne (1771–1859). *Histoire et Philosophie des Sciences, Sciences et Techniques en Perspective* **37**, 1–260.

Papperitz, Erwin (1909). Darstellende Geometrie. In *Encyklopädie der mathematischen Wissenschaften mit Einschluss ihrer Anwendungen*, volume III. Leipzig: Teubner, 520–595.

Parshall, Karen Hunfer and Rowe, David E. (1994). *The Emergence of the American Mathematical Research Community, 1876–1900: J.J. Sylvester, Felix Klein, and E.H. Moore*. Providence: American Mathematical Society.

Poncelet, Jean-Victor (1817). Philosophie mathématique. Réflexions sur l'usage de l'analise algébrique dans la géométrie; suivies de la solution de quelques problèmes dépendant de la géométrie de la règle. *Annales de Mathématiques Pures et Appliquées* **8**, 141–155.

Poncelet, Jean-Victor (1822). *Traité des Propriétés Projectives des Figures. Ouvrage utile a ceux qui s'occupent des Applications de la Géométrie Descriptive et d'Opérations Géométriques sur le Terrain*. Paris: Bachelier.

Poncelet, Jean-Victor (1827a). Note sur divers articles du bulletin des sciences de 1826 et de 1827, relatifs à la théorie des polaires réciproques, à la dualité des propriétés de situation de l'étendue, etc. *Annales de Mathématiques Pures et Appliquées* **18**, 125–142.

Poncelet, Jean-Victor (1827b). Note sur divers articles du bulletin des sciences de 1826 et de 1827, relatifs à la théorie des polaires réciproques, à la dualité des propriétés de situation de l'étendue, etc. *Bulletin des Sciences Mathématiques, Astronomiques, Physiques et Chimiques* **8**, 109–117.

Poncelet, Jean-Victor (1828). Mémoire sur la théorie générale des polaires réciproques. *Journal für die Reine und Angewandte Mathematik* **4**, 1–71.

Poncelet, Jean-Victor (1864). *Applications d'Analyse et de Géométrie qui ont servi de Principal Fondement au Traité des Propriétés Projectives des Figures*. Paris: Mallet-Bachelier.

Reich, Karin (2005). Karl Georg Christian von Staudt, 'Geometrie der Lage' (1847). In Grattan-Guinness, Ivor (ed.) *Landmark Writings in Western Mathematics*. Amsterdam, Oxford: Elsevier, 441–447.

Reye, Theodor (1866). *Geometrie der Lage*. Hannover: Carl Rümpler.

Reye, Theodor (1898). *Geometry of Position*. New York: Macmillan.

Rowe, David E. (1992). Klein, Lie, and the 'Erlanger programm'. In Boi, Luciano, Flament, Dominique and Salanskis, Jean-Michel (eds.) *1830–1930: A Century of Geometry*. Lecture Notes in Physics, volume 402. Berlin: Springer, 45–54.

Rowe, David E. (2001). Felix Klein as Wissenschaftspolitiker. In Bottazzini, Umberto and Daham-Dalmédico, Amy Daham (eds.) *Changing Images in Mathematics: From the French Revolution to the New Millennium*. London: Routledge, 69–92.

Schoenflies, Arthur Moritz (1909). Projektive geometrie. In *Encyklopädie der mathematischen Wissenschaften mit Einschluss ihrer Anwendungen*, volume III. Leipzig: Teubner, 389–480.

Scott, Charlotte Angas (1900). On Von Staudt's 'Geometrie der Lage'. *The Mathematical Gazette* **1**(19), 307–314.

Smith, David Eugene (ed.) (1906). *History of Modern Mathematics*. Fourth edition. New York: John Wiley and Sons.

Steiner, Jacob (1826). Einige geometrische betrachtungen. *Journal für die Reine und Angewandte Mathematik* **1**, 161–182.

Steiner, Jacob (1832). *Systematische Entwickelung der Abhängigkeit geometrischer Gestalten von einander, mit Berücksichtigung der Arbeiten alter und neuer Geometer über Porismen, Projections-Methoden, Geometrie der Lage, Transversalen, Dualität und Reciprocität, etc.* Berlin: Fincke.

Stolz, Otto (1871). Die geometrische Bedeutung der complexen elemente in der analytischen geometrie. *Mathematische Annalen* **4**, 416–442.

Terquem, Olry (1859). Sur diverses géométries. *Nouvelles Annales de Mathématiques* **18**, 445–446.

Tobies, Renate (1994). Mathematik als bestandteil der kultur – zur geschichte des unternehmens Encyklopädie der mathematischen Wissenschaften mit Einschluss ihrer Anwendungen. *Oesterreichische Gesellschaft für Wissenshaftgeschichte* **14**, 1–90.

Veblen, Oswald and Young, John Wesley (1910). *Projective Geometry*. Boston: Ginn and Company.

Voelke, Jean-Daniel (2008). Le théorème fondamental de la géométrie projective: évolution de sa preuve entre 1847 et 1900. *Archive for the History of the Exact Sciences* **62**, 243–296.

von Dyck, Walther (1904). Einleitender bericht über das unternehmen der herausgabe der Encyklopädie der mathematischen Wissenschaften. In *Encyklopädie der mathematischen Wissenschaften mit Einschluss ihrer Anwendungen*, volume I. Leipzig: Teubner, i–xx.

von Staudt, Karl Georg Christian (1847). *Geometrie der Lage*. Nürnberg: Friedrich Korn'schen Buchhandlung.

Witzschel, Benjamin (1858). *Grundlinien der Neueren Geometrie*. Leipzig: Teubner.

Zacharias, Max (1913). Elementargeometrie und elementare nichteuklidische Geometrie in synthetischer Behandlung. In *Encyklopädie der mathematischen Wissenschaften mit Einschluss ihrer Anwendungen*, volume III. Leipzig: Teubner, 859–1172.

Zech, Paul (1857). *Die Höhere Geometrie in ihrer Anwendung auf Kegelschnitte und Flächen zweiter Ordnung*. Stuttgart: E. Schweizerbart'sche Verlagshandlung und Druckerei.

10

Anachronism: Bonola and non-Euclidean geometry

Jeremy Gray
Open University and University of Warwick

Abstract: Roberto Bonola, in his celebrated history of non-Euclidean geometry, imported a distinction between elementary and advanced mathematics commonly drawn around 1900 back to the discoveries of the 1820s and 1830s. In so doing he fell into anachronism and misrepresented the subject, most obviously in his treatment of the work of Lobachevskii.

10.1 Introduction

A century of books and articles since Roberto Bonola's influential *La geometria non-euclidea: esposizione storico-critica del suo sviluppo* (1906) have proposed that non-Euclidean geometry is an identifiable mathematical field that it is meaningful to investigate. It has an extensive pre-history and a history of some duration, stretching across the nineteenth century and for some decades beyond. In this chapter I argue that Bonola's book gave a distorted account of its subject by seeing it through a contemporary divide of mathematics into elementary and higher mathematics.

Contemporary writers such as the Italians Gino Fano and Federigo Enriques, and the German Max Zacharias enable us to be precise about the distinction drawn between elementary and higher geometry around 1900. Elementary geometry was more-or-less synonymous with the content and methods of Euclid's *Elements* and analogous treatments of non-Euclidean geometry. Higher geometry took many forms, the one most

[a] From *Anachronisms in the History of Mathematics: Essays on the Historical Interpretation of Mathematical Texts*, edited by Niccolò Guicciardini

relevant here is Riemannian differential geometry.[1] By observing this distinction in the structure of his work, and by even denying himself the simpler approach due to Carl Friedrich Gauss (and also by paying too much attention to chronology and priority) Bonola gave a singularly poor account of the crucial discoveries of Nikolai Ivanovich Lobachevskii. He compounded his error by failing to appreciate the significance of the introduction of hyperbolic trigonometry by Lobachevskii and János Bolyai.

By importing a distinction drawn around 1900 back to the discoveries of the 1820s and 1830s Bonola fell into anachronism and misrepresented the history of non-Euclidean geometry.

10.2 Anachronism

Historians move between a selection of texts, which can include texts no-one has read for centuries and well-read, even canonical, texts, and they re-arrange pieces of these texts in pursuit of some (simple or elaborate) argument, until they publish texts of their own. Anachronistic views may affect historians' selection of texts, the refinement of the selection process as they begin to write, and the presentation of the final text, and will not always be visible on the surface. Why should we be concerned? What, more precisely, in the use of anachronism should concern us? After all, in the words of the invitation that has called us to the Bacon conference in Pasadena on April 2018: "It would be unjustified to deny the historian of mathematics such a possibility of translation, of familiarity with past texts: after all such possibility and familiarity are historical facts." But, as Niccolò Guicciardini continued, "However, we recognize that the greatest masters in the history of mathematics have achieved more convincing interpretations exactly because they taught us how to 'see the differences' between our mathematics and the mathematics of the past."[2] I shall argue that the effect of anachronism in Bonola's history of non-Euclidean geometry was to obscure the changes that take us from the mathematics of the past to our mathematics. More precisely, in a certain sense it allowed Bonola to over-estimate certain differences and under-estimate others to the detriment of a historical understanding of the discovery of non-Euclidean geometry. Our present concern is with

[1] Introduced by Bernhard Riemann (1854).

[2] I quote here from the email Guicciardini circulated among the invitees.

anachronism in the history of geometry, and here a small number of long words and phrases stand out: mathematician, Euclidean geometry, and non-Euclidean geometry. It will become clear that they change their meanings, so the possibility, even the advisability, of anachronism is present.

The word "mathematician" is not a problem in this context. Lobachevskii was a mathematician on any sensible definition of the term: a professor at a university paid to use his expertise for teaching and, increasingly, research. János Bolyai was a young pensioned-off military engineer who assisted his father in a Technical College. Gauss was professionally an astronomer, and also the head of a Government-directed national geographical survey – but of course his status derived from his high reputation as a mathematician as well. The people who came before them in the story – notably Girolamo Saccheri, Johann Heinrich Lambert, and Adrien-Marie Legendre, but there were many others – were mathematicians in an eighteenth-century sense of the term. The people who come after them in the story are overwhelmingly university figures – mathematicians in the modern sense of the term, and it is safe to conclude that in this case the actor's category of "mathematician" and the modern disciplinary one can be used with only mild, harmless, anachronism.

The same is not true of "Euclidean geometry" and "non-Euclidean geometry." We can, for example, regard Euclidean geometry as synonymous with the content of Euclid's *Elements*, together with any other valid deductions in that framework, in which case it is a largely formal, axiomatic affair. Or we might regard Euclidean geometry as the mathematical study of physical space, one form of which is embodied in the methods of Euclid's *Elements*. There were philosophies of geometry, such as Adrien-Marie Legendre's, that sought to find a unique set of acceptable axioms from which the content of Euclid's *Elements* could be recovered; and others, such as Jean le Rond d'Alembert's, that sought to bridge the passage from experience to formal knowledge.[3] The calculus-based methods of differential geometry were, of course, taken to be applicable to (and perhaps necessarily about) the study of physical space, which made them compatible with Euclidean geometry, but there were no axioms in differential geometry.

In the same spirit, is research into non-Euclidean geometry a matter of

[3] See the many editions of Legendre's *Éléments de Géométrie* and d'Alembert's essay Géométrie in the *Encyclopédie Méthodique: Mathématiques*.

axiomatics or a question of applicability? Could it be logically possible without being true? In what sense was it a geometry at all if it was not also true? What should we take non-Euclidean geometry to be, or to have been?

A reasonable answer to this last question is that Bolyai and Lobachevskii discovered a physically plausible geometry that was different from Euclid's but equally axiomatic in its formulation. This was to have the momentous consequence that Euclidean geometry was no longer obviously or inevitably true. However, and famously, their work convinced almost no-one in their lifetimes. Only when it was transformed into a successful new theory of geometry by a radically new idea – Riemann's great generalisation of Gauss's differential geometry – did a concomitant change of attitude about geometry occur. Indeed, there was more than one change: there was a return to axiomatic treatments of elementary geometry, appeals were made to the idea of groups of transformations, projective geometry was taken by some to be fundamental.

10.3 Bonola (1906)

Let us take a first look at Bonola's book, noting his selection and arrangement of the subject matter, and the sources he used. A more detailed examination will follow after a consideration of what elementary mathematics was considered to be.

Bonola's account set the history out in book form for the first time, and greatly influenced its subsequent treatment. His book is in two parts. The first part is chronological, and is organised in these five chapters:

1. Greeks and Arabs on the parallel postulate (also Wallis, and editions of the *Elements*)
2. Eighteenth-century forerunners (Saccheri, Lambert, Klügel, Legendre, and others)
3. The founders (Gauss, Schweikart, Taurinus)
4. The founders, continued, (Lobachevskii and Bolyai) – this includes a lengthy section on the spread of non-Euclidean geometry that draws on a wealth of sources
5. Later developments (chiefly Riemann and Helmholtz)

The second part, with which we shall not be concerned, contains two topics that did not fit in the first.[4]

Bonola did not write a preface setting out the aims, scope, and limitations of his book; it opens up with Euclid's definition of parallels. Nor did he define a field of elementary geometry. But he did make it very clear where he thought that advanced topics in geometry began. Chapter Five "Later Developments" of Bonola (1906) begins:

> To describe the further progress of Non-Euclidean Geometry in the direction of *Differential Geometry* and *Projective Geometry*, we must leave the field of Elementary Mathematics and speak of some of the branches of Higher Mathematics, such as the *Differential Geometry of Manifolds*, the *Theory of Continuous Transformation Groups*, *Pure Projective Geometry* (the system of STAUDT) and the *Metrical Gcometries* which are subordinate to it. As it is not consistent with the plan of this work to refer, even shortly, to these more advanced questions, we shall confine ourselves to those matters without which the reader could not understand the motive spirit of the new researches, . . .

Bonola was right that a number of difficult domains of mathematics grew up in the nineteenth century with which the history of non-Euclidean geometry is entangled. What he introduced over the next fifty pages does indeed illuminate these fields sufficiently to capture the "motive spirit" of those researches. The charge I shall level is that his division of mathematics into elementary and higher is anachronistic.

I do not mean that that distinction can never be made; there are often occasions where it can be drawn, whether the better to embark on the higher reaches or, as here, to give a reason for treating them only superficially. I mean that there was a specific view of what is elementary that is tied to the period around 1900 and that has shaped the analysis.

10.3.1 Bonola's sources

Bonola had a wealth of scholarship to draw on. The ongoing edition of Gauss's *Werke* had revealed the surprising degree to which Gauss had sympathised with the work of Bolyai and Lobachevskii, and had

[4] They are entitled: Statics and the law of the lever, and Clifford's surface. Horatio S. Carslaw, the English translator, added three further appendices, of which the last points to a strange weakness in Bonola's account: Constructions in non-Euclidean geometry, the independence of projective geometry from the parallel postulate, and the impossibility of proving the parallel postulate. The English edition also contains George B. Halsted's translations of Bolyai's *Appendix* (1831) and Lobachevskii's (1840).

even explored several of the same ideas himself. As the ideas of non-Euclidean geometry became accepted in the mathematical community in the 1870s mathematicians looked for precursors, some in the circles around Gauss, some elsewhere, and forgotten treatments by Saccheri and Lambert were rediscovered.[5]

The definitive source book for this material was published by Friedrich Engel and Paul Stäckel as *Die Theorie der Parallellinien von Euklid bis auf Gauss* in 1895, and Bonola relied on it. It begins with the first 32 propositions of Euclid's *Elements*, up to the point where the parallel postulate is used for the first time. It then looks at John Wallis's attempt on the parallel postulate, with excursions on Naṣīr al-Dīn al-Ṭūsī (whom Wallis mentioned) and editions of the *Elements* in the West. The next chapter is on Saccheri and gives a German translation of his recently rediscovered book *Euclides ab omni naevo vindicatus* (1733). Then comes Lambert and his *Theorie der Parallellinien* (1786). Then, after some remarks about d'Alembert, Fourier, Lagrange, Laplace, and Legendre, comes some extracts from Gauss's correspondence, the contribution of Ferdinand Karl Schweikart, and extracts from two books by his nephew, Franz Adolf Taurinus. With the exception of the contributions of Gauss and Taurinus, who used hyperbolic trigonometry, it is reasonable to label all of this material as "elementary" (on any commonly used sense of the term). The book ends with a 21-page bibliography of writings on the theory of parallels from 1482 to 1837.

10.3.2 Roberto Bonola (1874–1911)

Bonola was a mathematician in the circles around Federigo Enriques and was involved in his attempts to modernise the teaching of geometry. Enriques was one of the most eminent Italian mathematicians of his generation, internationally respected for his work with Guido Castelnuovo on algebraic surfaces. He had diverse interests in mathematics and philosophy, and advocated both a rigorous logical approach to geometry and a psychological grounding of the basic terms (such as the straight line) in everyday human experience.[6]

Bonola's first account of non-Euclidean geometry appeared in 1900 in the book *Questioni riguardanti la geometria elementare* edited by

[5] Bonola observed that Eugenio Beltrami and Corrado Segre had discussed Saccheri's work.
[6] See Enriques (1906) and, for example, Gray (2008).

Enriques.[7] It is a more straightforwardly mathematical survey of the subject than his later book. Six years later, when Bonola published that book, Enriques was, amongst other things, writing his long article for the *Encyklopädie der mathematischen Wissenschaften* on the foundations of geometry. It was Enriques who, at Bonola's request, wrote the introduction to the English translation of the book on non-Euclidean geometry in 1911, and he took the opportunity "to render homage to the memory of a friend, prematurely torn from life [. . .]" (Bonola, 1906, p. v).

In his introduction, Enriques drew attention to "the clear exposition of the principles of a theory now classical, but also a critical account of the developments which led to the foundations of the theory in question" (Bonola, 1906, p. v). But, in a manner not entirely consistent with this, he soon went on to praise "The admirable simplicity of the author's treatment, the elementary character of the construction he employs, [and] the sense of harmony" that accord with the artistic temperament and broad education of the author (Bonola, 1906, p. vi). This indicates that the book is not quite the straight-forward historical treatment that one might expect: what are these "elementary constructions" doing?

10.4 Elementary mathematics

We can say more about the concept of elementary geometry in Bonola's day and in his milieu. In Bonola's time, and in Italy especially, the school syllabus in mathematics emphasised Euclidean geometry; the same was true in England, and in both cases the idea of a classical heritage was part of the mix. The high status of elementary mathematics in Italy derived from its classical past, its internal charms (mathematical rigour and clarity of thought), and its role in intellectual culture. But it was also the case that many Italian mathematicians had participated in the struggles for unification, and after it had been achieved many of them played an unusually strong role in the Italian Parliament.[8] There they pushed successfully for the revival of Italian scientific life and the creation of new scientific institutions with a strong emphasis on mathematics. Francesco

7 As with Klein's activities in Germany, Enriques's book was part of a campaign to re-align the teaching of mathematics in schools with the needs and discoveries of higher mathematics. But even as it seeks reform, it opens the preface with the remark that the geometry of the Greeks, known to us by the name of Euclid, is "in itself so beautiful and harmonious that, after twenty centuries, we would not know how to replace it."

8 See Bottazzini and Nastasi (2013, Part III). Over a dozen eminent mathematicians were members of the Italian Senate at one time or another in the years after 1870.

Brioschi and Luigi Cremona in particular advocated a "return to Euclid" in the late 1860s, and the use of Euclid's *Elements* in the teaching of geometry in schools.[9] Enrico Betti and Brioschi then published a suitable annotated edition of the *Elements* for the purpose.[10] Their efforts succeeded in overcoming the opposition of some other mathematicians, and became established in Italian education, duly to become the orthodoxy that Enriques and his cohorts sought to reform.

Fano had studied under Corrado Segre between 1888 and 1892, and after spending a year with Felix Klein in Göttingen he became Castelnuovo's assistant in Rome in 1894.[11] In the introduction to his lithographed book on non-Euclidean geometry (Fano, 1898), Fano divided the approaches to the subject into three types, of which the first was the "elementary direction of Bolyai–Lobachevskii" that was completed by Eugenio Beltrami's account of geometry on a surface of constant negative curvature.[12]

He then explained in the opening chapter, which took the story as far as the work of Gauss, that to conduct any science it is necessary to specify some postulates or fundamental truths, which may be somewhat arbitrary provided they are consistent and independent. This, he said, is what Euclid had done, and his *Elements* had for centuries defined the field of geometry. However, an analysis of Euclid's postulates had led to criticism of the fifth or parallel postulate, and eventually an alternative postulate had been found which led to non-Euclidean geometry. This account closely follows the book by Engel and Stäckel.

A lengthy second chapter then describes the work of Bolyai and Lobachevskii in some detail and takes it as far as the discovery of the trigonometric formulae for triangles in the new geometry. In the third chapter Fano discussed the work of Beltrami, and only now did he introduce ideas from the differential geometry of surfaces such as the Riemannian metric, intrinsic curvature, and maps of surfaces. This division into chapters suggests a working definition of elementary geometry that would encompass the material in Euclid's *Elements*, trigonometry,

[9] Perhaps unwisely, they suggested following the example of English schools, see Bottazzini and Nastasi (2013, p. 272).

[10] *Gli Elementi d'Euclide con note aggiunte ed esercizi ad uso de' ginnasi e de' licei*, Firenze, 1868.

[11] Castelnuovo worked closely with Segre and later with Enriques.

[12] The other directions were the essentially analytic direction initiated by Riemann and the analytico-geometric direction of Cayley and Klein.

and *ad hoc* extensions to non-Euclidean geometry, but exclude Riemannian, and even Gaussian, ideas.

The impression is even clearer in Fano's next book, on descriptive geometry, which was published in a lithographed edition in 1902. Here Fano said that[13]

> Elementary geometry as it is studied in secondary schools occupies itself with certain geometric figures and certain properties of those figures. Among triangles, for example, it distinguishes between equilateral and isosceles triangles; among plane quadrilaterals between squares and rectangles, etc. It studies special properties of some of these figures, and also properties of circles and spheres [. . .] (Fano, 1902, p. 3).

It studies, he concluded, properties of figures that are independent of their position, orientation, and transformation by similarity. This is very much Euclidean geometry, although not necessarily as presented in Euclid's *Elements* and the name of Euclid is not mentioned.

Elementary geometry for Fano, then, was something like the content of Euclid's *Elements*, perhaps with the Euclidean parallel postulate replaced by an alternative, studied with the sorts of techniques used in Euclid's *Elements*. In particular, it has a strong axiomatic flavour, but in Fano's case it is compatible with elementary trigonometry. It is also the case that the methods of Cartesian or analytic geometry are excluded.

We can get an indication of what Enriques took elementary geometry to be by looking at his essay in the *Encyklopädie der mathematischen Wissenschaften*. Enriques's essay "Prinzipien der Geometrie" (The Principles of Geometry) (1907) is a much more sophisticated work that surveys geometry as the contemporary mathematician was meant to understand it. But it opens with some general, philosophical remarks about elementary geometry that recall how several basic concepts had been defined: point, straight line, and plane; segment, angle, and betweenness; congruence and motion of figures. This indicates both what he took to be the elementary concepts of geometry and what he thought might have to be done to make them rigorous.[14]

The parallel postulate came in for particular treatment. It is dealt with

13 "La Geometria Elementare, che si studia nelle scuole secondarie, si occupa di certe figure geometriche e di certe proprietà di queste figure. Essa distingue ad es fra i triangoli, i triangoli equilateri ed isosceli; fra i quadrangoli piani, i quadrati, i rettangoli, ecc.; e studia le proprietà speciali di ciascuna di queste figure; essa studia altresi le proprietà del cerchio, della sfera, ecc.ecc."

14 A rigorous treatment was famously to be given by Klein (1908) in volume two of his *Elementarmathematik vom höheren Standpunkt aus*) (*Elementary Mathematics from an advanced*

in two sections. The first follows the presentation of Engel and Stäckel closely and goes as far as the work of Bolyai and Lobachevskii and Riemann's spherical geometry. It concludes by mentioning the associated trigonometries. The second starts with the idea of surfaces of constant curvature and concludes with a discussion of the possible physical applicability of non-Euclidean geometry, so it is close to Fano's position.

In this section Enriques raised two questions. The first asks: how was the logical possibility of non-Euclidean geometry and therefore the independence of the parallel postulate from the other axioms and postulates of the *Elements* discovered? (The second asks about simple postulates equivalent to the parallel postulate; we shall not pursue that here except to note that they are all broadly elementary and synthetic.)

Enriques's answer to his first question is that Lobachevskii had shown that the hyperbolic trigonometry of triangles is a consistent set of formulae, and that this is the first proof of the consistency of non-Euclidean geometry. This was a minority view, and is not adequate as put, but it could be strengthened into a rigorous and convincing one.[15]

Max Zacharias (1913) had a somewhat similar understanding of elementary geometry in his 315-page article on the subject in the *Encyklopädie der mathematischen Wissenschaften.*[16] He admitted that it would be almost impossible to define such a diverse subject, and offered instead a mixture of elementary topics informed by contemporary reflections upon them such as those David Hilbert and Klein had provided.

The article opens with a survey of nineteenth-century work under the heading of "The parallel axiom, Euclidean and non-Euclidean geometry." This takes the story as far as the work of Bolyai, Lobachevskii, and Gauss before conceding that only with Riemann's account was the

Standpoint), but precisely because of its standpoint this work does not help at all to determine what the limits of elementary geometry were taken to be.

15 Enriques was not altogether persuaded of the contributions of Riemann and Beltrami, on the strange grounds that the representation of non-Euclidean geometry as the geometry on a surface of constant negative curvature only represented a piece of the plane; this is true of Beltrami's pseudo-sphere, but not of the Beltrami disc. He was much more convinced by Klein's projective representation using the Cayley metric, but it is hard to see how Enriques could accept this but not the Beltrami disc, especially after the insights of Henri Poincaré, which Enriques was to explain further on in the essay.

16 Heinrich Wilhelm Max Zacharias (1873–1962) graduated in mathematics from Berlin and became a Gymnasium teacher there. He received his doctorate in Rostock with a thesis on geometry under the direction of Otto Staude. He went on to write several books on mathematics, including treatments of both non-Euclidean and projective geometry, and translated Girard Desargues's *Brouillon Projet* into German for the Ostwald Klassiker series (information from Wikipedia). According to Wassily Hoeffding, who worked with him editing the *Jahrbuch*, by 1944 Zacharias was a fervent Nazi; see Hoeffding (2012, p. 593).

consistency of non-Euclidean geometry secured. It then turns to later attempts to defend the parallel postulate. Then Zacharias examined "Elementary Euclidean Geometry." He referred readers to Enriques's article in the *Encyklopädie der mathematischen Wissenschaften* for a survey of attempts to give accounts of the basic ideas and theorems of the subject that were complete and logically consistent. Then he took up the long-running question of whether motion of figures was fundamental to the definition of congruence. Those who thought it was he regarded as followers of Hermann von Helmholtz, those who did not he considered to follow Moritz Pasch's initiative, where, of course Hilbert's *Grundlagen der Geometrie* was situated.

The forty pages on the history of elementary non-Euclidean geometry are very similar to Bonola's account (Bonola's book was of course a source for Zacharias, but he cited many authors). It starts with Saccheri and Lambert, and moves on to Gauss, Schweikart, and Taurinus. Zacharias did not count Taurinus among the discoverers of non-Euclidean geometry precisely because Taurinus intended to establish the validity of the parallel postulate. The account of Lobachevskii's work is several pages long and fair, most aspects of it are summarised from the opening synthetic sections through the introduction of hyperbolic trigonometry to the final formulae for triangles in non-Euclidean geometry. As we shall see, this is more faithful, easier to follow, and less anachronistic than Bonola's treatment. Zacharias finished with a selection of later work that fitted reasonably under the umbrella of a synthetic treatment of elementary geometry.

On this evidence it would seem that there was more or less a consensus among Italian mathematicians around Segre and Enriques and German mathematicians around Felix Klein that elementary geometry was generally concerned with Euclidean geometry or variants of it obtained by varying the definition of parallel, together with the associated trigonometry. Advanced investigations of axioms systems and novel geometries, such as Hilbert had produced, were not taken to be elementary, nor were ideas of Gaussian curvature, and still less were Riemannian ideas considered elementary.

It should be briefly noted that other authors did not entirely agree with this in any case informal division of the subject. The *Encyklopädie der Elementar-Mathematik*, edited by Heinrich Weber and Josef Wellstein, was somewhat more cognisant of the difficulties in defining the fundamental notions of point, line, plane, and surface, and more willing to include some of Hilbert's ideas. It also regarded plane projective geometry as elementary. But it too included non-Euclidean geometry, although its treatment was much more thorough than the ones discussed above (and couched in terms of families of coaxial circles). One could say that the content of elementary geometry was the same, but the approach went deeper and broader.

The term "elementary" is in fact ambiguous, and we shall see that this ambiguity is a further reason for Bonola's use of misleading anachronism. As well as having the breadth just described, it has an aspect of being plausible, intuitive, well known, and taught in school. Around 1900 there was a palpable sense of excitement and paradox in the discovery that so many of the ideas one had learned in school and accepted as natural were in fact problematic, hard to define, and full of unexpected possibilities. It may be "obvious" what distances, lines, angles, and even axioms are, but it was also a topic of active research, as all the authors just mentioned also discussed at length in the work. Indeed, one could argue that what makes elementary geometry a term specific to the end of the nineteenth century is that it focuses on topics that had been taken as intuitive, even primitive, for many years but had recently been shown to be capable of precise definitions that yielded surprising consequences.

10.5 Bonola's "elementary" account

To judge by his book, Bonola took the crux of the topic under historical investigation to be the discovery by Bolyai and Lobachevskii that the axioms of Euclid's *Elements* that do not involve the parallel postulate are compatible with both the parallel postulate and the alternative proposed by Bolyai and Lobachevskii. The extensive pre-history was naturally the history of investigations of the parallel postulate (most but not quite all being flawed attempts at defending it), the subsequent history being the construction of adequate bases for the work of Bolyai and Lobachevskii, which were well known to be of a different kind from the offerings of Bolyai and Lobachevskii themselves.

Bonola's book, as its chapter headings indicate, is a simple-minded account: these authors did these things. It generally presents a fair summary of what every one did on their own terms, with occasional commentaries to avoid leaving a false mathematical impression (for example, when a use of the Archimedean axiom can be avoided). The book is, in fact, more of a historically minded survey article than a history of mathematics; Fano (1898) is very similar. Like him, Bonola assumed, for example, that the reader knows Euclidean geometry, and has some sense of Euclid's *Elements*, although he allowed that readers may need help to take the step from Bolyai and Lobachevskii to Riemann.

I shall concentrate on two sections in Bonola's account where I think the question of anachronism is most apparent: the pages on Lobachevskii and the chapter on Riemann. For the record, as is well known the approaches of Lobachevskii and Bolyai overlapped considerably, so Bonola's exposition of János Bolyai's work is likewise weakened, but Bonola gives more attention to some of the discoveries Bolyai made that Lobachevskii did not, notably the fact that in the new geometry it is sometimes possible to find a circle equal in area to a given square.[17]

10.5.1 Lobachevskii

Bonola's treatment of the work of Lobachevskii is very strange. Bonola summarised Lobachevskii's arguments about the nature and properties of parallelism in a page and a half and then, unlike Fano, omitted all mention of the nine sections of the numerous truly elementary arguments that show that if the angle sum of one triangle is greater than (resp. equal to, resp. less than) two right angles ($2R$) then so it is in every triangle; and that there cannot be a geometry (differing from Euclid's only over the parallel postulate) in which the angle sum of a triangle is greater than $2R$. Bonola might have pleaded that these results were first proved by Saccheri, and so he discussed them there, but the reader simply cannot discover that Lobachevskii knew them.

Much worse, Bonola's account obscures entirely how Lobachevskii introduced the formulae of hyperbolic trigonometry. The formulae of hyperbolic trigonometry are crucial to the success of Bolyai and Lobachevskii. It is by exploiting them that they were able to describe the implications of their new assumption about parallels, and it is the wealth of these implications that convinced them that their new assumption fits consistently into the rest of the assumptions of Euclid's *Elements* (minus the parallel postulate itself). Not to describe how Lobachevskii or Bolyai introduced them is a remarkable lapse.

17 An English translation with commentary on Bolyai's work may be found in Gray (2004).

Lobachevskii's argument goes in a small number of steps:

(i) there is a non-constant angle of parallelism function,
(ii) to each family of mutually parallel lines there is a family of disjoint orthogonal curves ('horocycles'),
(iii) given two horocycles in the same family there is a formula connecting lengths along the horocycle cut out by a pair of parallels and their separation.

The first step is Lobachevskii's basic assumption about parallels.

The second step, after Lobachevskii has proved that parallelism is an equivalence relation (forgive the anachronism!), is a minor anachronism itself: Lobachevskii never said orthogonal but had a clumsier way of saying something that immediately implies this.

The third step rests on a property of horocycles and their parallels that Lobachevskii should really have made explicit, but let that pass.

At this point, Lobachevskii could begin to replace classical geometrical concepts with formulae – as it turns out, those of hyperbolic trigonometry. It is this step, or, if you prefer, this sequence of three steps, that is the crucial novelty in his work.

Bonola's account fails completely to appreciate this. Rightly, it begins by saying "But the most important part of the *Imaginary Geometry* is the construction of the formulae of trigonometry" (Bonola, 1906, p. v). It continues with Lobachevskii's introduction of the concepts of horocycle and horosphere, although Bonola does not tell us what they are here (that information was given earlier, in his account of Gauss's work), and notes that they are employed by Lobachevskii to deduce the trigonometric formulae. But what then follows in a page is a proof that Lobachevskii's and Taurinus's formulae are equivalent, and a hint of Lobachevskii's way of deriving them (taking as given the formula for the angle of parallelism).

Taurinus, Bonola had already said, developed the formulae of hyperbolic trigonometry in the context of his explorations of the consequences of denying the parallel postulate, but he persuaded himself that he had in fact found a defence of Euclidean geometry. In 1824 he asked Gauss to endorse his work, but Gauss refused. Bonola opted to report on Taurinus's work after discussing Gauss's, so he had already presented the formulae by the time he turned to consider Lobachevskii, but it makes for an awkward exposition all the same. Needless to say, Lobachevskii certainly did not know anything about Taurinus's work. As for the introduction of the hyperbolic functions, Bonola (1906) had supplied a footnote on p. 82 to their introduction by Lambert. Note that Lambert did not use them in his investigations into non-Euclidean geometry.

We then get a mention of Lobachevskii's discovery that very small non-Euclidean triangles would be nearly indistinguishable from Euclidean ones. Bonola then ended his account of Lobachevskii's work by arguing that in his last work, the *Pangeometry*, Lobachevskii gave a treatment that was free of internal contradictions because of the analytic form in which it can be expressed. He even gives a page-long quotation from the *Pangeometry* to that effect; the opinion is presented as Bonola's, not merely as Lobachevskii's (Bonola, 1906, pp. 93–4).

The placing of key developments with their first discoverer (horocycles with Gauss, hyperbolic trigonometry with Taurinus) surely reflects a view of history that first discovery or use is important, and more important than later but independent rediscovery. This view risks implying that mathematics unfolds chronologically, a view Bonola sensibly denied elsewhere, because the deeper questions are about intention and effect. Bonola, intent on doing justice to Taurinus and keeping an obscure figure in the light, garbled his account of Lobachevskii's work. That said, chronological priority seems to have been important to Bonola. By adhering to it too tightly, as he did, he robbed Lobachevskii of his most profound insights, and made it easier for his whole account to slip into anachronism.

It is a grave failure not to indicate how hyperbolic trigonometry entered Lobachevskii's investigation of non-Euclidean geometry. That Bonola also endorsed Lobachevskii's claim that the new formulae in and of themselves conferred consistency on the new geometry only heightens the interest in seeing how they get into the subject at all, but that is exactly what Bonola's account fails to describe. Moreover, Lobachevskii's claim is worthy of scrutiny. The historian Morris Kline endorsed the general opinion of a century when he wrote that the acceptance of the validity of the new geometry by Gauss, Lobachevskii, and Bolyai "was an act of faith. The question of consistency of non-Euclidean geometry remained open for another forty years" (Kline, 1972, p. 880).

It would have been possible for Bonola to do what Zacharias was to do and explain what Lobachevskii (and Bolyai) did, in Bonola's case at the modest cost of repeating the formulae.[18] I suggest that the reason Bonola did not do that was that in the context of Euclidean geometry he regarded trigonometry as a dull addition. There is a charm to the austere world of axiomatic mathematics, and a sense that is fundamental; trigonometry reduces everything to mere calculation. (To be more precise, it is the trigonometry of triangles that is useful but dull; the trigonometric functions were regarded as very interesting and important.) On this view, hyperbolic trigonometry is an unremarkable addition to non-Euclidean geometry, relieved only by the novelty of its conclusion. It is the view that there are certain formulae, which are to be explained mathematically when they are first discovered, and which are then there to be used. Such arguments are necessarily technical, but elementary (in Bonola's sense), and the historian may omit them and go straight to the conclusions.

However plausible such a standpoint might be regarding ordinary or spherical trigonometry, which were both thoroughly understood disciplines, it is quite misleading when applied to the novel discipline of hyperbolic trigonometry, as I shall indicate in Section 10.6. But, by diminishing the importance of hyperbolic trigonometry Bonola made it easier for himself to see the work of Bolyai and Lobachevskii as largely axiomatic and therefore elementary, and easier to see the work of Riemann as belonging to the separate domain of advanced mathematics. It also allowed him to omit any discussion of Gauss's differential geometry.

After the discussion of the work of Bolyai and Lobachevskii Bonola gave an account of the posthumous reception of their work, noting the contributions of Guillaume-Jules Houël and Giuseppe Battaglini. Then comes the break, and the step up to Riemann and Helmholtz. Bonola made it clear that he can only give a superficial account here, because he must leave the field of "Elementary Mathematics." This highlights the problem. Bonola has a category of elementary mathematics, which contains work on alternatives to the parallel postulate conducted in the main with methods such as those found in Euclid's *Elements*. Nothing in Lobachevskii's work seems to rely on differential geometry, the novelty is another, if duller, branch of elementary mathematics: hyperbolic trigonometry.

[18] Curiously, Fano avoided Lobachevskii's argument and resolved the issue in a different way.

10.5.2 Distance

The very fact the Riemann rates a separate chapter, in which a considerable amount of mathematics is explained so that what is said about Riemann is intelligible, is significant. It comes about because there had been a change in the concept of geometry that was still difficult for people to understand fifty years after Riemann wrote. (It still is difficult to reach a wide audience with Riemann's ideas.) This is an opportunity to turn to another of Guicciardini's suggestions. As he put it in an email to invitees to the Pasadena meeting:

> It be might interesting to distinguish between anachronisms related to reading past European texts with which one imagines a form of continuity with present-day mathematics, and anachronism with respect to non-European texts often perceived to be from other worlds.

What is this "form of continuity" if the reformulation of geometry by Riemann, and the use of those ideas in the context of non-Euclidean geometry by Riemann and Beltrami, has to be described in so heavily mathematical a way that it seems like an entirely fresh start, disconnected from what had gone on before? Bonola marked the development of differential geometry and the re-presentation of non-Euclidean geometry in that context by an abrupt change of style; plainly he had no way to analyse this and could only record it as a change.

I shall argue that it is the idea that geometry is founded on a concept of distance that provides the form of continuity between Lobachevskii and Riemann, and that it is Bonola's conception of elementary geometry that obscured this from him.

Lobachevskii (2010, p. 3) had explained at the start of the *Pangeometry*, as he had in his earlier works, that the fundamental notions of geometry as hitherto understood were insufficient to resolve questions about parallels. This, he said, was because

> most of the definitions that are ordinarily given in the *Elements of geometry* [...] not only do not indicate the generation of the magnitudes that are being defined, but also, they even do not show that these magnitudes can exist.

Lobachevskii preferred to start with circles and spheres, and so with a primitive concept of distance, and define the plane as the points common to equal spheres with two fixed points as centres (and the straight line by means of equal circles in a plane).[19]

With this in mind we can look at Lobachevskii's remarks in the *Pangeometry* that Bonola quoted:[20]

> Having shown in what precedes in what manner we must calculate lengths of curves, areas of surfaces and volumes of bodies, we can now assert that *Pangeometry* is a complete geometric doctrine. A simple glimpse at Equations (19), which express the dependence that exists between the edges and the angles of rectilinear triangles, is sufficient to prove that from there, *Pangeometry* becomes an analytical method which replaces and generalises the analytical methods of ordinary geometry. It would be possible to start the exposition of *Pangeometry* from Equations (19), and even to try to substitute for these equations other equations that would express the dependencies between the angles and the edges of any rectilinear triangle. But in that case, we must prove that the new equations are compatible with the fundamental notions of geometry. Equations (19), having been deduced from these fundamental notions, are therefore necessarily compatible with them, and all other equations that we might substitute for them would lead to results that contradict these notions, unless these equations are consequences of Equations (19). Thus, Equations (19) are the basis of the most general geometry, since they do not depend on the assumption that the sum of the three angles in any rectilinear triangle is equal to two right angles.

In other words, Equations (19) are fine as pieces of analysis, but to be interpretable as statements in geometry they must be consistent with the fundamentals of geometry. However, they are geometrical statements because they were derived from those fundamentals. This returns us, of course, to the derivation, with the extra clarity that it involves an appreciation of the concept of distance.

The best examination of Lobachevskii's argument is due to Seth Braver (2011), which is a splendid, leisurely attempt to make Lobachevskii (1840) accessible to "any modern reader . . . undaunted by high school mathematics" (as it says on the back of the book). Braver proceeds by steadily amplifying the individual steps of Lobachevskii's argument, filling out points that Lobachevskii reasonably took for granted. When he reaches this point (on p. 110), however, he takes the unusual step of supplying what he calls a missing lemma, which says that parallels that cut out segments of equal length on one horocycle in the corresponding

[19] Jean-Baptiste Fourier had done this before, as Engel and Stäckel and Bonola both remark. It is unlikely that Lobachevskii knew of Fourier's opinion.

[20] Here in the recent translation by Lobachevskii (2010, p. 75).

orthogonal family cut out equal lengths on any other orthogonal horocycle. Lobachevskii assumed this because he knew, but could not express, a symmetry property of parallels that Braver has excavated from between the lines of the *Geometrische Untersuchungen.*

What is unsatisfactory to the modern mathematician in Lobachevskii's proof is not that it was insufficiently spelled out, but that it rests on a *primitive* notion of distance. In a world in which there has always been one geometry of that kind and now there are two and can apparently be no more this might be acceptable. It would be unreasonable to hold the new non-Euclidean geometry to higher standards than the familiar Euclidean geometry. But Riemann's approach took the ground from under both of these geometries. Distance could no longer be a primitive concept, and the intuitive character of elementary geometry was found to be unduly naive.

It is oddly difficult to say that distance is the fundamental concept in Euclid's *Elements*; straightness of lines has at least an equal claim. It is possible to copy a given line segment in different places, to draw circles (which by definition have equal radii), and to appeal to incidence properties to say that one line segment is less than, equal to, or greater than another. So equality of segments is much used, and we may regard distance as a synonym, but as much if not more of the first 31 propositions of Book I is concerned with angles between straight lines. Once the concept of parallel straight lines is introduced, Euclid proved, in Prop. 34, that opposite sides of a parallelogram are equal. This implies that two parallel lines are equidistant, but for whatever it is worth Euclid never says that, as surely he would have done if the *Elements* was based solely on the concept of distance.[21]

The imprecision of elementary geometry was much discussed by authors like Enriques and Zacharias, as it had to be after the thorough reworking of the subject by Mario Pieri, Hilbert, Klein and others. Axiomatisations of elementary geometry had an effect on histories of the subject. If the enquiries of everyone from, say, Wallis to Bolyai and Lobachevskii were taken to be into the status of the parallel postulate then the consistency of a system of non-Euclidean geometry requires the exhibition of a system of objects that demonstrably satisfies the requisite axioms. Such a system was provided by the various disc models in two

[21] Finally, as remarked earlier, in the terms that Fano used, the geometry in Euclid's *Elements* studies figures up to similarity, not size.

dimensions (Riemann's and Poincaré's disc with a novel metric in the spirit of differential geometry; the Beltrami–Klein projective model). It was then possible to take the points and geodesics in one of these spaces and show, item by item, that they obeyed the axioms. This was a process indicated by Henri Poincaré (1891); I am not sure where or even if this was done before in the numerous explorations of elementary non-Euclidean geometry. (What Hilbert did himself was much more abstract, because he proceeded without the axiom of continuity, also known as the Archimedean axiom.)

This line of argument emphasises the role of differential geometry (or, if you prefer, the Cayley–Klein projective metric) and marginalises the original argument of Lobachevskii. That argument is not so much flawed, although it is true that no amount of consistent conclusions can prove that the whole system is consistent, as unpersuasive. What has to be done is to make the idea of distance clearer and more flexible than Lobachevskii did — to say precisely what it is so that it can figure convincingly in proofs — and this has come to mean Riemannian differential geometry.

One could argue that a firmly metrical approach to geometry unites Bolyai, Lobachevskii, and Riemann and separates them both from Euclid. Equally, an axiomatic presentation of geometry unites both them and Euclid and separates them from Riemann. The looseness and hidden ambiguities of elementary geometry make it difficult to see where they belong, and by importing elementary geometry as a fundamental category of explanation Bonola hobbled his explanation.

10.6 The acceptance and rejection of non-Euclidean geometry

The saddest and most dramatic aspect of the whole story of Bolyai and Lobachevskii and non-Euclidean geometry is, of course, that their work was not accepted in their lifetimes, while steadily after the death of Gauss it became momentous. A rational rejection of their work would be easier to understand if it was in some way flawed. Évariste Galois's work met a similar fate, and there it is clear that the very large leaps in his argument and the entirely unfamiliar notion of a group that he introduced only obscurely form a large part of the explanation of its rejection. But if you believe that the accounts by Bolyai and Lobachevskii are fundamentally sound then that possibility is lost. And if you further believe that the work is a piece of *elementary* geometry then it would seem that there could have been only three possible responses: acceptance, rejection on specific grounds, and neglect.

Bonola, like many others at the time, found the accounts of Bolyai and Lobachevskii good enough, although in need of underpinning. This left him with the problem of explaining why it was not accepted. Bonola put this down to the domination of Kantian ideas about geometry, and the naive, pre-Kantian epistemology of both Bolyai and Lobachevskii, who advocated astronomical tests of the validity of geometry. This is an attractive idea, if under-developed. It would require a strong adherence in France and Germany in the 1830s and 1840s to Kantian philosophy that somehow weakened by the 1860s. However, there is little evidence of Kantianism in France. The German situation was more complicated, and Bonola's ideas had some force: Klein spoke of the "so-called non-Euclidean geometry" in order to keep Rudolf Hermann Lotze, the resident Kantian in his university, quiet; and Kantians spent the next fifty years trying to come to terms with the new geometry. However, I think it is unlikely that the rejection happened exclusively on such high intellectual grounds.

Equally, historians who point to a logical flaw in the works of Bolyai and Lobachevskii tend to exaggerate the effect of the flaw, which was only pointed out in the period when the work was being accepted and its initial rejection had become a puzzle. Rather, what seems to have happened is that Lobachevskii's "On the principles of geometry" was submitted to the Academy of Sciences in St. Petersburg for publication in 1832, only to be dismissed by Mikhail Vasilyevich Ostrogradskii as being mostly incomprehensible and the work of "a worse than mediocre geometer."[22]

Worse still, Lobachevskii's work in German and French was ignored. The public silence of Gauss did not help, as can be seen from the reactions to the discoveries in the Gauss *Nachlass*. It should also be noted that others who knew Gauss well also accepted non-Euclidean geometry; Friedrich Bessel was willing to contemplate its use in astronomy. This suggests that provisional acceptance of the new geometry was possible early on, pending further elucidation – quite a common situation in mathematics.

The implications of the new geometry go beyond the problems it might cause for Kantians; it is alarming to think that all our knowledge

[22] See Ostrogradskii (1963, p. 103).

of space might have been wrong, even if a remedy is to hand. Was it not easier, if you heard of the new ideas at all, to reject them or to put your faith in the idea that investigations into the parallel postulate had a tendency to prove inconclusive and therefore unproductive? But Bonola did not want to find the rejection of the work of Bolyai and Lobachevskii wholly, or even largely, irrational.

But if you believe that the work of Bolyai and Lobachevskii is elementary, then its rejection is difficult to accept. This leads us to ask: Was it elementary by the standards of the 1830s? The merit of the concept of elementary mathematics around 1900 was that it captured a distinction between the intuitive mathematics taught in school and the paradoxical advanced concepts that had been found necessary to provide rigour for those intuitions. Evidently, no such distinction could be drawn in the 1830s, so the questions can only mean: Was the work of Bolyai and Lobachevskii sufficiently plausible or convincing as an intuitive account of basic geometry?

If we look again at the *Geometrische Untersuchungen* we might conclude that it was not. The novel concept of parallel lines is unfamiliar; the introduction of the horocycles generalises a family of straight lines orthogonal to a family of parallel lines to a family of curves with unknown properties. There is little reason to believe that the methods of Euclidean geometry will illuminate these curves, and indeed they do not. Instead, they yield an argument about exponential growth (with, as mentioned above, a gap). Then comes the transition to three dimensions, and the use of the insight that the formulae of spherical trigonometry are independent of the parallel postulate. Finally, there is the high level of sophistication of the argument. All this is intended to open up the study of space, whether abstract or physical, and to propose that figures with familiar names might have unfamiliar properties. This requires a willingness to believe that on some unspecified surface, and also in three-dimensional space, there is a novel concept of distance.

Perhaps this should not have intimidated the best mathematicians of the day, had they read it carefully, but it is very likely that it was not an easy argument when it was first published. It is an advanced, speculative argument that may be plausible but only becomes so when the concept of distance has been re-analysed, and changed from a primitive, immutable given to anything expressed by a Riemannian metric. It requires not just Gauss's idea of a metric on a surface (generalised, in a way Gauss had not described, to three dimensions), but Riemann's interpretation of that idea as saying that any geometry is a surface (or higher-dimensional manifold) with a metric.

Like other authors of his day, Bonola recognised the validity of this argument by assigning the final defence of non-Euclidean geometry to advanced mathematics and then presuming that, once the reader knows that the new geometry will turn out all right in the end, the arguments that led to its discovery are elementary. As they are, once they are grounded in advanced ideas that had become very familiar to the leading mathematicians around 1900.

This anachronistic failure to see that Bolyai and Lobachevskii's arguments were not elementary in their day, but had become elementary (in the sense of unproblematic) in the context of later mathematics, might help to explain the rejection and later acceptance of their ideas. But the force of this point of view should not be exaggerated: such was the extent of the neglect that it remains speculative. That said, elementary models of non-Euclidean geometry such as the Poincaré disc, greatly help introduce the subject to beginners, but it is still the case that no model of non-Euclidean geometry is properly understood until the Riemannian perspective on geometry is opened up.

10.7 Conclusion

In the first decade of the twentieth century, elementary geometry had a double meaning. It referred both to an intuitive body of knowledge that was more or less the same as the contents of Euclid's *Elements*, and whatever results had been discovered since Euclid's time by similar methods. It also referred to the sophisticated knowledge that had only recently been found necessary to pump into the foundations of geometry in order to give it the required degree of rigour.

It would be unreasonable to ask of someone using elementary geometry in the first decades of the nineteenth century to employ the insights of Klein or Hilbert, so to treat the investigations of Bolyai and Lobachevskii under the heading of elementary geometry is to give it the first meaning, and this is what Bonola did.

This, I suggest, guided his selection of what topics to present and how to deal with them, and at crucial points to play up the continuities with Euclid. It led to omissions in the account of the work of Bolyai and Lobachevskii that did not fit the dichotomy of elementary and advanced mathematics, notably the transition to the methods of hyperbolic trigonometry and to neglect the questioning of the naive notion of distance.

Other writers, like Fano and Zacharias, did not omit the first of these points, but included it in straightforward summaries of Lobachevskii's work. Zacharias did not give it a critical examination, and after all he could see that Enriques had found it entirely acceptable. But he did not probe the introduction of novel methods, and that Bonola was not alone only indicates the force of the idea that there was an identifiable body of elementary geometry. Rather, he was a prisoner of his concept of elementary geometry, and his anachronistic assumption that that was what Lobachevskii and Bolyai were doing.

This also led Bonola to exaggerate the size of the step to Riemann. Had he appreciated that Lobachevskii had left the safe surroundings of elementary geometry when he placed much more reliance on the concept of distance, and that Bolyai had gone even further by attempting to introduce the methods of differential geometry (although not in a foundational way) he could have made more of the insights of Gauss and brought more of the continuities between Bolyai and Lobachevskii and Riemann. Instead, he depicted a stark conceptual break. Or, one might say, given the remarkable conceptual advances made by Riemann, Bonola diminished the conceptual advances made by the discovers of non-Euclidean geometry, Gauss included, and in that way gave a distorted account of what they had achieved.

Acknowledgements I would like to thank Umberto Bottazzini, Jemma Lorenat, and Niccolò Guicciardini for their helpful critical comments on earlier drafts of this chapter.

References

Beltrami, Eugenio (1868). Saggio di interpretazione della geometria non Euclidea, *Giornale di Matematiche* **6**, 284–312. (English translation in J. Stillwell, *Sources of Hyperbolic Geometry*, American Mathematical Society (1996), pp. 7–34.)

Bolyai, János (1832). Scientiam spatii absolute veram exhibens. Appendix to *Tentamen juventutem studiosam in Elementa Matheseos purae*

. . . *introducendi*, Maros-Vásérhely. (Translated G.B. Halsted as *The Science Absolute of Space*.)

Bonola, Roberto (1900). Sulla teoria delle parallele e sulle geometrie non euclidee. In Enriques (1900, pp. 143–222).

Bonola, Roberto (1906). *La geometria non-Euclidea: esposizione storico-critica del suo sviluppo*. Bologna: Zanichelli. (English translation by H.S. Carslaw, with a preface by F. Enriques, *History of non-Euclidean geometry*, Open Court, Chicago, 1912. Reprinted 1955 by Dover, New York.)

Bottazzini, Umberto and Pietro Nastasi (2013). *La patria ci vuole eroi: Matematici e vita politica nell'Italia del Risorgimento*. Bologna: Zanichelli.

Braver, Seth (2011). *Lobachevskii Illuminated*. Cambridge: Cambridge University Press.

d'Alembert, J. le Rond (1754). Géométrie, *Encyclopédie méthodique: Mathématiques*.

Engel, Friedrich and Paul Stäckel (1895). *Die Theorie der Parallellinien von Euklid bis auf Gauss*. Leipzig: Teubner.

Enriques, Federigo (ed.) (1900). *Questioni riguardanti la geometria elementare*. Bologna: Zanichelli.

Enriques, Federigo (1906). *Problemi della scienza*. Bologna: Zanichelli.

Enriques, Federigo (1907). Prinzipien der Geometrie, *Encyclopädie der mathematischen Wissenschaften*, III.1.1, 1–129.

Fano, Gino (1898). *Lezioni di geometria non euclidea*, lithographed notes. Rome: Luigi Cippitelli.

Fano, Gino (1902). *Lezioni di geometria descrittiva*, lithographed notes. Torino: Litografia G. Paris.

Gauss, Carl F. (1900). *Werke*, vols. IV and VIII. Göttingen: Königlichen Gesellschaft der Wissenschaften.

Gray, Jeremy J. (ed.) (2004). *János Bolyai, Non-Euclidean Geometry and the Nature of Space*. Cambridge MA: Burndy Library, MIT Press.

Gray, Jeremy J. (2008). *Plato's Ghost: The Modernist Transformation of Mathematics*. Princeton, New Jersey: Princeton University Press.

Hoeffding, Wassily (2012). *The Collected Works of Wassily Hoeffding*. New York: Springer.

Klein, Felix (1908). *Elementarmathematik vom höheren Standpunkt aus*, Vol. 2. Leipzig: Teubner. (English translation *Elementary Mathematics from an advanced Standpoint*, Vol. 2, New York: Dover, 2004.)

Kline, Morris (1972). *Mathematical Thought from Ancient to Modern Times*. New York: Oxford University Press.

Lambert, Johann H. (1786). Theorie der Parallellinien, *Leipziger Magazin für die Reine und Angewandte Mathematik*, 137–164, 325–358. (Reprinted in Engel and Stäckel, 1895.)

Legendre, Adrien-Marie (1860). *Éléments de géométrie*, 8th ed. Paris: Firmin Didot Frères.

Lobachevskii, Nikolai I. (1840). *Geometrische Untersuchungen zur Theorie der Parallellinien*. Berlin: in der Fincke'schen Buchhandlung. (Reprinted Mayer and Müller, 1887; translated G.B. Halsted as *Geometric Researches in the Theory of Parallels*, Open Court Publishing Company, 1914.)

Lobachevskii, Nikolai I. (2010). *Pangeometry*, edited and translated by Athanase Papadopoulos. Zürich: European Mathematical Society.

Ostrogradskii, Mikhail Vasilyevich (1963). *Life and Work, Scientific and Pedagogic Heritage*, G. Gnedenko and I.B. Pogrebitskii (eds.) Moscow: Academia Nauk. In Russian.

Poincaré, Henri (1891). Les géométries non euclidiennes. *Revue générale des sciences pures et appliquées* **2**, 769–74. (Modified reprint in Poincaré, 1902.)

Poincaré, Henri (1902). *Science et Hypothèse*. Paris: Flammarion. (English translation by W.J. Greenstreet, *Science and Hypothesis*, Walter Scott Publishing Co., 1905; reprinted by Dover, New York, 1952.)

Riemann, Georg Friedrich Bernhard (1854). Ueber die Hypothesen, welche der Geometrie zu Grunde liegen. *K. Ges. Wiss. Göttingen 13*, 1–20. Reprinted in *Gesammelte Mathematische Werke, Wissenschaftliche Nachlass und Nachträge, Collected Papers*, R. Narasimhan (ed). New York: Springer, (1990), pp. 304–319.

Saccheri, Giovanni G. (1733). *Euclides ab Omni Naevo Vindicatus: sive conatus geometricus quo stabiliuntur prima ipsa universae geometriae principia*. Milan: Montano. Translated by G.B. Halsted as *Euclid Freed of every Flaw*, Open Court Publishing Company, 1920. New edition published as *Euclid Vindicated from Every Blemish*, edited and annotated Vincenzo De Risi, translated G.B. Halsted and L. Allegri. New York: Springer (2013).

Zacharias, Max (1913). Elementargeometrie und elementare nicht-euklidische Geometrie in synthetischer Behandlung. *Encyclopädie der mathematischen Wissenschaften*, III.1.2, 862–1175.

11

Anachronism and incommensurability: words, concepts, contexts, and intentions

Joseph W. Dauben
Herbert H. Lehman College and The Graduate Center, City University of New York

> The past is a foreign country: they do things differently there. –
> L.P. Hartley

Abstract: Historians are constantly confronted with the twin problems of translating texts and interpreting their meanings. Translation may be as seemingly straightforward as explaining a technical or abstract concept in one's own language, or as fraught with challenges as capturing the sense of a single line of Haiku in Japanese, or how concepts such as 率 *lü* should be translated when they appear in ancient Chinese mathematical works. When mathematicians like Georg Cantor or Abraham Robinson demonstrate the consistency of concepts that, since the paradoxes of Zeno and Democritus, have been assumed to be paradoxical notions like infinitesimals or the actual infinite, how should the works of earlier mathematicians be regarded who either used such concepts or believed to have proven their impossibility? Is it anachronistic to use nonstandard analysis or transfinite numbers to "rehabilitate" or explain the works of Leibniz, Euler, Cauchy, or Peirce, for example, as recent mathematicians, historians, and philosophers of mathematics have attempted? At the other extreme, chronologically, how may ideas readily accepted in the West – like incommensurable numbers, parallel lines, and similar triangles – but foreign to traditional Chinese mathematics have adversely affected the interpretations of ancient Chinese mathematical works? Is the problem of anachronism in the history of mathematics something that can be avoided, or is it an inescapable effect of incommensurability between our contemporary point of view and times past, which are always destined to be foreign, and for which translations can only approximate but never make up for differences and nuances of context?

11.1 Introduction

In his most recent book, *Isaac Newton and Natural Philosophy*, Niccolò Guicciardini raises the question of anachronism in the history of mathematics by asking whether it is fair or anachronistic to judge the priority dispute between Newton and Leibniz over creation of the calculus on terms used today – namely, who was the first to publish the "calculus" as this discipline is understood nowadays?[1] I am going to approach this same question of anachronism from several different but related perspectives, first in the case of modern theories of the infinite and infinitesimals as they have been applied to the concept of continuity that was a key element in the philosophy of Pragmatism as developed by the American polymath Charles Sanders Peirce, and then in terms of ancient Chinese mathematics, where anachronism figures in different ways in explanations both Chinese and Western historians of mathematics have given to the methods used to extract square roots, to determine the heights of distant objects, or to establish the Chinese equivalent of the Pythagorean theorem.

11.2 Transfinite set theory, nonstandard analysis, and Charles Sanders Peirce

Beginning with the case of nonstandard analysis, this was a rigorous theory of infinitesimals devised by Abraham Robinson (1918–1974) in the mid-1960s. Robinson's early training in mathematics was at the Hebrew University in Jerusalem, under Abraham Fraenkel (1891–1965).

[1] Guicciardini writes: "It is often said that in his youth Newton 'discovered calculus'. However, it is not trivial to say exactly what is meant by it. Historically speaking, it is not a good idea to establish the meaning of the above expression too strictly, but rather leave the historical actors directly involved to speak and see what *they* meant by calculus, and how *they* practised it. Calculus as we know it today (and there is still ambiguity, sometimes dissent, about what it should be and how it should be taught) consists of a series of notations, concepts, methods and theorems that have been achieved over a long period of time; certainly it was not invented by a single man. Rather, we should conceive the discovery of calculus as a long process, at least spanning the period beginning with Pierre de Fermat and Johannes Kepler and culminating with the work of Euler" (Guicciardini, 2018, p. 49).

(a) (b) (c)

Figure 11.1 (a) Robinson at the Hebrew University, Jerusalem, 1957–1958. (b) Robinson's MAA lecture, *Nonstandard Analysis*, filmed in 1970 and reproduced here by permission of the Mathematical Association of America. (c) Robinson celebrating his Brouwer medal with Arend Heyting, Leiden, April 26, 1973. Parts (a) and (c) included courtesy of the estate of Renée Robinson.

Robinson spent World War II doing aeronautical research at the Royal Aircraft Establishment in Farnborough, England, and later received his Ph.D. in mathematics from Birkbeck College, University of London, with a thesis on the metamathematics of algebra in 1949. Having taught at the Royal College of Aeronautics in Cranfield, Robinson was offered a position teaching applied mathematics at the University of Toronto in the early 1950s, after which he spent five years (1957–1962) back at the Hebrew University, where he assumed the chair in mathematics previously held by Fraenkel in the Einstein Institute.

It was while Robinson was on a sabbatical leave, replacing Alonzo Church for a year at Princeton University in 1960–1961, that Robinson first had the idea of a rigorous theory of infinitesimals. At the time he had been thinking about nonstandard models of arithmetic – and also talking with Kurt Gödel at the Institute for Advanced Study about Leibniz. All of a sudden, as he was walking into Fine Hall (home of the Mathematics Department at Princeton), the idea of creating a nonstandard model of the real numbers – including infinitesimals – flashed into his mind (Dauben, 1995, pp. 276–287).

In 1967 Robinson accepted a position in the Mathematics Department at Yale University, and shortly thereafter, in 1970, the first meeting devoted to nonstandard analysis was held at the Mathematisches Forschungsinstitut in Oberwolfach, Germany. This was a clear sign that the subject had become one of international importance. In 1973, the L.E.J. Brouwer medal was awarded to Robinson by the Dutch Mathematical Society, and by then Gödel was hoping that Robinson might be his successor at the Institute for Advanced Study. Tragically, Robinson's life was cut short by pancreatic cancer, of which he died at age fifty-five in 1974.

Robinson's legacy – especially the creation of nonstandard analysis – has had both its proponents and detractors. What is of interest to us here as historians of mathematics is the means it has offered in the eyes of some for rationally reconstructing – and even radically revising – past accounts of the history of mathematics. But isn't doing so obviously anachronistic? To what extent might taking this approach be justified? One philosopher of mathematics who turned to nonstandard analysis in this regard almost immediately after Robinson had published his rigorous theory of infinitesimals was Imre Lakatos (1922–1974):

> Robinson's work revolutionizes our picture of this most interesting and important period [the pre-Weierstrass era]. It offers. . . a rational reconstruction of the discredited infinitesimal theory which satisfies modern requirements of rigour and which is no weaker than Weierstrass's theory. This reconstruction makes infinitesimal theory an almost respectable ancestor of a fully-fledged, powerful modern theory, lifts it from the status of pre-scientific gibberish and renews interest in its partly forgotten, partly falsified history (Lakatos, 1978, Vol. 2, p. 44).

Robinson summarized his own take on the significance of nonstandard analysis for the history of mathematics more modestly:

> Nevertheless we venture to suggest that our approach has a certain natural appeal, as shown by the fact that it was preceded in history by a long line of attempts to introduce infinitely small and infinitely large numbers into Analysis (Robinson, 1965, p. 184).

It may seem ironic that the figure most closely associated with the first successful attempt to introduce infinitely large numbers into mathematics was Georg Cantor (1845–1918), who nevertheless was a vigorous opponent of infinitesimals. Born in St. Petersburg in 1845, for reasons of his father's health the family moved to Frankfurt, Germany, in 1854. The young Georg went to school in Darmstadt, graduating from the Realschule and then the Höhere Gewerbeschule, a trade school, before completing his studies in mathematics at the University of Berlin in 1866 (with semesters spent at the Polytechnicum in Zürich and another summer term at Göttingen). Cantor devoted his dissertation at Berlin to number theory, which he wrote under the direction of Ernst Kummer.

III. Abhandlungen zur Mengenlehre.

1. Über eine Eigenschaft des Inbegriffes aller reellen algebraischen Zahlen.

(a) (b) (c)

Figure 11.2 (a) Georg Cantor, 1870s. (b) Richard Dedekind, 1870. (c) Page from Cantor (1874, p. 258).

And then, in 1870, Cantor accepted a position at the University of Halle where he would spend the rest of his career as a mathematician.

It was Cantor's colleague at Halle, Eduard Heine, who interested him in a long-standing problem in the theory of trigonometric series. In the early 1870s, Cantor managed to prove the uniqueness of representations of arbitrary functions over domains even if a finite number of points in their domain were excepted, and soon he was able to generalize this result to include infinite sets of exceptional points so long as they were distributed in certain specified ways. This remarkable discovery prompted Cantor to investigate what he called "derived sets" of various species, which in turn led him to develop a theory of real numbers based on convergent sequences, and, within a decade, his theory of transfinite numbers.

Meanwhile, Cantor's colleague, Richard Dedekind, for different reasons, was also interested in a rigorous theory of the continuum of real numbers. Dedekind had reached the conclusion that "The line is infinitely richer in point-individuals than is the domain of rational numbers in number-individuals" (Dedekind, 1872, p. 9).

But exactly how much richer no one could say until Cantor, in 1874, published his famous proof in the *Journal für die reine und angewandte Mathematik* that the set of all real numbers is *non*-denumerably infinite.

The paper in which Cantor published this ground-breaking discovery bore a rather strange – not to say misleading – title: "Über eine Eigenschaft des Inbegriffes aller reellen algebraischen Zahlen" (On a Property of the Set of All Real Algebraic Numbers). This was no doubt in anticipation of opposition to his discovery that Cantor expected from his older colleague and former teacher at the University of Berlin, Leopold Kronecker. Cantor had been a student of Kronecker at the University of Berlin, and Kronecker's opposition to irrational numbers, for instance, was already well known. As Heine admitted in a letter to Hermann A. Schwarz in 1870:

Meine kleine Arbeit, "Über trigonometrischen Reihen" von der ich in diesem Augenblick die Correctur in Händen habe, und die nun mehr, nach vielen Verhandlungen mit Kronecker, der mich veranlassen wollte sie zurückzuziehen (der Nähere unten) im laufenden (71te.) Bande des Journals s. 353 erscheint hatte mir viele Freude gemacht .... (E. Heine to H.A. Schwarz, May 26, 1870; Dauben, 1979, Appendix B, p. 308).

(My little work "On trigonometric series," of which I have at this moment the publisher's proofs in my hands, and which, moreover, after many negotiations with Kronecker who wanted to persuade me to withdraw it (details below), appears in the current (71st) volume of the journal on p. 353, has made me very happy.)

Kronecker had objected to the appearance of irrational numbers in a paper Heine had written and tried to prevent its publication, as likewise happened to another paper of Cantor's a few years later on the invariance of dimension (Dauben, 1979, p. 66). Had Cantor entitled his paper as blatantly as "The Nondenumerability of the Real Numbers," that would certainly have aroused Kronecker's opposition to publication of what was clearly the most startling result Cantor had achieved – much more so than his announcement that the algebraic numbers were denumerably infinite.

In 1883 Cantor published his first paper on point-set theory, which introduced his transfinite ordinal numbers. At first, Cantor used an argument involving nested intervals to prove that assuming the real numbers were denumerable led to a contradiction. Some ten years later, having defined his transfinite cardinal numbers in the meantime, he could now express this result algebraically in the form of his famous Continuum Hypothesis:

$$1883: P(R) > P(N)$$
$$1895: 2^{\aleph_0} = \aleph_1$$

Throughout this period Kronecker was a relentless critic of both the new analysis championed by his colleague at the University of Berlin,

Carl Weierstrass, as well as Cantor's new set theory and the accompanying transfinite numbers. Kronecker was determined to show that: "... the results of modern function theory and set theory are of no real significance" (L. Kronecker to G. Mittag-Leffler, in a letter of January 26, 1884, quoted in Schoenflies (1927, p. 5); see also Dauben (1979, p. 135)). Nevertheless, Cantor's work was staunchly defended and promoted by the Swedish mathematician, Gösta Mittag-Leffler, who arranged for the French translations of Cantor's most important papers to be published in his influential international journal, *Acta Mathematica*.

Cantor published a comprehensive introduction to his transfinite set theory in 1895, his "Beiträge zur Begründung der transfiniten Mengenlehre" (Contributions to the Founding of Transfinite Set Theory), which appeared in two parts in *Mathematische Annalen* in 1895 and 1897. The first of Cantor's two papers began with three aphorisms, each a subtle hint to the defense he was prepared to give his revolutionary new ideas, including this line from Seneca, quoted by Alexander von Humboldt in volume four of his *Cosmos*:

> Veniet tempus, quo ista quae nunc latent in lucem dies extrahat (Seneca, *Quaestiones Naturales*, 7, 25).
>
> (The time will come when these things which are now hidden from you will be brought into the light.)

To justify his introduction of transfinite numbers, Cantor considered them as new irrationals:

> The transfinite numbers themselves are in a certain sense *new irrationals*, and in fact I think the best way to define the *finite* irrational numbers is entirely similar; I might even say in principle it is the same as my method described above for introducing transfinite numbers. One can certainly say: the transfinite numbers *stand or fall* with the finite irrational numbers; they are alike in their most intrinsic nature [*innerstes Wesen*]; for any numbers like these are definite, delineated [*abgegrenzte*] forms or modifications (αφωρισμενα) of the actual infinite (Cantor 1887/1888; quoted from Cantor, 1932, pp. 395–396).

With these examples of Cantor and Robinson in mind, how do Cantor's transfinite set theory and Robinson's nonstandard analysis raise questions about anachronism in the history of mathematics? Robinson himself considered such historic figures as Leibniz, l'Hôpital, Lagrange, d'Alembert, and Cauchy as fully justified by his theory of nonstandard analysis (Robinson, 1966, p. 260–82), and Cantor believed that transfinite set theory could serve to correct serious matters of doctrine as professed by the Roman Catholic Church and even remedy errors in the philosophy of St. Thomas with respect to matters of the infinite and questions about eternity (Dauben, 1978, p. 296).

With respect to whether such claims are anachronistic, the position advanced by Nick Jardine in his article, "Uses and Abuses of Anachronism in the History of the Sciences," is applicable. Jardine (2000, p. 261) argues that under certain circumstances anachronisms may be "entirely legitimate". Jardine is concerned with complex matters of "the social conditions of production of actions and representations," but here the matter is more straightforward – nothing in the use of nonstandard analysis to explore the efficacy of earlier appeals to infinitesimals is contrary to the basic intuitions of those predecessors. And this is exactly the point made by Hilary Putnam in his rational reconstruction of the analysis the American philosopher and logician Charles Sanders Peirce (1839–1914) gave to account for the essence of continuity. In this case, Peirce's analysis of continuity depended, in its most essential character, on infinitesimals.

Peirce was something of a child prodigy. His father, the Harvard

(a) (b)

Figure 11.3 (a) C.S. Peirce, ca. 1867. (b) Georg Cantor, ca. 1895. Part (a) reproduced courtesy of Houghton Library. Part (b) courtesy of the Archive of the Martin Luther University Halle-Wittenberg.

University mathematician, Benjamin Peirce, educated his son personally in mathematics. For much of his early career Peirce was employed by the US Coast and Geodetic Survey. While teaching logic at Johns Hopkins (where he was a member of the faculty from 1879 until 1884), Peirce did the unthinkable – he divorced his first wife and married a French actress. As a consequence, he lost his job at Johns Hopkins and retired to a farmstead home in Milford, Pennsylvania, that he named Arisbe. There he subsisted on odd writing jobs and the support of benefactors, including the publisher Paul Carus, the philanthropist George Arthur Plimpton, and financial donations from close and concerned colleagues like William James and Josiah Royce. When Peirce died in 1914 at age 74, his massive collection of disorganized papers and correspondence went to the Philosophy Department at Harvard University. When Harvard set about to organize and edit Peirce's papers, the philosopher Josiah Royce described the situation thusly:

> We have just received at Harvard the extant logical manuscripts of Charles S. Peirce, a gift from his widow, and, as I hope, a real prize. I look forward to some arrangement for editing them. They are certainly fragmentary but also certainly inclusive of some valuable monuments of his unique and capricious genius (Houser, 1992, p. 1249).

A major problem that anyone working on Peirce must face is the obscurity of much of his writing. As the logician Christine Ladd-Franklin, who had been among Peirce's students at Johns Hopkins, once said of Peirce's writing:

> Many of his contributions to the philosophical dictionary were of the purely cabalistic type. The second part of the article on Symbolic Logic, for instance, was finally, against the urgent advice of Professor Couturat, who himself had contributed the admirable first part, sent to the printer, though it is doubtful if any one will ever be able to read it. But it will never be known what reams of closely written matter were excluded! (Franklin, 1916, p. 721).

Initially the job of editing the Peirce papers was entrusted to two graduate students, Paul Weiss and Charles Hartshorne, who between 1931 and 1935 published six volumes of Peirce's *Collected Papers*. Two more volumes appeared in 1958, edited by Arthur Burks.[2] Unfortunately,

[2] Subsequently, a massive project was launched in 1975 – the chronological edition of Peirce's works. Unfortunately, Peirce's mathematics has not been systematically included due to the four volumes (in five books) of Peirce's mathematical works edited by Carolyn Eisele and published by Mouton (Peirce, 1976). As of 2019, the Peirce Edition Project has brought the

this edition is especially problematic with respect to the treatment it gives of Peirce's mathematics. As Weiss and Hartshorne readily admit:

> We were young men who knew nothing about editing, nothing about publishing; but also, we were given no help [from the Harvard Philosophy Department]. We were not encouraged in our work (Bernstein, 1970, p. 180).

Before turning briefly to underscore the problems of working with the *Collected Papers*, one positive result should be mentioned: namely, they preserve parts of manuscripts that are now missing from Peirce's originals. For example, the paper on which Peirce wrote "Multitude and Number" (CSP MS 25) now in the Houghton Library archives at Harvard University is dry and brittle, and over the years pieces have broken off, as in Figure 11.4. Fortunately, the printed version preserves such now missing parts of what Peirce wrote.[3]

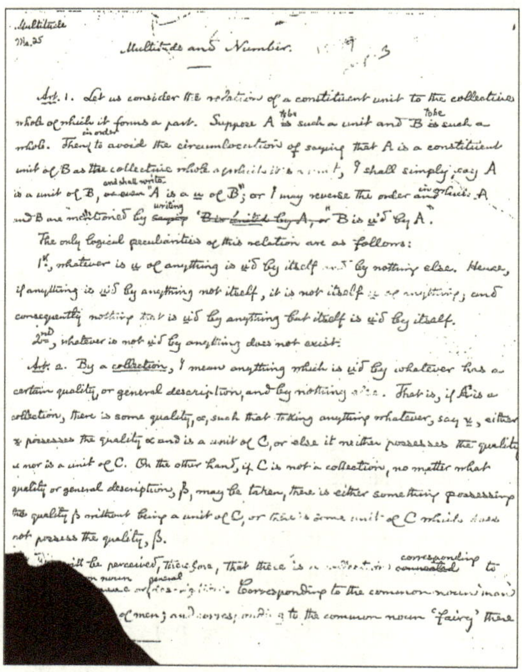

Figure 11.4 CSP MS 25: 1 (Houghton Library, MS Am 1632 (25)).

edition as far as 1892 with publication of volume 8 in 2009. Volume 9 is expected to continue the chronological edition as far as 1892–1893.

[3] Figures 11.4, 11.5, 11.6, 11.7, and 11.9 are from CSP MS AM 1632 (25), in the collection of Peirce manuscripts preserved in the archives of the Houghton Library, and are reproduced courtesy of the Houghton Library, Harvard University, Cambridge, MA. Here, in the figures and text, this manuscript is referred to as CSP MS 25.

However, in preparing Peirce's papers for publication, Hartshorne and Weiss edited with a heavy hand, literally, and blue-penciled out entire paragraphs that they chose to omit from the printed edition. They must have done so assuming the parts they marked for deletion were either unnecessary or redundant. Rather than prepare a clean typescript to convey to the printer, they apparently sent Peirce's original papers, edited as they felt best, directly to the press. For example, among paragraphs omitted from MS 25 devoted to "Multitudes and Numbers," one includes the remarkable statement: "I borrow from Georg Cantor the following definitions . . . " (Figure 11.5, CSP MS 25: 5).

The *Collected Papers* thus leave the impression that everything in MS 25 on infinite multitudes is due entirely to Peirce, but on Peirce's

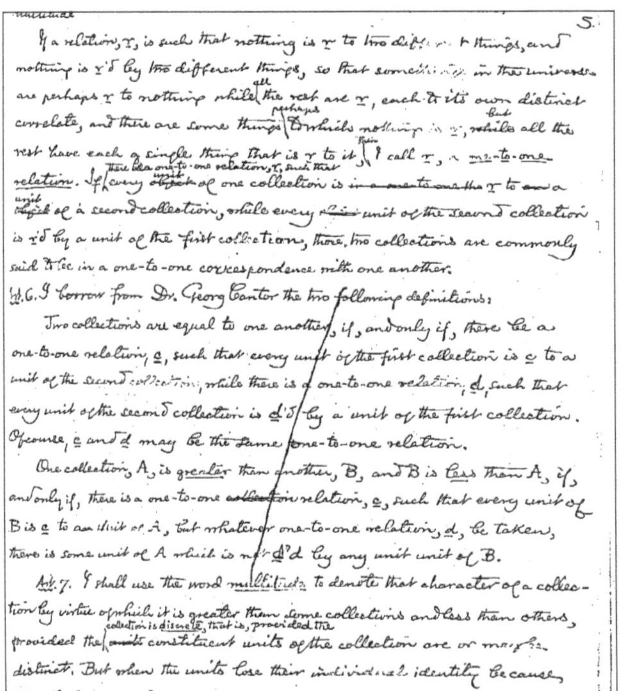
5.

If a relation, r, is such that nothing is r to two different things, and nothing is r'd by two different things, so that some things in the universe are perhaps r to nothing while all the rest are r, each to its own distinct correlate, and there are some things perhaps to which nothing is r, but all the rest have each a single thing that is r to it, I call r, a one-to-one relation. If there be a one-to-one relation, r, such that every unit of one collection is r to a unit of a second collection, while every unit of the second collection is r'd by a unit of the first collection, those two collections are commonly said to be in a one-to-one correspondence with one another.

Art. 6. I borrow from Dr. Georg Cantor the two following definitions:

Two collections are equal to one another, if, and only if, there be a one-to-one relation, c, such that every unit of the first collection is c to a unit of the second collection, while there is a one-to-one relation, d, such that every unit of the second collection is d'd by a unit of the first collection. Of course, c and d may be the same one-to-one relation.

One collection, A, is greater than another, B, and B is less than A, if, and only if, there is a one-to-one relation, c, such that every unit of B is c to a unit of A, but whatever one-to-one relation, d, be taken, there is some unit of A which is not d'd by any unit of B.

Art. 7. I shall use the word multitude to denote that character of a collection by virtue of which it is greater than some collections and less than others, provided the collection is discrete, that is, provided the constituent units of the collection are or may be distinct. But when the units lose their individual identity because the collection exceeds every possible existence of the universe, the word multitude ceases to be applicable. I will take the word multiplicity to mean the greatness of any collection discrete or continuous.

Figure 11.5 CSP MS 25: 5 (Houghton Library, MS Am 1632 (25)).

own testimony, unfortunately cancelled from the published edition, this was certainly not the case.

Further along, on page 14, Peirce cancelled the word "number" and replaced it with the word "multitude," an important indication of the evolution of his thinking. But again, if one is only working with the published edition of Peirce's *Collected Papers*, there is no indication that Peirce made this change in nomenclature. Worse yet, given the importance of diagrammatic reasoning for Peirce, what is one to make of the meaning of the following sentence as it appears in the *Collected Papers*:

> The denumerable multitude added to itself gives itself . . . as made plain by zigzagging through two denumerable series (Peirce, 1931–1936, 4.193).

A mathematician might guess at what "zigzagging" means here, but if the editors had included Peirce's diagram (Figure 11.6), this would have all been much clearer (see MS 25: 32).

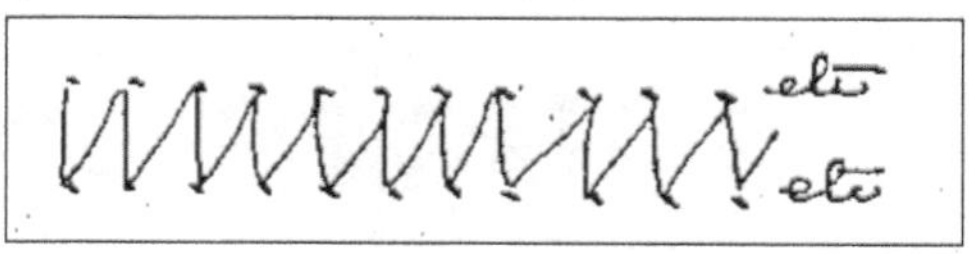

Figure 11.6 Detail of CSP MS 25: 32 (Houghton Library, MS Am 1632 (25)).

Elsewhere Hartshorne and Weiss did include diagrams, one for example by which Peirce proved that even a denumerable number of denumerable multitudes is denumerable – basically a simple extension of his zigzag method (Figure 11.7). But the transcription of Peirce's text as it appears in the *Collected Papers* is *not* what Peirce wrote! The *Collected Papers* version (Figure 11.8) says: "Let ℵ denote the denumerable multitude . . . " (Peirce, 1931–1936, 4.196). Figure 11.7 reproduces the original page, and enlarged in Figure 11.9, this makes clear what Peirce actually said.

Given that Peirce was the "Father of Semiotics," he was always very careful about symbols. Here the editors not only misunderstood what Peirce meant by his trefoil symbol 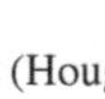, they conflated it with Cantor's transfinite cardinal $\aleph_0$ in a way that was sure to mislead the unsuspecting reader. Peirce's "post numerals" as he referred to the trefoils were generated by power sets; Cantor's transfinite alephs, however, were the result

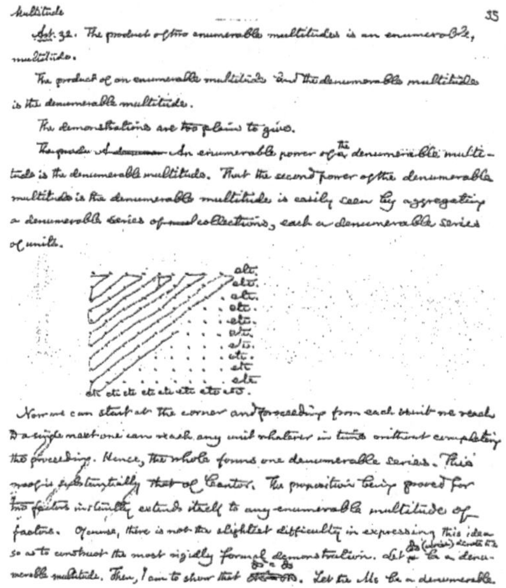
multitude 35

Art. 32. The product of two enumerable multitudes is an enumerable multitude.

The product of an enumerable multitude and the denumerable multitude is the denumerable multitude.

The demonstrations are too plain to give.

An enumerable power of the denumerable multitude is the denumerable multitude. That the second power of the denumerable multitude is the denumerable multitude is easily seen by aggregating a denumerable series of collections, each a denumerable series of units.

Now we can start at the corner and proceeding from each unit we reach to a single next one can reach any unit whatever in time without completing the proceeding. Hence, the whole forms one denumerable series. This proof is substantially that of Cantor. The proposition being proved for two factors instantly extends itself to any enumerable multitude of factors. Of course, there is not the slightest difficulty in expressing this idea so as to construct the most rigidly formal demonstration. Let ℵ (aleph) denote the denumerable multitude. Then, I am to show that $ℵ^2 = ℵ$. Let the Ms be a denumerable

Figure 11.7 CSP MS 25: 32 (Houghton Library, MS Am 1632 (25)).

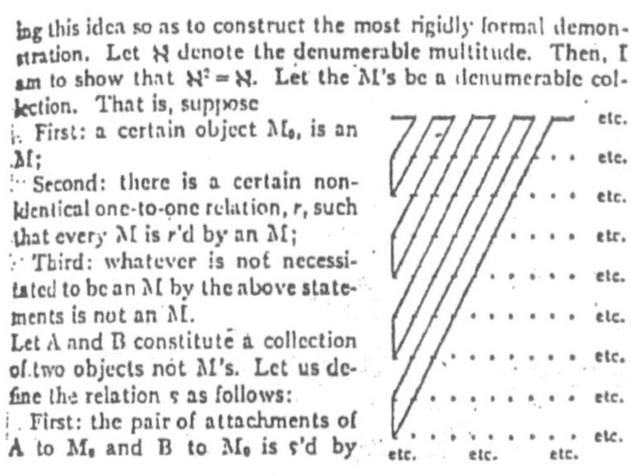
ing this idea so as to construct the most rigidly formal demonstration. Let ℵ denote the denumerable multitude. Then, I am to show that $ℵ^2 = ℵ$. Let the M's be a denumerable collection. That is, suppose

First: a certain object M_0, is an M;

Second: there is a certain non-identical one-to-one relation, r, such that every M is r'd by an M;

Third: whatever is not necessitated to be an M by the above statements is not an M.

Let A and B constitute a collection of two objects not M's. Let us define the relation ς as follows:

First: the pair of attachments of A to M_0 and B to M_0 is ς'd by

Figure 11.8 (Peirce, 1931–1936, 4.196). *Collected Papers of Charles Sanders Peirce*, Volumes I–VI, edited by Charles Hartshorne and Paul Weiss. Cambridge, MA: The Belknap Press of Harvard University Press, Copyright ©1931, 1932, 1933, 1934, 1935, 1958, 1959, 1960, 1961, 1963, 1986 the President and Fellows of Harvard College.

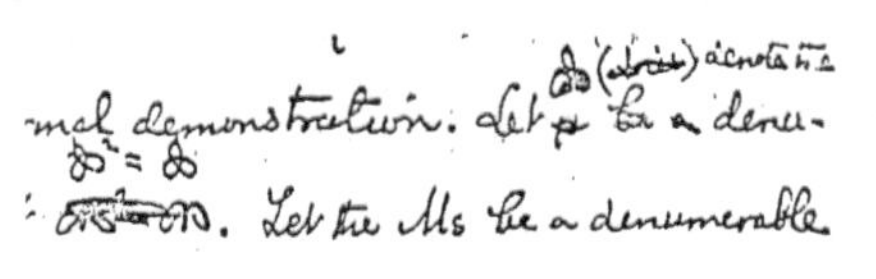
-mal demonstration. Let ℵ (aleph) denote the denu-
$ℵ^2 = ℵ$. Let the Ms be a denumerable

Figure 11.9 CSP MS 25:32 (Houghton Library, MS Am 1632 (25)).

of an entirely different method that he called principles of generation, a much weaker means of defining transfinite cardinal numbers in terms of number classes of well-ordered sets. How does all of this relate to Peirce's understanding of continuity?

In a paper of 1880, "On the Logic of Number," he begins with what he calls a simple system of quantities, but provides a flawed definition of continuity:

> A system of simple quantity is either continuous, discrete, or mixed. A continuous system is one in which every quantity greater than another is also greater than some intermediate quantity greater than the other (Peirce, 1881, p. 86).

Whereas Cantor had approached the problem of continuity from his work on trigonometric series, and Dedekind from having to teach the matter rigorously, Peirce came to his understanding of the infinite through logic, and in 1880 was only just beginning his study of continuity. It was from what De Morgan called the "Syllogism of Transposed Quantity" (which is true only in finite cases) that Peirce began his thinking about the infinite and continuity:

> Every Texan kills a Texan, Nobody is killed by but one person, Hence, every Texan is killed by a Texan (Peirce, 1931–1936, 3.288).

In pursuing his study of quantity both finite and infinite, Peirce decided that a purely logical definition of continuity was required. As he explained in an article in *The Monist* in 1897:

> A perfectly satisfactory logical account of the conception of continuity is required. This involves the definition of a certain kind of infinity; and in order to make that quite clear, it is requisite to begin by developing the logical doctrine of infinite multitude. This doctrine still remains, after the works of Cantor, Dedekind, and others, in an inchoate condition. For example, such a question remains unanswered as the following: Is it, or is it not, logically possible for two collections to be so multitudinous that neither can be put into a one-to-one correspondence with a part of the whole of the other? To resolve this problem demands, not a mere application of logic, but a further development of the conception of logical possibility (Peirce, 1897, 205–206).

But what did Peirce mean here by the need to define "a certain kind of infinity" before the concept of continuity could be accounted for logically? What did this have to do with "logical possibility"? And what was "inchoate" about the works of Cantor and Dedekind? To illustrate his idea of continuity, Peirce introduced a procedure he called "interpolation" on the unit interval. At Step 1, there was only 1 point; at Step 2, there were two points, at Step 3, four points, then eight, and so on:

	0	.01	.1	...
Step 1			.1	
Step 2		.01		.11
Step 3	.001	.011	.101	.111
Step 4	.0001 .0011	.0101 .0111	.1001 .1011	.1101 .1111

Here, Peirce said, one had a "premonition of continuity" whereby the continuum would "stick together" (Peirce, 1976, vol. 3, pp. 87–8). Sometimes he used alternatively evocative language: "glued together" or "welded." Just as irrational numbers were interpolated between the rational numbers, he imagined infinitesimals must be interpolated between irrational numbers to "glue" the continuum together.

As Peirce said in a letter of 1906, by a "Leibnizian infinitesimal" of the first order he meant an assumed quantity "smaller than any finite quantity," the first quantity after .1, .01, .001, . . . As he explained in letters to the philosopher Josiah Royce and the mathematician C.J. Keyser: "such supermultitudinous collections stick together by logical necessity. Its constituent individuals are no longer distinct, or independent. They are not subjects but phases expressive of the properties of the continuum" (Peirce, 1976, vol. 3, p. 95). Peirce went so far as to argue that his infinitesimals were substantiated by nature, since he took their existence to be necessary for physics. He also believed physics, and even how matter acted on the brain to produce thought, required infinitesimally subtle monads (Peirce, 1976, vol. 3, pp. 898 and 857). Several decades earlier Cantor had said much the same thing, arguing that transfinite set theory would be useful in elucidating certain distinctions between physical matter and the aether.[4]

With all of the above in mind, can modern mathematics – to wit, nonstandard analysis – be of any help in explaining what Peirce had in mind here? The philosopher Hilary Putnam believes the answer is an emphatic "Yes!" According to Putnam, Peirce's "metaphysical speculations" involving infinitesimals were "characterized by enormous originality and profundity in conception combined with precision in technical detail" (Putnam, 1995, p. 2).

4 For discussion of Cantor's "World Hypotheses," which he introduced as a set theoretic approach to mathematical physics, see Dauben (1979, pp. 291–294).

As his contribution to the Sesquicentenary Celebration of Peirce's birth held at Harvard in 1989, Putnam addressed the matter of "Peirce's Continuum." Putnam noted that the draft of one of Peirce's Cambridge Conferences lectures of 1898 seemed "bizarre":

> In MS 439 (the draft of the third Cambridge Conferences Lecture on which I am relying), Peirce said a number of things that sound completely bizarre to today's mathematical sensibility, and then challenged anyone who might think he is crazy to show that he had actually contradicted himself. What is worse, it looks very easy to show that he did indeed contradict himself (Putnam, 1995, p. 6).

Especially "bizarre" was the idea that any point could "burst into a discrete multitude of points" at any time, exploding with points flying off in all directions. According to Peirce, a point:

> . . . might burst into any discrete multitude of points whatever, and they would all have been one point before the explosion. Points might fly off, in multitude and order like all the real irrational quantities from 0 to 1; and they might all have had that order of succession in the line and yet all have been at one point in the line. Men will say this is self-contradictory. It is not so. If it be so, prove it. The apparatus of the logic of relatives is a perfect means of demonstrating anything to be self-contradictory that really is so, but that apparatus not only refuses to pronounce this self-contradictory, but it demonstrates, on the contrary, that it is not so (Peirce, MS 439, pp. 27–28; quoted from Putnam, 1995, pp. 6–7).

The question Putnam could not help but ask: "Had Peirce then simply lost his marbles in 1898?" (Putnam, 1995, p. 7). Putnam soon explains that, despite the apparent ambiguity of what Peirce had written, this all made logical sense within Peirce's frame of pragmatic reference:

> . . . if we suppose that what we ordinarily call a "point" can in some sense have *parts* – and I am going to argue that this is what Peirce believed – then the "contradiction" between Peirce's statements also disappears . . . How can I justify putting the notion of a "point part" into Peirce's mouth when he did not explicitly use any such notion? (Putnam, 1995, p. 8).

As Putnam went on to explain, Peirce believed in infinitesimals:

> Indeed, he said things in a number of places in these lectures which directly imply the existence of infinitesimals, including geometric infinitesimals, that is to say, line intervals whose length is not zero but is less than any positive real length whatsoever (Putnam, 1995, p. 8).

Here, the crucial factor for Putnam is the fact that nonstandard analysis can capture the essence of what Peirce had in mind in saying that the continuum was "glued together":

In nonstandard analysis we say that two points P and Q whose distance is infinitesimal are "identical modulo the infinitesimals," and we symbolize this by using the wiggly equals sign: $P \approx Q$. If P is a point, the collection of all points Q such that $P \approx Q$ is called the monad of P. It can be shown that every point has to lie in the monad of a "standard" point, that is, a point whose distance from the origin is a (standard) real number (Putnam, 1995, p. 9).

Thus, on Putnam's reconstruction, he says there is no contradiction with respect to nonstandard analysis:

What Peirce was doing, then, was first imagining a transformation as the result of which these point parts fly apart, without changing their relative order, that is, they are mapped onto distinct points (distinct monads) having the order type of the real numbers between 0 and 1; and Peirce was saying that even before the transformation, that is, even when the point parts were parts of a single line (were "at" a single point) they had that same order. On my proposed reconstruction, he was quite right in claiming that there is no contradiction in this point of view (Putnam, 1995, pp. 9–10).

Peirce regarded the collection of whole numbers as a "potential collection":

In the Eighth of his Cambridge Conferences Lectures, Peirce repeatedly referred to infinite sets as potential aggregates: for instance, "We have a conception of the entire collection of whole numbers. It is a potential collection, indeterminate and yet determinable, and we see that the entire collection of whole numbers is more multitudinous than any whole number" (MS 948, 13) (Putnam, 1995, p. 15).

For Peirce, the line is a collection of "*possibilia*," all of which Putnam characterizes as "Peirce's daring metaphysical hypothesis":

The reason that the line is a collection of points that "lack distinct individuality" is that it is a collection of *possibilia*, and possibilia are not fully determinate objects for Peirce. To say that the line is a collection of possibilia is to say that one can construct things that stand in a certain triadic relation, the relation "Proceeding to the right from A you reach B before you reach C" (MS 948, 15). *What answers to our conception of a continuum is a possibility of repeated division which can never be exhausted in any possible world, not even in a possible world in which one can complete abnumerably infinite processes.* That is what I take Peirce's daring metaphysical hypothesis to be (Putnam, 1995, p. 17).

As Putnam concludes, for Peirce, metaphysically, "possibility outruns actuality":

The Peircean picture is that the multitude of possibilities is so great that as soon as we have a possible world in which some of these possibilities are realized – say, a possible world in which some abnumerable multitude of the division are made – then we immediately see that there is a possible world in which still more divisions can be made, and hence there is no possible world in which all these non-exclusive possibilities are all actualized. We might summarize this by saying that the metaphysical picture is that possibility intrinsically outruns actuality, not just because of the finiteness of human powers, or the limitations imposed by physical laws (Putnam, 1995, p. 19).

It is on this interpretation of Peircean infinitesimals that Putnam believes he has shown Peirce's stand to be a consistent one:

My aim was not, of course, to expound the whole metaphysics of Peirce's Cambridge Conferences Lectures, but to reconstruct one key element, which, if it could not be shown to be consistent, would have represented a fatal flaw in the whole edifice. And that, I believe, has been done (Putnam, 1995, p. 19).

In fact, Putnam's vision of what Peirce meant is quite similar to the now well-established model of infinitesimals created by Abraham Robinson in the 1960s, as reimagined in a revisionist calculus textbook by H. Jerome Keisler wherein he imagines microscopes applied to the continuum of real numbers. This approach, as Keisler emphasizes, places: "the intuitive ideas of the founders of the calculus on a mathematically sound footing, and is easier for beginners to understand than the more common approach via epsilon, delta definitions" (Keisler, 1976, Preface). One has only to look at an explanatory page (see Figure 11.10) to grasp the gist of Keisler's inspiration. The "microscopes" can be applied to the continuum of real numbers, at any level of refinement one wishes. Between any two distinct numbers of the continuum, no matter how close, more and more can be identified until the continuum is "glued" together, in Peirce's sense of continuity, or "blows up" if you prefer!

At the end of his book introducing nonstandard analysis in a comprehensive way in 1966, Abraham Robinson devoted a final chapter to addressing the ways in which nonstandard analysis might serve historically to rehabilitate the long maligned concept of infinitesimals. Specifically, Robinson was concerned with history of the calculus. He was certain that prejudice against infinitesimals had led to an overly severe critique

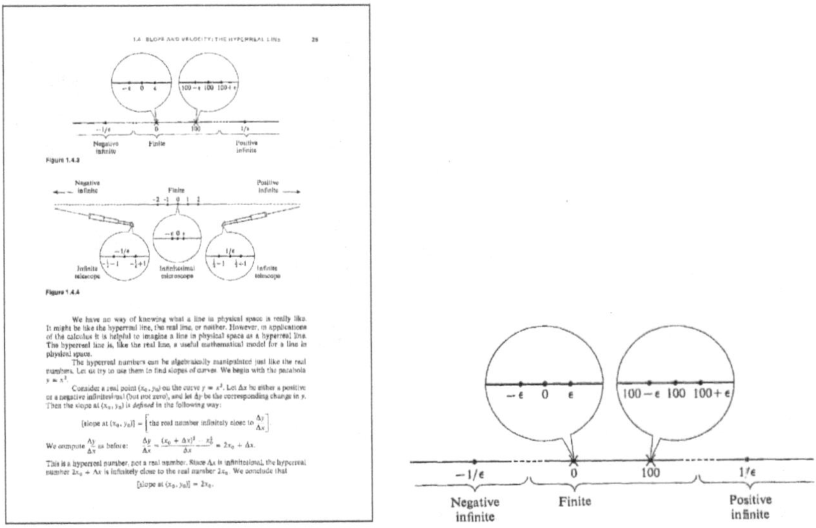

Figure 11.10 The circles represent "infinitesimal microscopes" which are powerful enough to show an infinitely small portion of the hyperreal line. The set R of real numbers is scattered among the finite numbers. About each real number c is a portion of the hyperreal line composed of the numbers infinitely close to c (shown under an infinitesimal microscope for $c = 0$ and $c = 100$). The numbers infinitely close to 0 are the infinitesimals (Keisler, 1976, p. 25). Reproduced by permission of H. Jerome Keisler.

of Leibniz and his successors, and a rather more lenient treatment of those who preferred the doctrine of limits than was warranted:

For over half a century now, accounts of the history of the Differential and Integral Calculus have been based on the belief that even though the idea of a number system containing infinitely small and infinitely large elements might be consistent, it is useless for the development of Mathematical Analysis. In consequence, there is in the writings of this period a noticeable contrast between the severity with which the ideas of Leibniz and his successors are treated and the leniency accorded to the lapses of the early proponents of the doctrine of limits (Robinson, 1966, p. 260).

Having established a theory of certain non-archimedean fields containing infinitesimals in a rigorous way, Robinson offered brief reassessments of how the history of the calculus might be "redrawn":

Since (as we believe) we have shown that the theory of certain types of non-archimedean fields can indeed make a positive contribution to classical Analysis, it seems appropriate to conclude our work with a number of remarks which

(a) (b) (c)

Figure 11.11 (a) Robinson (1966). (b) Abraham Robinson. (c) Euler (1748).

attempt to indicate in what ways the picture sketched above should be supplemented or even redrawn. Our comments will by necessity be fragmentary since a complete history of the Calculus would be beyond the scope of this book (Robinson, 1966, p. 261).

Nevertheless, in a series of brief vignettes, Robinson suggested how his own work might be used to reinterpret and thereby rehabilitate work done by such of his predecessors using infinitesimals as Leibniz, l'Hôpital, Lagrange, d'Alembert, and Cauchy. Interestingly, he did not include Leonhard Euler, although Robinson had borrowed the frontispiece from Euler's *Introductio in Analysin Infinitorum* (1748) for his own introduction to *Non-standard Analysis* (Figure 11.11(a)).

Robinson was careful to emphasize that his rational reconstruction of a truly infinitesimal calculus should not be misconstrued to affirm the actual existence of infinitesimals:

Returning now to the theory of this book, we observe that it is presented, naturally, within the framework of contemporary Mathematics, and thus appears to affirm the existence of all sorts of infinitary entities. However, from a formalist point of view we may look at our theory syntactically and may consider that what we have done is to introduce new deductive procedures rather than new mathematical entities (Robinson, 1966, p. 282).

Nevertheless, he was happy to conclude by quoting his former teacher at the Hebrew University, Abraham Fraenkel, that perhaps one day a new "Cantor" would appear – someone who would make infinitesimals both respectable and useful in applications – as indeed Robinson himself had done. According to Fraenkel, the test of the need for infinitesimals was their applicability to the differential and integral calculus, but in this they had failed completely:

Bei dieser Probe hat aber das Unendlichkleine restlos versagt .... Gewiss wäre es an sich denkbar (wenn auch aus guten Gründen äusserst unwahrscheinlich und jedenfalls beim heutigen Stand der Wissenschaft in ungreifbarer Ferne liegend), das ein zweiter Cantor dereinst eine einwandfreie arithmetische Begründung neuer unendlichkleiner Zahlen gäbe, die sich als mathematisch brauchbar erwiesen und inrerseits vielleicht einen einfachen Zugang zur Infinitesimalrechnung eröffnen könnten (Robinson, 1966, p. 279).

However, the infinitely small has failed this test completely ... Certainly it would be conceivable (although, for good reasons, extremely unlikely, and in any case far into the future from now given the present state of knowledge), that a second Cantor would someday give a correct arithmetical justification of new, infinitesimal numbers which would prove mathematically useful, and perhaps, on the other hand, could provide easy access to the calculus.

Here Fraenkel's intuition about the future was as prescient as Robinson's was clear about the usefulness of nonstandard analysis in justifying intuitions that went back to Newton and Leibniz, who initiated modern interest in infinitesimals but could not provide the rigorous justification that Robinson had at last achieved. As Imre Lakatos put it, Robinson's nonstandard analysis raises the theory of infinitesimals "from the status of pre-scientific gibberish and renews interest in its partly forgotten, partly falsified history" (Lakatos, 1978, vol. 2, p. 44). Robinson believed that not only was his approach a natural one, given the many mathematicians who had relied on their intuitions to posit the usefulness of infinitesimals for mathematics, it also served to demonstrate why those intuitions had not misled his predecessors into making grievous errors in using infinitesimals despite their inability to secure the foundations that Robinson at last provided. This, in Jardine's parlance, is a perfectly acceptable sort of anachronism whereby just as Putnam was able to justify the essential correctness of Peirce's intuition about how infinitesimals served to "glue together" the points of a line to make it continuous, so too Robinson succeeded in showing the reasonableness of "redrawing" the early history of the calculus to reinstate past views that, cast in the light of nonstandard analysis, could be seen more clearly. In these cases at least, an anachronistic explanation nevertheless served to clarify, not confound, what had confused earlier defenders of theories based on infinitesimals like the calculus.

11.3 Anachronisms and ancient Chinese mathematics

In contrast to the extremes of infinitely large transfinite numbers and infinitesimals, as well as examples of anachronistic treatments of recent mathematics in the West, anachronism is just as problematic, perhaps even more so, with respect to the interpretation and understanding of mathematics in ancient China. Consider the following examples in the mathematician's "tool kit," as Niccolò Guicciardini has referred to the various methods mathematicians use in the course of their day-to-day activities. The three examples discussed in what follows are all related to ancient Chinese mathematics involving the third-century mathematician 劉徽 Liu Hui (fl. 263 CE).

11.3.1 Ancient Chinese surveying methods: the double-difference method

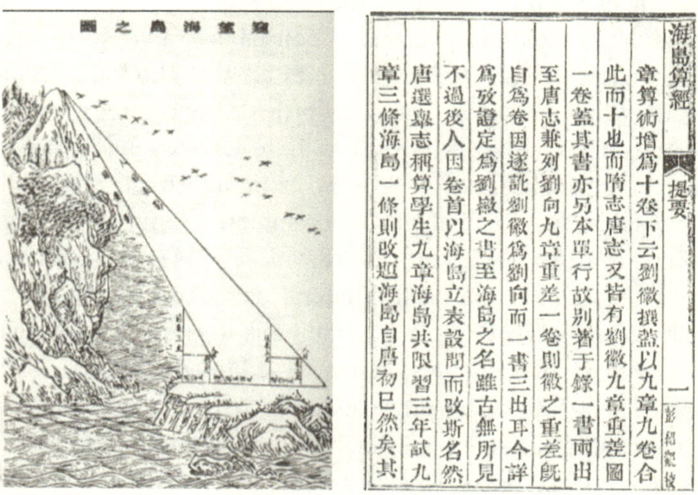

(a) (b)

Figure 11.12 (a) *Sea Island Mathematical Manual*. (b) *HDSJ*.

劉徽 Liu Hui is best known as the commentator on the ancient classic text, 九章算術 *Jiuzhang suanshu* (Nine Chapters on the Art of Mathematics), but he was also the author of another brief work, one that he wrote in 263 CE and designed to serve as a companion to the *Nine Chapters*. That work, known as the《海島算經》*Haidao suanjing* (Sea Island Mathematical Manual) (*HDSJ*), was written, Liu Hui says, because an important mathematical method was *missing* from the received version of the *Nine Chapters*. What Liu Hui added was a new chapter devoted to the method of double differences, basically a surveying technique to determine – according to the first problem in the book – the height of a distant island that cannot be measured directly (Figure 11.12(a,b)). The work itself was lost at an early date, and is known from a version hand-copied into the Ming dynasty encyclopedia, 永樂大典 *Yongle dadian* (Yongle encyclopaedia) (1403–1408), and from another hand-copied version in the Qing encyclopedic work, 四庫全書 *Siku quanshu* (Complete Library of the Four Branches of Literature) (1772–1778).

Somewhat later, another version edited by the mathematician 李黃 Li Huang (1746–1812) was published in 1820: 海岛算經細草圖說 *Haidao suanjing xicao tushuo* (Detailed solutions [of the problems] of the *Sea Island Mathematical Manual* with figures and explanations). The first problem in all editions is to ascertain the height of a far-off island using a technique known as the "double-difference method." The basic idea is easy to follow from the diagram (11.14) featuring two poles of known height (AG, EK), which are erected a known distance apart (KG). The top tips of the two poles are aligned by observation with the top of the island. From the other known quantity, the difference between the distances of the two points of observation on the ground from their respective poles (BC), one is asked to determine the height of the island.

Later mathematicians of the Qing Dynasty discussed the problems of the *HDSJ*, including 李鏐 Li Liu (1839–1901) in his 海岛算經緯筆 *Haidao suanjing weibi* (Main points of the *Sea Island Mathematical Manual*) and 沈欽裴 Shen Qinpei (fl. 1807–1823), 重差圖說 *Zhongcha tushuo* (Figures and explanations for the double-difference [method]). In 1920 the Belgian Jesuit priest Louis van Hee (1873–1951) published a translation of the *Sea Island* treatise in *T'oung Pao* (the oldest Western journal devoted to Sinology) (Figures 11.13(a)–(c)). Van Hee included a diagram (Figure 11.13(c)) from which he points out, as did the earlier Chinese commentators already mentioned, that the solution follows directly from the similarity of the triangles, two of which are oblique. I'll get to the anachronistic nature of this remark, and the reason for it, in a moment.

Van Hee published a more detailed version of the *Haidao suanjing* in *Quellen und Studien* more than a decade later, in 1932. Meanwhile, the eminent historian of Chinese mathematics 李儼 Li Yan (1892–1963),

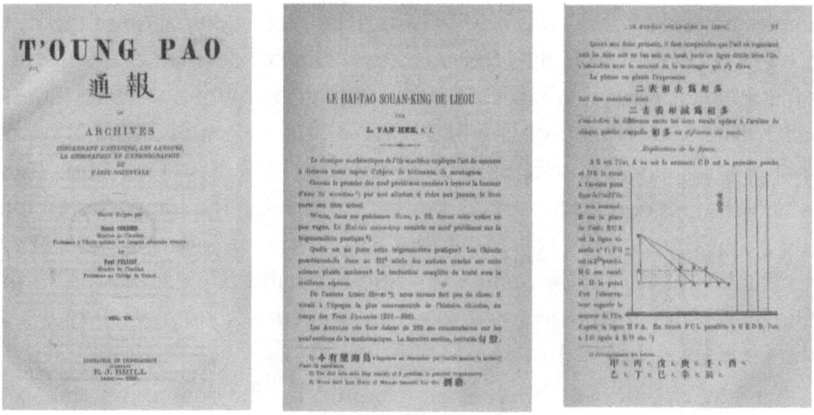

(a) (b) (c)

Figure 11.13 (a) *T'oung Pao*. (b) and (c), pages from van Hee (1920).

had published his own commentary on the double-difference method (Li Yan, 1926), followed by a similar discussion of the method in *A Concise History of Ancient Chinese Mathematics* written with his student, 杜石然 Du Shiran, first published in Chinese in two volumes (Li Yan and Du Shiran, 1963/1964) and later, in 1987, as a single-volume English translation by John Crossley and Anthony W.-C. Lun. The English translation includes a diagram with the same sort of explanation of the problem as given by van Hee.

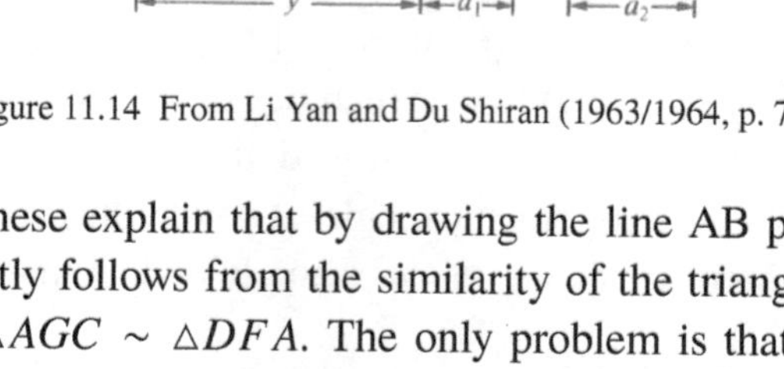

Figure 11.14 From Li Yan and Du Shiran (1963/1964, p. 76).

As all of these explain that by drawing the line AB parallel to EH, the result neatly follows from the similarity of the triangles, $\triangle ABC \sim \triangle DEA$ and $\triangle AGC \sim \triangle DFA$. The only problem is that the concepts of parallel lines and oblique triangles were *not* subjects of traditional, ancient Chinese mathematics.

The famous Chinese mathematician 吴文俊 Wu Wenjun (Wu Wen-tsun) (1919–2017) explained this first in a Chinese article (Wu Wenjun, 1977) (translated into English as "The out–in complementary principle" in 1983) wherein he pointed out that all previous interpretations of the double-difference method were wholly anachronistic because they violated basic premises of ancient Chinese mathematics, which he sought to restore. At virtually the same time, another Chinese historian, 白尚恕 Bai Shangshu (1921–1995), reiterated the old interpretation of the *chong-cha* double-difference method, and so far as I know, he was the last to do so (Bai Shangshu, 1982).

In essence, what Wu Wenjun pointed out was that the solution of the problems of the ancient text of the *Haidao suanjing* could be explained simply and straightforwardly without reference to parallel lines or similarity of oblique triangles, but on a transparent application of a basic principle of ancient Chinese mathematics, what he called the "out–in complementary principle." This amounts to the following.

How can one prove that the two areas A and B of a given rectangle are equivalent, equal in area?

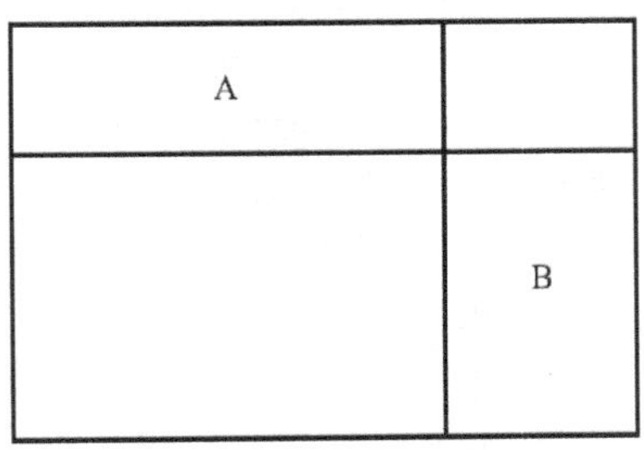

Figure 11.15.

The two areas will obviously be equal if they lie on the diagonal of the rectangle as follows: see Figure 11.16.

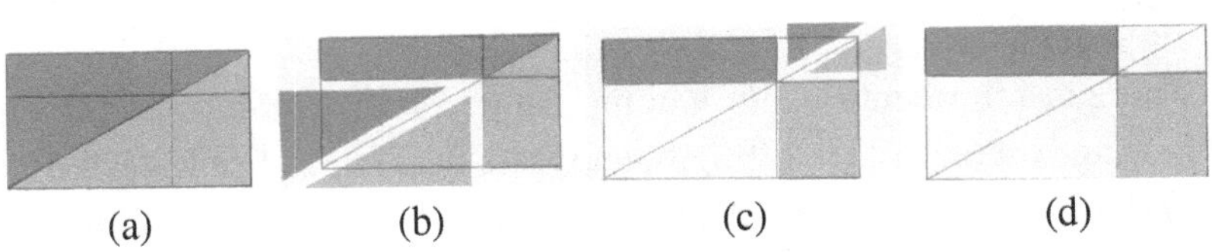

Figure 11.16.

From Figure 11.16(a) it is clear that the diagonal divides the rectangle into two equal areas, one below the diagonal, the other above. By subtracting the equal areas below and above the diagonal as in parts (b) and (c) of the figure, what remains by subtracting equal areas from equal areas is the result, namely that the remaining two areas, A and B, must also be equal, QED: A = B (part (d) of the figure). In essence this is a "proof without words," in so far as one can immediately see, visually from the diagrams, the correctness of the reasoning here.

The same principle can then be used in numerous applications, for example to prove the *gou-gu* property of right triangles (the Chinese equivalent of the "Pythagorean" theorem, to use another anachronism here in terminology). Again we proceed following the idea of the out-in complementary principle, but using a diagram from the Song dynasty version of the ancient classic text, 周髀算經 *Zhoubi suanjing* (Mathematical classic of the Zhou gnomon) (ZBSJ), traditionally used to explain the claim that given any right triangle, the sum of the areas (Figure 11.17(a)) on either side of the right angle subtending the hypotenuse is equal in area to the square on the hypotenuse itself (Figure 11.17(b)).

In this case, in the spirit of the out-in complementary principle, the portions of the dark and light gray squares lying "outside" the hypotenuse square are considered to be "out," whereas those portions coinciding with the hypotenuse square are considered to be "in" (Figure 11.17(a)). By simply moving the "outside" areas to equivalent places "inside" the hypotenuse area (Figure 11.17(b)), we again have a "proof without words" of the *gou-gu* or "Pythagorean theorem."

As Liu Hui says, those parts of the diagram "inside" don't move, while those "outside" can be moved inside, whereby the correctness of the assertion that the sum of the two areas on the sides of the right triangle subtending the hypotenuse, namely the two squares in Figure 11.17(a), can be suitably rearranged to comprise the area of the square on the hypotenuse itself, as in Figure 11.17(b), exactly the transition depicted in Figure 11.18(a–e). Voilá – the out–in complementary method has again done its job.

There is a thirteenth-century commentary on the *Haidao suanjing* by 楊輝 Yang Hui (ca. 1238–1298) that appears in Chapter 1 of his 算法通變本末 *Suanfa tongbian benmo* (Alpha and Omega of Variations on Methods of Computation, 1274) entitled 乘除通變算寶 *Chengchu tongbian suanbao* (Precious Reckoner for Variations of Multiplication and Division). There he notes how obscure and abstruse the double-difference method was – as presented in the *Haidao suanjing*:

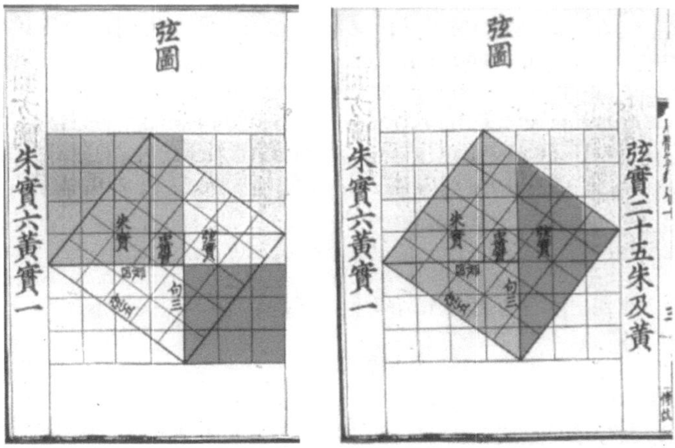

(a) (b)

Figure 11.17.

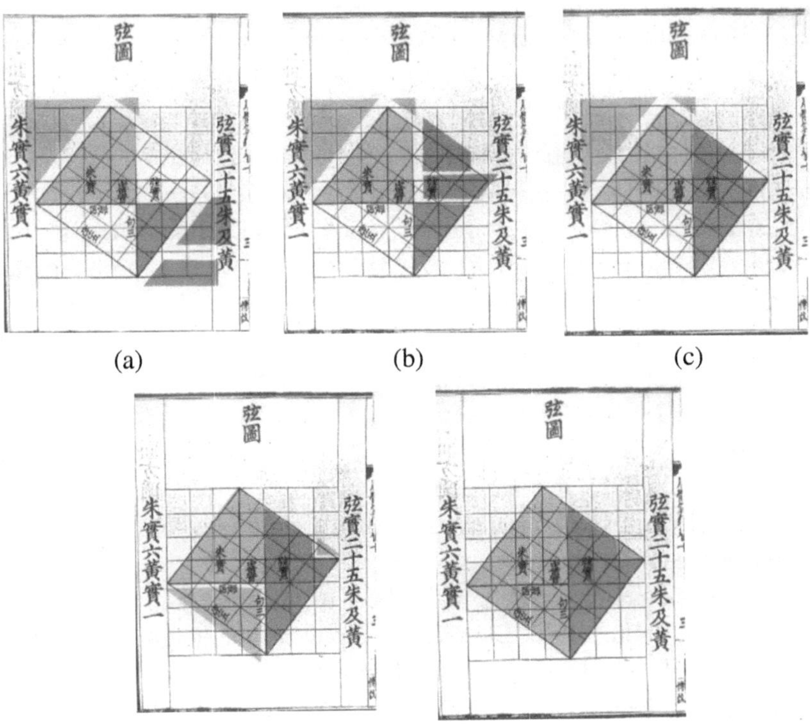

(a) (b) (c)

(d) (e)

Figure 11.18.

> . . . the setting of the methods and problems of the *Hai Tao* [HDSJ] is obscure and extremely abstruse. None could unravel its mystery. Although Li Shun-fêng [Li Chunfeng] commented on it, he only laid down the methods and did not explain their origins, while the *I Ku Kên Yüan* [*Yigu genyuan*] does not contain detailed explanations of the working. Even after following the techniques of his mathematical exercises, one still does not understand their purport (Lam Lay Yong, 1977, p. 14). (Lam follows the Wade–Giles system of transliterating Chinese characters; pinyin equivalents are supplied in square brackets.)

But with the help of a diagram he's made, Yang Hui says that he has figured it out:

> Actually, this is an extension of the *kou ku* [*gougu*] chapter of the *Chiu Chang* [*Jiuzhang* (*suanshu*)]. This book dates back for more than a thousand years. Li Shun-fêng of the T'ang dynasty supplemented this book with the 'working', but no one has yet stated nor explained clearly the purport of the method. The author has placed the small diagram on the island in the sea before him, and came to understand a little of the method employed by his predecessors. If the complete method is handed down, is not the secret purport being slighted? And if it is not handed down, then there is nothing to further the good work of his predecessors (Lam Lay Yong, 1977, pp. 179–80).

The double-difference method also appears in an even earlier work than the *Haidao suanjing*, namely in the 周髀算經 *Zhoubi suanjing* (*ZBSJ*), the earliest of the *Ten Classics* of ancient Chinese mathematics. In that text there is a received diagram. The dating of this work is problematic, but according to the commentary on the text by the third-century mathematician 趙爽 Zhao Shuang, it is possible to reconstruct the diagram and provide a proof based on the method of double differences. As Christopher Cullen (1996, p. 219) notes in his translation and commentary on the *Zhoubi*, "Once again it is clear that a Chinese mathematician of this era did not think in terms identical to our own. Whereas we would automatically use similar triangles in such a situation, Zhao reaches an equivalent result by different means."

Wu Wenjun explains exactly what these means were, and does so with reference to Figure 11.20. ED and GF represent the heights of the two gnomons, DH and FI the lengths of their respective shadows cast by the sun at A. The height of the sun AB may now easily be determined by appeals to the double-difference method, by observing the equivalence of the following areas due to the diagonals AGI and AEH.

Applying the out–in complementary method, this argument requires no appeals to parallel lines or oblique triangles. This is very similar, in

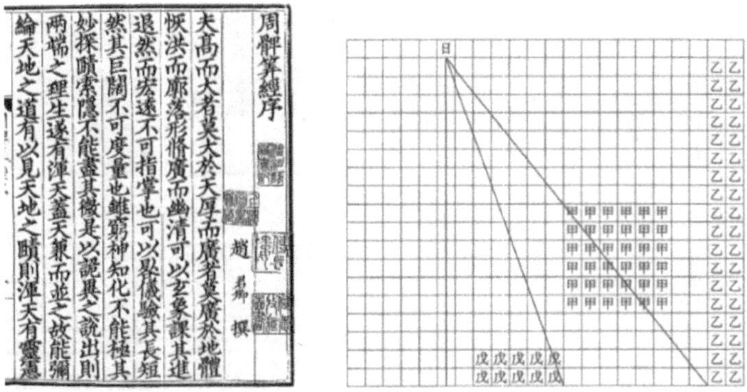

(a) 周髀算經 *Zhoubi suanjing* (b) 日高圖 *ri gao tu*

Figure 11.19 (a) Mathematical Classic of the Zhou Gnomon. (b) Height of the Sun diagram (Guo Shuchun and Liu Dun, 2001, p. 73).

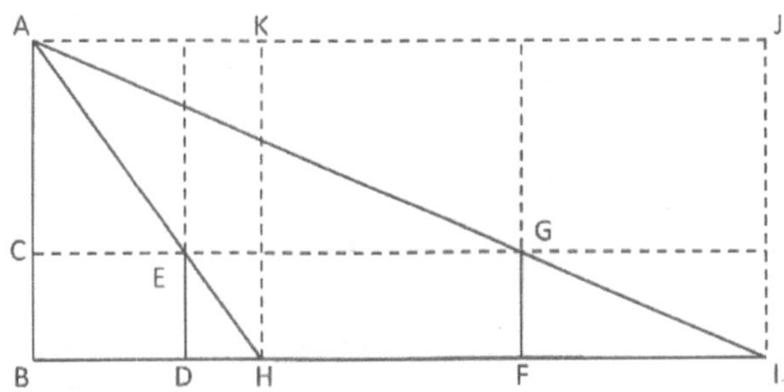

□JG = □GB;
□KE = □EB.
Therefore □JG − □KE = □GB − □EB.
Since □JG = AC × FI and □KE = AC × DH,
then (FI − DH)(AC) = ED × DF, and thus
AC = (ED × DF)/(FI − DH).

Figure 11.20.

fact, to the diagram and approach taken by Yang Hui, who also discusses the *Haidao suanjing*, with diagrams clearly referring in exactly these same terms to the out–in complementary method in his 續古摘奇算法 *Xugu zhaiqi suanfa* (Continuation of ancient mathematical methods for elucidating the strange properties of numbers) (1275); see Figure 11.21.

Yang Hui's work was subsequently lost, and therefore was unknown to writers of the Ming and early Qing dynasties. It was rediscovered by Mikami Yoshio 三上義夫 (1875–1950) thanks to a copy made in

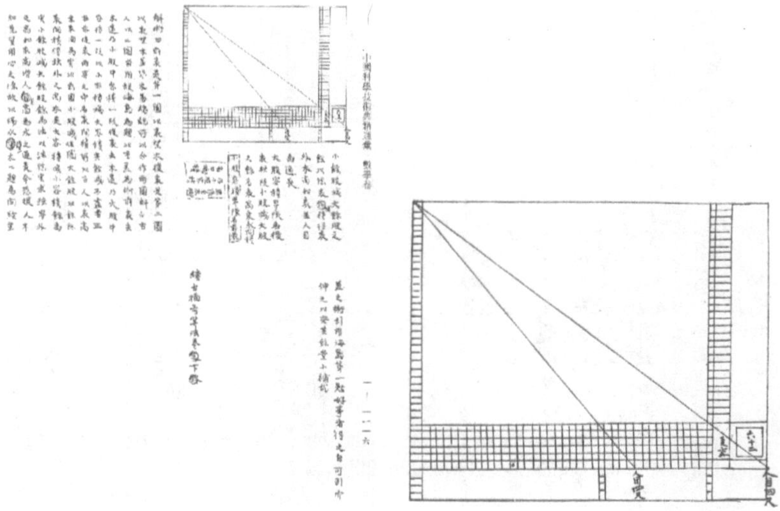

(a) Li Yan's copy, *YHSF* (b) Figure enlarged

Figure 11.21 Li Yan's copy of the *YHSF*, from Guo Shuchun (1993, vol. 1, 1116).

1662 by the Japanese mathematician 関孝和 Seki Kōwa (Seki Takakazu) (1642–1708), who copied it from an edition reprinted in Korea in 1433, of which the Chinese historian of mathematics Li Yan also obtained a copy. Another copy of the Korean reprint of the *Yanghui suanfa*, purchased by a Chinese diplomat to Japan, 楊守敬 Yang Shou-Jing (1839–1915), is now in the Beijing Library (Lam Lay Yong, 1977, pp. xiii and xv–xvii). The diagrams are of considerable interest here because they serve to corroborate Wu Wenjun's rehabilitation of the ancient basis upon which the correctness of the double-difference method was established using the out–in complementary principle.

As Wu Wenjun pointed out as early as 1977:

> If done in the usual manner according to Euclidean geometry, an auxiliary line GM′ should naturally be drawn parallel to AH to make the proving plain, as shown in the diagram on the right (see Figure 11.22). The rest can then be proved by making use of the similar triangles and the theory of proportion. In fact the proving of the formula has been so traced by historians of mathematics in China and elsewhere in recent times, including Li Huang of the Qing Dynasty (1644–1911). But his is surely not the original method of Liu Hui; *it is in fact totally out of accord with the spirit of ancient Chinese geometry* (emphasis added) (Wu Wenjun, 1977, p. 70).

Wu Wenjun (1982a, p. 18) also saw a serious anachronistic flaw in

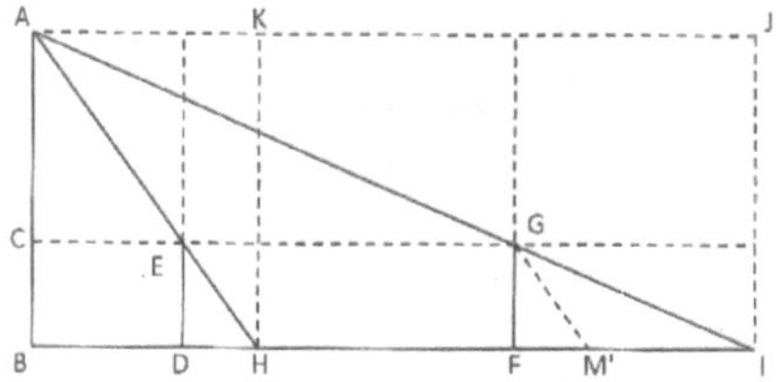

Figure 11.22 Diagram illustrating the "Western" approach to proving the double-difference rule by appeals to parallel lines and similar triangles.

treating ancient mathematics on terms only accepted much later. He was explicit about the problem of ascribing a knowledge of the calculus of the seventeenth century to Archimedes:

> It causes the glorious ancient achievements to vanish without trace. This is just as when post-seventeenth century calculus, a modern weapon, is used to prove Archimedes' formula about the area of a parabolic arc. This is of course very easy to do today, it is a simple exercise even for those who have just studied a little calculus. But if we use [calculus] to demonstrate that Archimedes' formula and theorem were correct, it is not only completely insignificant, it even turns history upside down and devalues the major contribution of Archimedes (quoted from Hudeček, 2014, p. 122).

The same could be said of the ancient Chinese surveying technique whereby the height of a distant island might be found using the double-difference method. To invoke parallel lines and use appeals to similar triangles to prove the correctness of the method introduces methods foreign to the spirit of ancient Chinese mathematics, and thereby not only obscures but totally misrepresents the ways in which ancient Chinese mathematicians thought about what they were doing. The out-in complementary method was a powerful, flexible, and transparent means of proving results involving areas and their equivalents, and the inspiration of the method, as well as its many applications, are wholly unappreciated if one uses foreign concepts to "explain" the results Liu Hui obtained in the *Haidao suanjing*.

11.3.2 Chinese algorithms for calculating square roots

A second problem in our "tool kit" from ancient China concerns a problem that appears in one of the earliest mathematical documents only recently discovered (excavated in 1983–1984), the 算數書 *Suan shu shu* (A book on numbers and computations). This is a work written on bamboo strips from the early Han dynasty, ca. 186 BCE. One of the problems asks for the determination of the square root of a given number, in this case the side of a square given its area, see Figure 11.25.

Problem 53: 方田 ***Fang Tian*****: square fields** (Given) a field of 1 *mu*, how many (square) *bu* are there? (The answer) says: 15 15/31 (square) *bu*. The method says: a square 15 *bu* (on each side) is deficient by 15 (square) *bu*; a square of 16 *bu* (on each side) is in excess by 16 (square) *bu*. (The method) says: combine the excess and deficiency as the divisor; (taking) the deficiency numerator multiplied by the excess denominator and the excess numerator times the deficiency denominator, combine them as the dividend. Repeat this, as in the "method of finding the width" (Dauben, 2008, p. 152).

Figure 11.23 Cover image (left) of Peng Hao (2001) reproduced by permission of Peng Hao and China Science Press, Beijing, PRC.

This problem is solved in the *Suan shu shu* by the method of excess and deficiency, which gives a fairly good approximation. Given that a field of one *mu* is equivalent to a field of 240 square *bu* (standard measures at the time the *Suan shu shu* was written), the "method" of the problem outlines how it would have been solved by placing counting rods on a counting board. The procedure begins by trial and error. If the numbers placed in the top row on a counting board represent the first two closest guesses for the side of a square approximating the area of 240 square *bu*, a square 15 *bu* on a side will fall short by 15 square *bu* (240 – 225 = 15), whereas a square 16 *bu* on a side will be in excess of one square *mu* by 16 square *bu* (256 – 240 = 16). These are then referred to as the deficiency and excess numerators and denominators, respectively, and would have appeared as follows on the counting board:

deficiency numerator	不足子	*bu zu zi*	15	16	赢子	*ying zi*	excess numerator
deficiency denominator	不足母	*bu zu mu*	15	16	赢母	*ying mu*	excess denominator

The method by which this problem is solved is explained later in the

Suan shu shu, in Problem 65, "Finding the Length," where according to the method of excess and deficiency used there, the following would be computed: $((15 \times 16) + (16 \times 15))/(16 + 15) = 480/31 = 15\ 15/31$ *bu*, as called for in the solution to this problem (Dauben, 2008, pp. 152–3; 162–3).

By the time Liu Hui wrote his commentary on the *Nine Chapters* some 500 years later, he approached this same problem of determining the square root of a given number quite differently:

> [Problem 4.12]: Suppose there is a [square] area of 55,225 *bu*. The question: What is the [length of the] side? The answer: 235 *bu* (Dauben *et al.*, 2013, Vol. 2, p. 373).

In the *Nine Chapters*, the method for finding square roots does not proceed as in the *Suan shu shu* by an approximation based upon a direct application of the method of excess and deficiency, but has been highly developed into an algorithm with its own technical designation as the 開方術 *Kai fang shu* or "Method of Finding the Square Root" (lit. "Method of Opening the Square" (Qian Baocong, 1963, vol. 1, p. 150)). This latter method allows determination of the square root by an iterative procedure based upon successive approximations by completing squares. The earliest surviving diagram from the *Yongle dadian* (1403–1408) capturing the essence of the method employed in the *Nine Chapters* reflects the algorithm that is applied, repeatedly, until either it comes to an end or one has reached whatever degree of exactness one may wish to determine the square root: see Figure 11.24.

Without going into all of the details, the geometry motivating the method makes clear the procedure involved. Given a square area of the sort depicted in the diagram, the first step is to determine a first approximation. Since the given area is 55,225 square *bu*, the side of the square designated here as 甲 *jia* (Figure 11.25) must lie somewhere between 200 *bu* and 300 *bu*, since:

$$200^2 = 40,000 < 55,225 < 90,000 = 300^2.$$

This in turn makes clear what the second approximation must be, once the first approximation (the *jia* square) is subtracted from the given area, i.e. $55,225 - 40,000 = 15,225$ square *bu*.

What remains is the gnomon surrounding the *jia* area, namely the area in white in Figure 11.25, after subtracting the 40,000 square *bu*

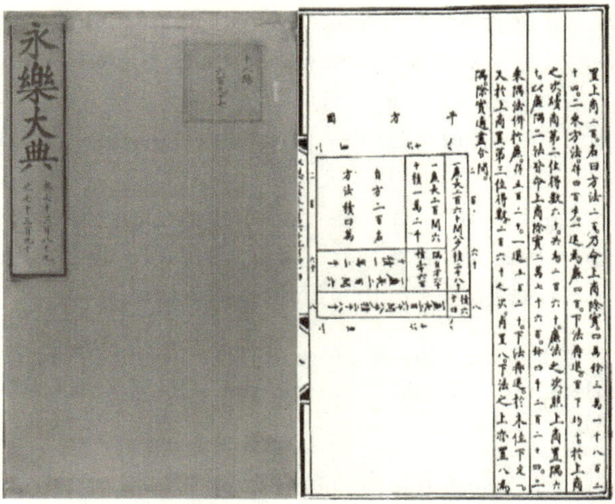

Figure 11.24 Oldest surviving diagram showing the derivation of the square root extraction method, from the *Yongle dadian* (Lam Lay Yong and Ang Tian Se, 2004, p. 76). This diagram, explaining the algorithm for extracting square roots, is from Yang Hui's 詳解九章算法 *Xiangjie jiuzhang suanfa* (1261), and appears in Chapter 15,344 of the 永樂大典 *Yongle dadian* (1403–1408). For details, see Lam Lay Yong and Ang Tian Se (2004, pp. 106–7).

representing the *jia* area. At the second step in this approximation of the square root, it is necessary to inspect the area of this remaining gnomon of 15,225 square *bu*. Consider now the two rectangular areas on either side of the *jia* square in Figure 11.26(b) plus the square completing this gnomon and designated by 乙 *yi* below in Figure 11.26(c). These must constitute an area that does not exceed 15,225 square *bu*, and by inspection this turns out to be 30 *bu* (40 *bu* being too large). Since each side of the two rectangular areas is determined by the length of the *jia* square, each of the two must be equal in area to (200)(30) square *bu*, and the square *yi* completing this gnomon is $30^2 = 900$ square *bu*. Thus in all this gnomon has a total area of $2(200)(30) + (30^2) = 12,900$ square *bu*. Subtracting this amount from the 15,225 square *bu* previously remaining leaves yet another gnomon of 2325 square *bu*, the white portion of the original square remaining in Figure 11.26(c). The algorithm continues, this time to determine the side of the last remaining smallest square at the bottom right of Figure 11.26(d). Here, the remaining white gnomon in Figure 11.26(c) is comprised of two rectangles of length 230, and widths equal to the side of the smallest light gray square in Figure 11.26(d). Upon

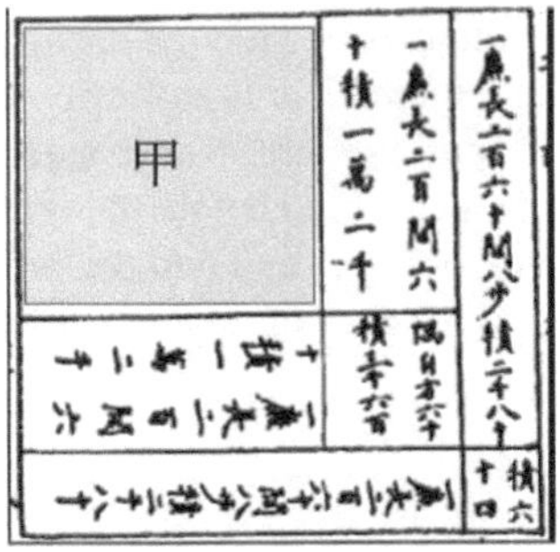

Figure 11.25.

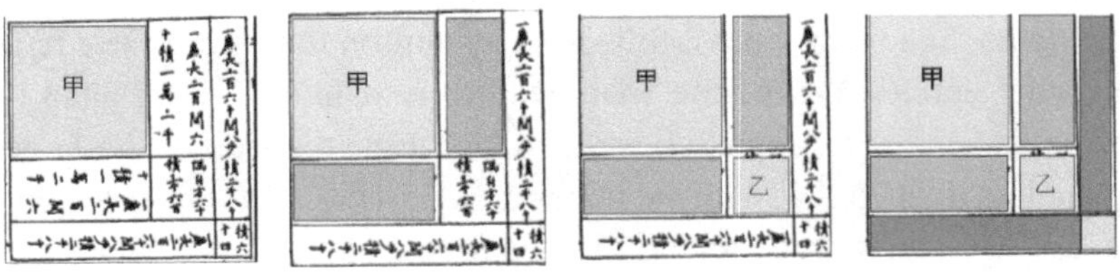

(a) (b) (c) (d)

Figure 11.26.

inspection, if the length of the side of this last smallest square is 5 *bu*, this will determine the area of this last gnomon as $2(230)(5) + 25^2 = 2325$ square *bu*. Clearly this means the end of this iterative process, having reached the final result exactly, namely the square root of 55,225 square *bu* is 235 *bu*.

Not all cases in reality turn out so neatly. What happens if one begins with an area that is not an exact square? Liu Hui addressed this possibility in a comment that appears in his remarks on Problem 4.16 in the *Nine Chapters*:

> If extracting the root does not come to an end, then the root cannot be extracted [evenly], and the side should be used to represent the root [当以面命之 *dang yi mian ming zhi*] (Dauben *et al.*, 2013, vol. 2, p. 397).

But what exactly does this mean? One eighteenth-century commentator, 李黄 Li Huang (1746–1812), in a commentary on the *Sea Island Mathematical Manual*, 海岛算經細草圖說 *Haidao suanjing xicao tushuo* (Detailed solutions [of the problems] of the *Hiadao suanjing* with figures and explanations) (1896), explained the meaning as follows:

What the sentence means is to use the side [of the square area which has been cut] as the denominator, the remainder area as the numerator. For instance, there is an area in 10 [square] *bu*, cut a square, whose side is 3 *bu*, and take it away, with the remainder area 1 [square] *bu*. Denote the remainder as $\frac{1}{3}$ *bu*. Multiply the denominator [by the side opened] and add the product into the numerator, which gives back its original number of area (Li, 1820/1896; quoted from Xu Yibao, 2005, p. 68).

In the case of $\sqrt{10}$, according to Li this meant one should take the side of the (largest) square (less than 10), namely 3, to be the denominator, with the remainder as the numerator. Thus for $\sqrt{10}$, the square root should be taken as $3 + \frac{1}{3}$. Mikami Yoshio and Qian Baocong both adopted this reading of the text, although this gives a solution that is a bit too large. But what about areas whose roots lead to remainders *larger* than the opened side? For example, consider $\sqrt{75}$, where the side of the largest square less than 75 is 8, with a remainder of $75 - 64 = 11$. On Li Huang's reading, the square root should then be $8\frac{11}{8}$, which is not even close to what the square root should be. This led 李儼 Li Yan (1892–1963) in his 中國數學大綱 *Zhongguo shuxue dagang* (Outline of the History of Mathematics in China) (Li Yan, 1958, p. 103) to simply say that the remainder in this case is 11, with no further explanation. In so doing, Li Yan relies on a passage in the *Zhoubi suanjing*, where 有奇 *you ji* (there is something more) was used to represent the concept of remainder. Also, in the 夏侯楊算經 *Xiahou Yang suanjing* (Xiahou Yang's Mathematical Classic), another ancient work comprising the "Ten Classics," Li Yan (1958, p. 103) finds another example where the square root of 52,290 is simply expressed explicitly as 723 with a remainder (奇 *ji*) of 171. As my former student and colleague 徐义保 Xu Yibao (1965–2013) expressed this situation in his PhD thesis devoted to *Concepts of Infinity in Chinese Mathematics*: "In view of these examples, Li Yan suggests that '*yi mian ming zhi*' has the same meaning as '*you ji*' or '*ji*.' This interpretation avoids using a fraction, and accordingly does not face the problems we have just discussed with Li Huang's understanding" (Xu Yibao, 2005, p. 71).

The Chinese historian of mathematics 李继闵 Li Jimin offers another reading of the phrase "*yi mian ming zhi*" and argues that it should be taken as tantamount in meaning to asserting the existence of irrational numbers. He first made this suggestion in a contribution entitled 《九章算術》中的比率理論 *Jiuzhang suanshu zhong de bilü lilun* (Theory of Rates in the *Nine Chapters on the Art of Mathematics*) published in a collection of studies devoted to the *Nine Chapters* and Liu Hui's commentary edited by Wu Wenjun (Li Jimin, 1982) and then again somewhat later in his own study of Liu Hu《九章算术》及其刘徽注研究 *Jiuzhang suanshu* jiqi Liu Hui zhu yanjiu (Study of the *Nine Chapters on the Art of Mathematics* and the Commentary by Liu Hui) (Li Jimin, 1990). Subsequently, the Taiwanese mathematician and member of the Academia Sinica, 李國偉 Lih Kowei, has offered a very different take on all this:

> This statement [that ancient Chinese mathematicians had an understanding of irrational numbers] represents some people's preferences, highlighting the highly optimistic evaluation of ancient Chinese calculations. We know that "irrational numbers" in the West began with the discovery of incommensurable magnitudes by the Greeks. However, after more than two thousand years of its historical development, it was not until the nineteenth century that the true foundations of mathematics were found. An over-estimation of the nature of China's ancient calculations may lead people to neglect to understand this history (Lih Kowei, 1994, pp. 49–54).

Lih Kowei argues that without a theory of magnitudes and proof of the impossibility of expressing certain magnitudes as incommensurable (as the side to the diagonal of a square), it cannot be said that ancient Chinese mathematics introduced a concept of irrational numbers.

Xu Yibao continues Lih Kowei's line of reasoning – that there is never any demonstration in ancient Chinese mathematics that there exist entities that cannot be expressed as ratios of numbers:

> For Chinese mathematicians, all entities such as lines, areas, and volumes can be measured out according to various units, and therefore can be represented by a combination of whole numbers. This fundamental belief in no way disposed them to discover incommensurability or irrational numbers. This would have required the realization that there were some entities that could not be described as a ratio of whole numbers, but no such statements are to be found in any of the currently-known treatises or commentaries on ancient Chinese mathematics (Xu Yibao, 2005, pp. 88-9).

Nevertheless, Chinese mathematicians were adept in offering approximations to incommensurable ratios like that of the circumference and diameter of the circle:

> On the other hand, Chinese mathematicians had little difficulty in providing fine approximations for the ratio of the circumference of a circle to its diameter or of the square root of a non-square number. It was the lack of the concept of geometrical magnitude which prevented them from confronting the logical

dilemma that first confounded the Pythagoreans, and in turn led to the discovery of incommensurable magnitudes and their arithmetic equivalent, irrational numbers. But ancient Chinese mathematicians never asked the question which plagued the Greeks about commensurability, and were never forced to confront the fact that there was no unit common to both the side and diagonal of the square (Xu Yibao, 2005, p. 89).

If Chinese mathematicians never confronted issues like the incommensurability of the side and diagonal of the square, then suggesting that ancient Chinese mathematics included the concept of irrational numbers is simply anachronistic. It reads our own understanding of a matter that so consumed the Greeks back into another past time and place, so that it seems almost unthinkable that Chinese mathematicians could not have been troubled by similar concerns that suggested, in some respects at least, mathematics was not as perfect or precise as one might wish.

11.3.3 The Chinese 勾股 **Gou-Gu** *(Pythagorean) theorem: proofs and diagrams*

One last example from the tool kit of mathematics remains to be considered here, at least briefly, raising again the question of anachronism in the history of mathematics. For the history of Chinese mathematics, publication of Volume 3 of *Science and Civilisation in China* (*SCC*), written by Joseph Needham (1900–1995) with the help of his Chinese collaborator, 王鈴 Wang Ling (1918–1994), marked a turning point. Needham met Wang Ling in China during World War II, when Needham was based in Chongqing from 1942 to 1946 as director of the Sino-British Science Co-operation Office (British Council Scientific Office in China). He subsequently arranged for Wang Ling to go to Trinity College, Cambridge, where he received his PhD in 1956 with a thesis devoted to a study of the *Nine Chapters* and history of mathematics in the Han Dynasty. Meanwhile, he was also working with Needham to produce the earliest volumes of *SCC*. One of the most controversial elements of Volume 3 was the diagram from the *Zhoubi suanjing* and a purported proof that it was said to have given of the *gou-gu* (Pythagorean) theorem.

At the time, an American graduate student, Arnold Koslow, was in King's College, Cambridge, doing a PhD in Philosophy. Koslow had previously studied Chinese at Columbia University in New York City, and had been put in touch with Joseph Needham, who invited Koslow to

(a)

(b)

Figure 11.27 (a) Wang Ling (far left) and, next to him, Arnold Koslow at the VIIIth International Congress of the History of Science, Excursion to Vinci, Italy, September, 1956. Photo courtesy of Arnold Koslow. (b) Wang Ling and Needham working in one of Needham's rooms at Gonville and Caius College, Cambridge. Photo reproduced courtesy of the Needham Research Institute (NRI), Cambridge, UK.

proofread Volume 2 of *SCC*. Needham was impressed by the corrections Koslow had suggested, and invited Koslow to collaborate further in reading the page proofs for Volume 3 of *SCC* when they began to arrive from Cambridge University Press. Needham also invited Koslow to accompany him and Wang Ling to the VIIIth International Congress for History of Science when it was held in Italy in 1956 (Dauben, 2018, pp. 113–20).

However, as Christopher Cullen (1996) notes in his study of the *Zhoubi suanjing*, the diagram Needham offered for the right-triangle theorem in *SCC* Volume 3 was clearly wrong. The proof appears as part of a dialogue where the master explains what some have taken to be a proof of the Pythagorean relation. Figure 11.28(b) is the diagram presumably reflecting Shang Gao's argument in a translation of the relevant passage in the *Zhoubi suanjing* that Needham credits to "A. Koslow." But the discrepancy between this diagram and the text is immediately obvious. The diagram specifically identifies the central inner square as yellow (中黄實 *zhong huang shi*) and also refers to a vermillion area (left of the central yellow square) (朱實 *zhu shi*).

However, the translated text mentions neither a central inner yellow square nor a vermillion area to the left of the central yellow square:

(4) Thus, let us cut a rectangle (diagonally), and make the width (*kou*) 3 (units) wide, and the length (*ku*) 4 (units) long. The diagonal (*ching*) between the (two) corners will then be 5 (units) long. Now after drawing a square on this diagonal, circumscribe it by half-rectangles like that which has been left outside, so as to form a (square) plate. Thus the (four) outer half-rectangles of width 3, length

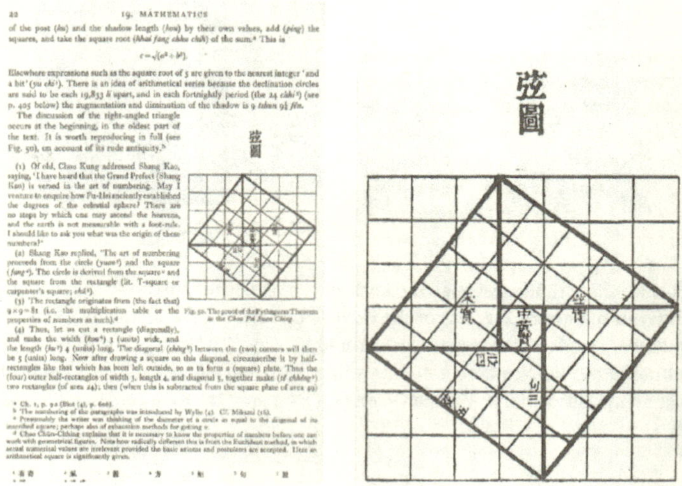

(a) (b)

Figure 11.28 (a) The page in question from the SSC. (b) The "hypotenuse diagram" (弦圖 *xiantu*), enlarged. Reproduced courtesy of the NRI.

4, and diagonal 5, together make (*tê chhêng*) two rectangles (of area 24); then (when this is subtracted from the square plate of area 49) the remainder (*chang*) is of area 25. This (process) is called "piling up the rectangles" (*chi chü*) (Needham, 1959, pp. 22–3).

Koslow, based solely on the Shang Gao dialogue, reconstructed the diagram so that it would better fit the passage, and with his reconstructed diagram, it is clear how neatly it served to demonstrate the equivalence of the *gou-gu* areas (Figure 11.29(b), "Figure 4") with the hypotenuse area (as in Figure 11.29(c), "Figure 5"). Figure 11.29(a) is Koslow's original page of reconstructed diagrams that he worked out with his new translation and then submitted to Needham in 1954. Needham and Wang Ling both agreed that what Koslow had suggested was better than their original translation.

Subsequently, Christopher Cullen, in an article written on the occasion of the International Congress of Mathematicians held in Beijing in 2002, revisited this matter and came to the startling conclusion that Koslow had invented the proof and misread the original text:

This version, due to Arnold Koslow, is quoted in Joseph Needham, *Science and Civilisation in China* (Cambridge, 1959) vol. 3, pp. 22–3. I offer this translation as representative of the lengths to which one may be tempted to go

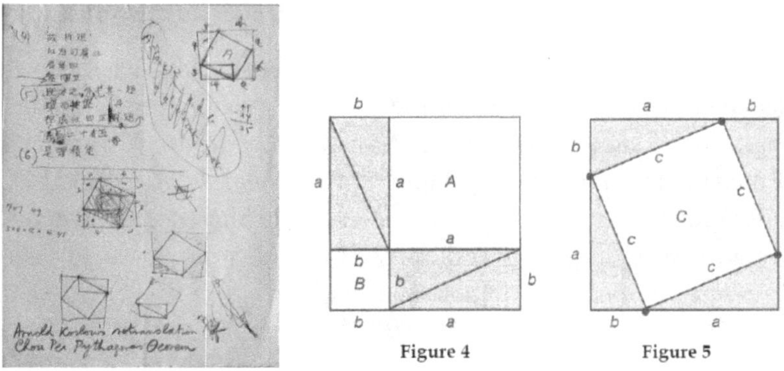

(a) (b) (c)

Figure 11.29 (a) Arnold Koslow's original sketches of the "proof" of the *gou-gu* theorem, as he reconstructed the diagram based on his revised translation of the original text, Archives of the NRI, Cambridge, UK. NRI*SCC*2/65/2/8, courtesy of the NRI. (b) and (c) Reconstructed diagrams showing equivalence of the areas *A* and *B* on the two legs of the right triangle ("Figure 4"), and the square *C* on the hypotenuse ("Figure 5") (Koslow, 2018, p. 82).

if one is determined to translate this text as if it was intended to convey a proof. I deliberately do not cite other examples (of which there are plenty, some by scholars whose other work I respect a great deal), and obviously each of them must be judged on its own merits. But I am regretfully convinced that all such efforts amount to making hamburgers without any ground beef to put inside the bun. One may admire the ingenuity of the attempt while declining to eat the result (Cullen, 2002, p. 786).

Arnold Koslow, a noted philosopher of science, is a colleague of mine at the Graduate Center of the City University of New York, in the PhD Program in Philosophy. It was therefore natural that I should ask him about this, and in reply he sent me the following explanation of what had happened. Although Needham accepted Koslow's translation, he said there was not enough money to change the diagram. Since Volume 3 of SSC was already in page proof, it could not be helped. In short, "New Translation, Wrong Diagram." As Arnold Koslow summarizes the situation as it actually happened:

And so, the new translation was printed with the old diagram (Volume 3 appeared in 1959). None of us were happy about this compromise, but Wang Ling liked the translation, Joseph was extraordinarily generous in his praise, and it would appear in this magnificent volume. I felt at the time that people could easily reconstruct the diagram from the translation (Koslow, 2018, p. 74).

In this case, anachronism has resulted not from a misunderstood word or technical term drawing upon modern understanding to "explain" a foreign, much earlier piece of mathematics, but from a diagram that appeared in the Song dynasty printing of the *Zhoubi suanjing* in 1213 CE and subsequently used by Joseph Needham to illustrate an argument provided several millennia earlier by Shang Gao purported to justify the *gou-gu* theorem. The "hypotenuse diagram" itself was based on the third-century commentary on the *ZBSJ* by Zhao Shuang, who refers to a "hypotenuse diagram" that he in turn used to clarify the meaning of the argument made earlier by Shang Gao. This is doubly anachronistic, not only because the passages attributed to Shang Gao were written centuries earlier, probably in the Western Han dynasty, but also because Shang Gao's references in the text (as Koslow noticed) bore no relation to the diagram, which did reflect Zhao Shuang's explanation of the *gou-gu* theorem, but not Shang Gao's.[5] Thus we have a clear case of anachronism on several levels for which the diagram later printed in the Southern Song text (and later versions) of the *Zhoubi suanjing* bore no relation to the text as published by Needham and Wang in SSC Volume 3.

11.4 Conclusion: hedgehogs and foxes

As the British historian Isaiah Berlin (1909–1997) famously categorized historians as either hedgehogs or foxes, to this point we have been playing the fox, examining a variety of examples across a broad historical spectrum of different times and cultures. It is now time to be the hedgehog, and zero in on the problem that serves as a leitmotif for all of the essays in this volume of Bacon Symposium lectures – anachronism in the history of mathematics. I want to argue here that anachronism can be both constructive or confusing, or even downright contradictory. In cases where it can clarify our understanding it may be taken to be a constructive intrusion in an historical analysis of things past; in others, it can in fact play an opposite, destructive role in distorting completely our proper understanding of the past.

From the examples I've given, I think it is fair to conclude that some

[5] For details concerning the problematic dating of texts and diagrams related to the *ZBSJ*, see Qu Anjing (1997) as well as the discussion in Cullen (1996, pp. 172–3).

uses of anachronisms are commensurable with past mathematical practice or thinking, as with Hilary Putnam's rational reconstruction of the Peircean continuum with respect to nonstandard analysis. On the other hand, there are uses of anachronism that are clearly *in*-commensurable with past mathematical thought. Using Cantorian alephs, for example, to represent Peirce's thought regarding the infinite does not help to clarify but only confuses.

Other concepts are so complex, with multiple meanings, that even the translation of a single Chinese character, for example, is fraught with anachronistic traps for the unwary. Take the case of 率 *lü*, a fundamental concept of ancient Chinese mathematics that appears in numerous, often quite different contexts. 率 *lü*, which in the *Nine Chapters* is used to express among other things the relation of the diameter and circumference of a circle, is often translated as "ratio," "rate," or "proportion." To avoid any anachronistic connotations using a term with very specific meanings in English for *lü*, Jean-Claude Martzloff (1997, p. 196) translates *lü* as "model," and explains: "we wish to highlight the fact that this term relates to certain types of numbers defined for use as a norm, model or standard." The most exhaustive treatments of the problem of translating *lü* are those provided by Karine Chemla throughout her commentary on the French translation of the *Nine Chapters*. In particular, in her glossary of terms, she devotes four pages to this subject (Chemla and Guo, 2004, pp. 956–9). She takes this up elsewhere as well, in particular in her introduction to the first chapter (Chemla and Guo, 2004, pp. 135–6). She also enumerates and analyzes the various ways in which the concept is used, and this is probably why she did not opt for any one translation for *lü*. See also Guo Shuchun (1984) and Li Jimin (1982). In the course of her translation of the *Nine Chapters* into French, rather than translate the word, she simply lets *lü* stand for itself as the transliteration of the character 率.

Anachronism arises in another way, graphically, in the interpretations various authors have offered to prove the validity of the double-difference method in the *Haidao suanjing* by introducing diagrams and arguments wholly foreign to ancient Chinese mathematics. Likewise, the assertion that ancient Chinese mathematicians had discovered the concept of irrational numbers, as Li Jimin argued in wanting to read their understanding of incommensurability into the Chinese realization that roots for some numbers could not be extracted evenly, simply contradicts the facts. As

Lih Kowei – a prominent Taiwanese mathematician and member of the Academia Sinica – underscored, without a theory of magnitude and an actual proof or any plausible argument that there exist some magnitudes – like the side and diagonal of a square or the diameter and circumference of a circle that cannot be expressed as a ratio of whole numbers – there is no basis for asserting an appreciation in ancient Chinese mathematics of either incommensurable or irrational entities.

Likewise, using an unsuitable diagram from a later work to illustrate an earlier argument, as in the case of the *Zhoubi suanjing* and the proof it presents of the *gou-gu* theorem, was anachronistic in making use of a pre-existing diagram, which may have been convenient, but only served to obscure the thought underlying the original argument establishing the *gou-gu* result presented by Zhao Shuang.

Like Nick Jardine (2000), who sought to find a place for anachronism in historical writing, especially in the history of science, so too more recently Naomi Oreskes (2013) defends her support of presentism in the history of science, but includes anachronism in her list of sins historians are often accused of committing. She then asks a pertinent question:

> Virtually all professional historians reject inevitability: we know that there were times when other options seemed possible and desirable, and those who desired them were not necessarily knaves or fools. So presentism is conflated, if not equated, with Whiggism, triumphalism, moralism, teleology, anachronism, and the general belief that the present is superior to the past. Presentism is bad history. But need it be? Is it possible to be a good presentist? (Oreskes, 2013, p. 602).

Oreskes draws heavily on examples from the history of geology and echoes a distinction first made by the historian of geology Stephen J. Gould between substantive and methodological uniformitarianism. Oreskes modifies this to consider the distinction between substantive versus methodological presentism. As she points out:

> Historians have a problem that is in some ways the inverse of uniformitarianism in geology: the problem of presentism. We might call it anti-uniformitarianism. For just as geologists have been taught to trust the present as an explanatory resource, historians have been taught to suspect it... Most important, we tend to believe that over-reliance on comparisons with the present will lead us to fail to understand the past on its own terms (Oreskes, 2013, p. 599).

Perhaps here the history of mathematics is sufficiently different, ontologically and epistemologically, from the other sciences that it deserves special attention. In fact, it is the only example among the sciences of a body of knowledge that is not intrinsically, systematically subject to falsification on Popperian terms, or to the anomalies that doom the physical and life sciences to endless cycles of normal science supplanted by revolutionary new theories that prove incommensurable in various ways with previous knowledge and attempts to explain it. This is not to say that mathematics does not experience revolutions, but these are not revolutions that either falsify its past or lead to the rejection of entire bodies of previous knowledge.[6] When revolutions occur in mathematics, they encompass the well-established knowledge of theorems, proofs, and algorithms that drive the everyday work of the mathematician. As Hermann Hankel (1869, p. 34) noted: "In most sciences one generation tears down what another has built and what one has established another undoes. In mathematics alone each generation adds a new storey to the old structure" (quoted from Dauben, 1984, p. 49).

In this sense, the history of mathematics deals with a subject whose past is always present, and presentism and anachronism may therefore be less problematic than for the other sciences rife with clearly discredited theories wherein the language of present theories is clearly "incommensurable" with the concepts and language used to account for natural phenomena in the past.

Oreskes (2013, p. 602) makes an important point about the problem with amateur accounts of the history of science, or those given by scientists themselves, that she notes are "often whiggish to an extreme, looking for the predecessors who 'got it right' and helped to build the path to our current truths." This is what Ivor Grattan-Guinness has so colorfully described as the temptation to which mathematicians usually fall prey in writing about fields to which they may have made major contributions themselves, as simply recounting what they see (with emphasis added) as "the royal road to *me*" (Grattan-Guinness, 1990, p. 157). As he went on to explain: in the process, this results in "an account of how a particular modern theory arose out of older theories instead of an account of those older theories in their own right" (Grattan-Guinness, 1990, p. 157).

Euclid, according to Proclus, when asked if there was any easy way

6 For a contrary view, see the well-known arguments advanced by Michael Crowe (1975) in his article, "Ten 'Laws' Concerning Patterns of Change in the History of Mathematics." For more recent discussions of the Crowe–Dauben debate, see, among others, Corry (1993), François and Van Bendegem (2010), and Sialaros (2018).

to learn geometry, replied that there was no shortcut, no "royal road" (Proclus, as translated by Morrow, 1970, p. 57). Often, as in the examples we've explored here, anachronistic reconstructions or explanations of mathematics only lead us down the wrong roads if what we are seeking is the understanding of what past mathematicians themselves intended and managed to accomplish on their own terms. Sometimes an anachronistic idea from the present can help to illuminate and clarify the past – but unless judiciously chosen and deftly used, more often than not, anachronisms in mathematics are simply intrusive spoilers that prevent us from appreciating the usually winding roads that lead to the present, often in wonderfully surprising and remarkable ways. It is those true "Eureka" moments that we most hope to capture, as historians of mathematics, as best and as authentically as we can.

References

Bai, Shangshu 白尚恕 (1982). 劉徽《海島算經》造術的探討 Liu Hui *Haidao suanjing* zao shu de tantao (An Exploration of Liu Hui's *Haidao suanjing*), 科技史文集 *Keji shi wenji* (*Annals of the History of Science and Technology*) (Shanghai) **8**, 79–87.

Bernstein, Richard (1970). Paul Weiss's recollection of editing the Peirce Papers. *Transactions of the Charles S. Peirce Society* **6** (3–4) (Summer–Fall), 161–183.

Butterfield, Herbert (1965). *The Whig Interpretation of History*. New York: W.W. Norton & Company.

Cantor, Georg (1874). Über eine Eigenschaft des Inbegriffes aller reellen algebraischen Zahlen. *Journal für die reine und angewandte Mathematik* **77**, 258–262.

Cantor, Georg (1887/1888). Mitteilungen zur Lehre vom Transfiniten. *Zeitschrift für Philosophie und philosophische Kritik* **91**, 81–125 and 240–265.

Cantor, Georg (1895/1897). Beiträge zur Begründung der transfiniten Mengenlehre. *Mathematische Annalen* **46**(4), 481–512; **49**(2), 207–246.

Cantor, Georg (1932). *Gesammelte Abhandlungen Mathematischen und Philosophischen Inhalts*. E. Zermelo (ed). Berlin: J. Springer. Reprinted Hildesheim: Olms, 1966; reprinted Berlin: Springer, 2013.

Chemla, Karine, and Guo Shuchun (2004). *Les Neuf Chapitres: Le Classique Mathématique de la Chine Ancienne et ses Commentaires*. Paris: Dunod.

Corry, Leo (1993). Kuhnian issues, scientific revolutions and the history of mathematics. *Studies in History and Philosophy of Science* **24**(1), 95–117.

Crowe, Michael (1975). Ten 'laws' concerning patterns of change in the history of mathematics. *Historia Mathematica* **2**, 161–166. Reprinted in Gillies (1992, pp. 15–20).

Cullen, Christopher (1996). *Astronomy and Mathematics in Ancient China: the Zhou bi suan jing*. Needham Research Institute Studies **1**. Cambridge, UK: Cambridge University Press.

Cullen, Christopher (2002). Learning from Liu Hui? A different way to do mathematics. *Notices of the AMS* **49**(7) (August), 783–790.

Dauben, Joseph W. (1974). Denumerability and dimension: The origins of Georg Cantor's theory of sets. *Rete* **2**, 105–134.

Dauben, Joseph W. (1977). C.S. Peirce's philosophy of infinite sets. *Mathematics Magazine* **50**(3) (May), 123–135.

Dauben, Joseph W. (1978). Georg Cantor: The personal matrix of his mathematics. *Isis* **69**, 534–550.

Dauben, Joseph W. (1979). *Georg Cantor: His Mathematics and Philosophy of the Infinite*. Cambridge, MA: Harvard University Press. Reprinted with a paperback edition by Princeton University Press, 1990. Chinese translation by Zheng Yu-xin, Nanjing: Nanjing University Press, 2008.

Dauben, Joseph W. (1981). Peirce's critique of Cantorian set theory. In *Proceedings of the C.S. Peirce Bicentennial International Congress, Amsterdam, The Netherlands*. Lubbock, Texas: Texas Tech University Press, pp. 93–98.

Dauben, Joseph W. (1984). Conceptual revolutions and the history of mathematics. In *Transformations and Tradition in the Sciences*, Everett Mendelsohn (ed). Cambridge, UK: Cambridge University Press. Reprinted in (and quoted from) Gillies (1992).

Dauben, Joseph W. (1995). *Abraham Robinson, The Creation of Nonstandard Analysis: A Personal and Mathematical Odyssey*. Princeton: Princeton University Press. Paperback edition, 1998; Chinese translation by Wang Qian, Beijing: Science Press, 2005.

Dauben, Joseph W. (1996). Charles S. Peirce: Evolutionary pragmatism and the history of science. *Centaurus* **38**, 22–82.

Dauben, Joseph W. (2008). 算數書 Suan shu shu (A Book on Numbers and Computations). English Translation with Commentary. *Archive for History of Exact Sciences* **62**(2), 91–178.

Dauben, Joseph W. (2018). The history of Chinese mathematics – the founding of the IHNS (CAS) – and the true story of a wrong diagram in SCC Volume 3. *Chinese Annals of History of Science and Technology* **2**(2), 86–128.

Dauben, Joseph W., Guo Shuchun, and Xu Yibao (2013). 九章筭術 *Nine Chapters on the Art of Mathematics: A Critical Edition and English Translation based upon a New Collation of the Ancient Text and Modern Chinese Translation*. In Three Volumes. (Library of Chinese Classics.) Shenyang: Liaoning Education Press.

Dedekind, Richard (1872). *Stetigkeit und die Irrationale Zahlen*. Braunschweig: Vieweg.

Euler, Leonhard (1748). *Introductio in Analysin Infinitorum*. Lausanne: M.-M. Bousquet.

François, Karen, and Jean Paul Van Bendegem (2010). Revolutions in mathematics, more than thirty years after Crowe's 'Ten Laws': A new interpretation. In *Philosophy of Mathematics: Sociological Aspects and Mathematical Practice*, Benedikt Löwe and Thomas Müller (eds). London: College Publications (Texts in Philosophy 11), pp. 107–120.

Franklin, Christine Ladd (1916). Charles S. Peirce at the Johns Hopkins. *The Journal of Philosophy, Psychology and Scientific Methods* **13**(26), 715–722.

Gillies, Donald, editor (1992). *Revolutions in Mathematics*. Oxford: Oxford University Press.

Grattan-Guinness, Ivor (1990). Does the history of science treat of the history of science? The case of mathematics. *History of Science* **28**, 149–173.

Guicciardini, Niccolò (2018). *Isaac Newton and Natural Philosophy*. London: Reaktion Books.

Guo, Shuchun 郭书春 (1984).《九章算术》和刘徽注中之率概念及其应用试析 *Jiuzhang suanshu* he Liu Hui zhu zhong zhi lü gainian jiqi yingyong shixi (An Analysis of the Concept of lü and its Applications in the *Nine Chapters on the Art of Mathematics* and Liu Hui's Commentary), 科学史集刊 *Kexueshi jikan* (*Collected Papers of History of Science*) **11**, 21–36.

Guo, Shuchun 郭书春, editor (1993). 中国科学技术典籍通汇：数学卷 *Zhongguo kexue jishu dianji tonghui: Shuxue juan* (*A Compendium of Chinese Classics of Science and Technology: Mathematics Section*). Zhenzhou: Henan jiaoyu chubanshe.

Guo, Shuchun 郭书春, and Liu Dun 刘钝, editors (2001). 算經十書 *Suanjing shishu* (*Ten Books of Mathematical Classics*). Beijing: Jiuzhang chubanshe.

Hankel, Hermann (1869). *Die Entwickelung der Mathematik in den letzten Jahrhunderten*. Antrittsvorlesungen. Tübingen: Fues'sche Sortimentsbuchhandlung.

Houser, Nathan (1992). The fortunes and misfortunes of the Peirce Papers. In *Signs of Humanity, vol. 3*, Michel Balat, Janice Deledalle-Rhodes, and Gérard Deledalle (eds). Berlin: Mouton de Gruyter, pp. 1259–1268.

Hudeček, Jiři (2014). *Reviving Ancient Chinese Mathematics: Mathematics, History and Politics in the Work of Wu Wen-tsun*. London: Routledge.

Jardine, Nicholas (2000). Uses and abuses of anachronism in the history of the sciences. *History of Science* **38**, 251–270.

Keisler, H. Jerome (1976). *Elementary Calculus: An Infinitesimal Approach*. Boston: Prindle, Weber, and Schmidt. Revised 1986, 2012.

Koslow, Arnold (2018). The Chinese version of the Pythagorean theorem and the case of the missing diagram: A memoir. *Chinese Annals of History of Science and Technology* **2**(2), 72–85.

Lakatos, Imre (1978). Cauchy and the continuum: the significance of non-standard analysis for the history and philosophy of mathematics (1966). In *Mathematics, Science and Epistemology (Philosophical Papers, Volume 2)*, John Worrall and Gregory Currie (eds). Cambridge, UK: Cambridge University Press, pp. 43–60.

Lam, Lay Yong 蓝丽蓉 (1977). *A Critical Study of the Yang Hui Suan Fa. A Thirteenth-century Chinese Mathematical Treatise*. 楊輝算法. Singapore: Singapore University Press.

Lam, Lay Yong 蓝丽蓉 and Ang Tian Se 洪天赐 (2004). *Fleeting Footsteps. Tracing the Conception of Arithmetic and Algebra in Ancient China*, revised edition. Singapore: World Scientific. (First edition, 1992.)

Li, Jimin 李继闵 (1982). 《九章算術》中的比率理論 *Jiuzhang suanshu* zhong de bilü lilun (Theory of Rates in the *Nine Chapters on the Art of Mathematics*). In 《九章算術》與劉徽 *Jiuzhang suanshu* zhong yu Liu Hui (The *Nine Chapters on the Art of Mathematics* and Liu Hui). Wu Wenjun (吳文俊), ed. Beijing: Beijing shifan daxue chubanshe, 1982: 228–45.

Li, Jimin 李继闵 (1990).《九章算术》及其刘徽注研究 *Jiuzhang suanshu* jiqi Liu Hui zhu yanjiu (*Study of the* Nine Chapters on the Art of Mathematics *and the Commentary by Liu Hui*). Xi'an: Shaanxi kexue jishu chubanshe. Reprinted by the Jiuzhang chubanshe in Taipei, 1992.

Li, Yan 李儼 (1926). 重差術源流及其新注 Chong-cha shu yuanliu jiqi xin zhu (New Commentary and Research into the Origin of the *Chong-cha* Rule), 學藝雜誌 *Xue yi zazhi* (*Arts and Sciences Magazine*) **7**(8), 1–15.

Li, Yan 李儼 (1958). 中國數學大綱 *Zhongguo shuxue dagang* (*Outline of the History of Mathematics in China*). Beijing: Zhongguo Quinghua Chubanshe.

Li, Yan 李儼 and 杜石然 Du Shiran (1963/1964). 中国古代数学简史 *Zhongguo gudai shuxue jian shi* (*A Concise History of Ancient Chinese Mathematics*). In two volumes. Beijing: Zhonghua xhuju. Translated as *Chinese Mathematics. A Concise History* by John N. Crossley and Anthony W.-C. Lun. Oxford: Oxford University Press, 1987.

Lieb, Irwin C. (1970). Charles Hartshorne's recollection of editing the Peirce Papers. *Transactions of the Charles S. Peirce Society* **6**(3–4) (Summer–Fall), 149–159.

Lih, Kowei 李國偉 (1994). «九章算術»与不可公度 *Jiuzhang suanshu* yu buke gongdu (*Nine Chapters on the Art of Mathematics* and the Incommensurable Magnitudes). 自然辯證法通訊 *Ziran bianzhengfa tongxun* (*Journal of Dialectics of Nature*) **2**, 49–54.

Martzloff, Jean-Claude (1997). *A History of Chinese Mathematics*. Stephen S. Wilson, translator. Berlin: Springer. Revised edition, 2004.

Morrow, Glenn R. (1970). *Proclus. A Commentary on the First Book of Euclid's Elements*. Princeton, NJ: Princeton University Press.

Needham, Joseph (1959). *Science and Civilisation in China. Vol. 3, Mathematics and the Sciences of the Heavens and the Earth*, with the collaboration of Wang Ling. Cambridge, UK: Cambridge University Press.

Oreskes, Naomi (2013). Why I am a presentist. *Science in Context* **26**(4), 595–609.

Peirce, Charles S. (1881). On the logic of number. *American Journal of Mathematics* **4**, 85–95. Reprinted in Peirce (1931–1936, vol. 3, 3.252–3.288).

Peirce, Charles S. (1897). The logic of relatives. *The Monist* **7**, 161–217. Reprinted in Peirce (1931–1936, vol. 3, 3.456–3.552).

Peirce, Charles S. (1931–1936). *The Collected Papers of Charles S. Peirce*. In Six Volumes. Charles Hartshorne and Paul Weiss (eds). Cambridge, MA: The Belknap Press of Harvard University Press.

Peirce, Charles S. (1976). *The New Elements of Mathematics by Charles S. Peirce*. C. Eisele (ed). The Hague: Mouton & Co.

Peng, Hao 彭浩 (2001). 張家山漢簡《算數書》註釋 Zhangjiashan hanjian Suan shu shu zhushi (*Commentary on the* Suan shu shu *Han Bamboo Slips from Zhangjiashan*). Beijing: Kexue chubanshe.

Putnam, Hilary (1995). Peirce's continuum. In *Peirce and Contemporary Thought. Philosophical Inquiries*, Kenneth Laine Ketner (ed). New York, NY: Fordham University Press, pp. 1–22.

Qian, Baocong 錢宝琮, editor (1963). 算經十書 *Suanjing shishu* (*Ten Books of Mathematical Classics*). Beijing: Zhonghua shuju.

Qu, Anjing 曲安京 (1997). On hypotenuse diagrams in ancient China. *Centaurus* **39**, 193–210.

Robinson, Abraham (1965). On the theory of normal families. *Acta Philosophica Fennica* **18**, 159–184. Reprinted in Robinson (1979, vol. 2, 62–87).

Robinson, Abraham (1966). *Non-Standard Analysis*. Amsterdam: North-Holland Publishing Co.

Robinson, Abraham (1979). *Selected Papers of Abraham Robinson*. H.J. Keisler, S. Körner, W.A.J. Luxemburg, and A.D. Young, eds. *Volume 1: Model Theory and Algebra*; *Volume 2: Nonstandard Analysis and Philosophy*; *Volume 3: Aeronautics*. New Haven, CT: Yale University Press.

Schoenflies, Arthur (1927). Die Krisis in Cantor's mathematischem Schaffen. *Acta Mathematica* **50**, 1–23.

Sialaros, Michalis, editor (2018). *Revolutions and Continuity in Greek Mathematics*. Berlin: Walter de Gruyter.

Sinkevich, Galina I. (Галина И. Цинкевич) (2012). Георг Кантор и польская школа теории множеств (*Georg Kantor and the Polish School of Set Theory*). St. Petersburg: St. Petersburg State University of Architecture and Civil Engineering.

van Hee, Louis (1920). Le Hai-Tao Souan-King de Lieou. *T'oung Pao* **20**(1), 51–60.

van Hee, Louis (1932). Le classique de l'île maritime, ouvrage chinois du IIIe siècle. *Quellen und Studien zur Geschichte der Mathematik, Astronomie und Physik* **2**(3), 255–280.

Wu, Wenjun (Wu Wen-tsun) 吴文俊 (1977). 出站互补原则 Chuzhan hubu yuanzi (The out-in complementary principle). In 中国古代科学技术成果 *Zhongguo gudai kexuejishu chengguo* (*Achievements in Science and Technology of Ancient China*), pp. 80–100; English translation, Beijing: Foreign Language Press, 1983: pp. 66–89.

Wu, Wenjun (Wu Wen-tsun) 吴文俊 (1982a). 我國古代測望之學重差理論評介兼評數學史研究中某些方法問題 Woguo gudai ce wang zhi xue zhongcha lilun pingjie jian ping shuxue shi yanjiu zhong mou xie fangfa wenti (Comments on the ancient Chinese surveying theory of *chongcha* and comments on some methodological questions in research on the history of mathematics), 科技史文集 *Keji shi wenji* (*Annals of the History of Science and Technology*) (Shanghai) **8**, 10–30.

Wu, Wenjun (Wu Wen-tsun) 吴文俊 (1982b). 《海島算經》古證探源 *Haidao suanjing* gu zheng tan yuan (Exploring the Origins of Ancient Proofs in the *Haidao suanjing*). In 九章算術與劉徽 *Jiuzhang shuanshu* yu Liu Hui (*Liu Hui and the* Nine Chapters on the Art of Mathematics). Beijing: Beijing Normal University Press, pp. 162–180.

Xu, Yibao 徐义保 (2005). *Concepts of Infinity in Chinese Mathematics*. PhD Dissertation, The City University of New York.

Index

Aaboe, Asger Hartvig, 9, 10
Abel, Niels Henrik, 237, 238, 240
Abel-Rémusat, Jean Pierre, 52
Acta Eruditorum, 199
aethereal bodies, 120
al-Dīn al-Ṭūsī, Sharaf, 152
Alexandria, library of, 107
algebra, history of, 19, 78
algebraic curves, 19–22
algorithm, 62–64, 73, 75, 76, 79, 154
  Babylonian, 74
anachorism, 85–87, 91, 98
anachronism
  approach to Chinese language, 54, 56
  conceptual, 42, 43
  emergence of, 3, 84, 156
  in values, 56
  textual, 44, 53, 54, 61, 64, 66, 73
  translation, 11
anachronistic assessment, 49
anachronistic expectation, 71
analysis, 20, 225, 233, 234, 236, 237, 243, 244, 247
  indeterminate, 134, 135, 138, 141, 156
  rigor, 234
analytic geometry, 19, 20
analytical mechanics, 17, 198, 203, 204, 207, 208, 213, 227
anatopism, 84
anatropism, *see* anatopism
ancient texts
  commentaries on, 44, 46
anharmonic ratio, *see* cross ratio
Apollonius, 10–13
  Conics, 10
Archimedes, 108
architecture, 109
Archytas of Tarentum, 125–127, 176
Aristarchus of Samos, 108
Aristotle, 7, 115, 116, 121, 170–174
  Metaphysics, 116, 170
  Physics, 170
arithmetic, 111, 112
Arnold, Vladimir, 219, 220
artillery construction, 122, 123, 127
assignment of variables, 74, 75
astronomy, 107, 109, 111–114, 116, 120, 121
Austin, John Langshaw, 25
axiom of continuity, 300
axioms
  geometry, 253, 256, 283
  projective geometry, 266

Babylonian mathematics, 9, 10, 44, 85
Babylonian programming, 74
Bachet de Méziriac, Claude-Gaspard, 151
Bai Shangshu, 331
Barbeau, Edward, 244, 245
Barozzi, Francesco, 189
Battaglini, Giuseppe, 295
Bell, Eric Temple, 255
Beltrami, Eugenio, xxiii, 288, 297
Berlin Academy, 234
Berlin, Isaiah, 348
Bernoulli, Jacob, 17, 198, 200, 208, 213, 215
Bernoulli, Johann, xxi, 15, 17, 197, 198, 200–209, 211–215, 218, 220
Bernoulli, Nicolaus, 236
Bernoulli's equation, 201, 205, 206, 208, 218, 219
Bertrand, Joseph, 134
Bessel, Friedrich, 301
Betti, Enrico, 288
Bhāskara, 89, 90, 92, 93
Bibliothèque Royale, 46
Biernatzki, Karl, xx, 134–135, 138–157

Biot, Edouard, xviii, 44–46, 48, 50–52, 54, 56, 57, 59, 60, 62, 63, 67, 75
Biot, Jean-Baptiste, 44
Blay, Michel, 17, 197, 213, 219
Bloch, Marc, 33
Boethius, 181
Bolyai, János, xxiii, 282–285, 290, 292, 293, 295, 299–304
Bolza, Oskar
  Vorlesungen über Variationsrechnung, 228
Bonola, Roberto, xxiii, 281, 282, 284, 291–298, 300–304
  La geometria non-euclidea: esposizione storico-critica del suo sviluppo, 281
  Questioni riguardanti la geometria elementare, 286
Borel, Émile, 240
Borrel, Jean (Johannes Buteo), 183, 184
Bos, Henk, xviii, xxi, 8, 18, 19, 21–24, 26, 31, 217
Bossuet, Jacques, 3
Boyer, Carl, 92
  The Concepts of the Calculus, 92
Brahmagupta, 87, 89, 92
  Brāhmasphuṭasiddhānta, 87
Braver, Seth, 298
Bressoud, David, 92, 99
Brioschi, Francesco, 288
Brollo, Basilio, 51

calculating surface, 62–65, 67, 70, 72, 75
calculus, 86, 91, 92, 98, 99, 101, 197, 198, 233, 242, 245
  Leibnizian, 213, 217
calculus of variations, 197, 223, 224, 228, 229
California Institute of Technology, xvii
Calinger, Ronald, 240
Callet, Jean-François, 245
Callippus, 116
Cambridge school, 24
Campanus, Johannes (Giovanni Campano, Campano da Novara), 181, 182
canonical transformations, 227–229, 232
Cantor, Georg, 310–315, 320, 321
Cantor, Moritz, xx, 135, 136, 139, 141, 144, 159, 160
  Vorlesungen über Geschichte der Mathematik (Lectures on the History of Mathematics), 155
Carathéodory, Constantin, 228–231
  Variationsrechnung und Partielle Differentialgleichungen Erster Ordnung, 228
Carnot, Lazare-Nicolas-Marguerite, 265
  Géométrie de position, 265
Carrus, Sauveur, 273, 274
Carus, Paul, 315
Castelnuovo, Guido, 286, 288
Cauchy, Augustin-Louis, 2, 8, 236–238, 242, 245, 313, 326
  Cours d'analyse de l'Ecole Royale Polytechnique, 236
Cayley, Arthur, 263, 265
central force field, 199, 200, 205
central force motion, 198, 218–220
  polar coordinates, 203
central forces, 205, 210, 213, 219, 220
Cesàro, Ernesto, xxii, 234, 239, 240
Charlier, Carl Vilhelm Ludvig, 227
Charpentier, Jacques, 176
Chasles, Michel, xxii, 251, 253, 255, 256, 258, 259, 261, 263, 265, 266, 268, 273, 274
Chemla, Karine, 134, 154, 349
Chimisso, Cristina, 29, 30
Chinese remainder theorem, *see* dayan rule
chronology, 135, 156
Church, Alonzo, 309
Cicero, 114
  Academica, 169
City University of New York, 347
Clairaut, Alexis-Claude, 197
Clark, Jonathan, 5, 16
Clebsch, Rudolf Friedrich Alfred, 254, 263
Collège de France, 45, 52
Collingwood, Robin, 14
Columbia University, 345
commensurability, 119, 120
Complete Treatise, *see* Suanfa tongzong
complex analysis, 238
conic section, 85, 202, 203, 212
conservation of mechanical energy, 205
construction of equations (Cartesian theory), 21, 24, 26, 34
continuity, 297, 315, 320, 321, 324, 325
continuity, principle of, 258, 273
Cooke, Roger L., 255
  The History of Mathematics: a Brief Course, 255
Coolidge, Julian Lowell, 255
  A History of Geometrical Methods, 255
Courant, Richard, 267
Cramer, Gabriel, 21
Cremona, Luigi, 264, 265, 288
  Elementi di geometria projettiva, 264
cross ratio, 256, 258, 267, 272, 274
Crossley, John, 330
Cullen, Christopher, 334, 345, 346
Cuomo, Serafina, 108
  Ancient Mathematics, 108

d'Alembert, Jean le Rond, 197, 217, 283, 286, 313, 326

dayan rule, 135, 138–157
de Guignes, Chrétien-Louis-Joseph, 51
De Morgan, Augustus, 237, 320
de Nemore, Jordanus
  Arithmetica (De elementis arismetice artis), 181
Dedekind, Richard, 144, 147, 311, 320, 321
definitions, 241, 245, 252
demarcation
  problem of, 109
Descartes, René, 15, 19, 21–24, 31, 208, 217
  Dioptrique, 23
  Géométrie, 18–20, 23, 31, 199, 204, 215
  analytic geometry, 15
Dewey, John, 33
diagram, 108
dialectic, 113
dialectic (in Plato's philosophy), 110, 113, 115
Dickson, Leonard Eugene, 155
dictionary, 78
  Chinese, 52, 53, 57
  de Guignes, 53, 57
  Morrison, 53
Dictionary of Scientific Biography, 167
differential equations, 199–205, 210, 213, 215, 217
  ordinary, 198, 213, 217
  partial, 197, 228, 229
  theory of, 219
Diogenes Laertius, 168, 191
Dirichlet, Peter Gustav Lejeune, 141, 145, 147
disciplinary boundaries, 3, 26, 28, 33
distance, 255, 297–300, 302–304
Divakaran, P.P., 86, 87, 99, 100
domestication (in translation theory), 14, 25, 218, 220
doppelverhältnis, *see* cross-ratio
double-difference method, 329, 331, 332, 334, 349
Du Shiran, 330
duality, theory of, 258
Dutch Mathematical Society, 309
Dutton, Denis, 25

Ecole Polytechnique, 44
elementary mathematics, 284, 288, 296
Encyclopaedia Britannica, 253
Encyklopädie der Elementar-Mathematik, 291
Encyklopädie der mathematischen Wissenschaften mit Einschluß ihrer Anwendungen, 255, 270, 273, 287, 289, 290
Engel, Friedrich, 286, 289, 290
Engel, Friedrich and Paul Stäckel
  Die Theorie der Parallellinien von Euklid bis auf Gauss, 286
Enriques, Federigo, 281, 286, 288, 289, 291, 299, 304
  Prinzipien der Geometrie (Principles of Geometry), 289
Epicurus, 113, 114
  Letter to Herodotus, 113
  Letter to Pythocles, 114
epistemological values, 78
epistemology, 125, 127
Erlangen Program, 263–265, 269
ETH (Eidgenössische Technische Hochschule Zürich), 310
ethics, 121, 127
Euclid, xxi, xxiii, 10–13, 108, 113, 168, 183–185, 189–191, 238, 284, 289, 293, 299, 300, 303, 351
  Data, 10
  Elements, 10, 20, 113, 176, 178–184, 190, 281, 283, 284, 286, 288–290, 292, 293, 299, 303
Eudoxus of Cnidus, xxi, 115, 116, 168, 176, 177, 192
Euler, Leonhard, xxi, xxii, 8, 15, 21, 141, 151, 197, 212, 216, 217, 220, 223–227, 230–241, 244, 245, 247, 326
  De seriebus divergentibus, 243
  Institutiones calculi differentialis, 243
  Introductio in Analysin Infinitorum, 234, 326
  Mechanica sive motus scientia analytice exposita, 200
  Mechanica, sive motu scientia analytice exposita, 215, 216
  Methodus Inveniendi Lineas Curvas Maximi Minimive Proprietate Gaudentes, 217, 224, 225, 229–231, 233, 247
  Theoria motus corporum solidorum seu rigidorum, 217
Euler–Lagrange equation, 224, 232
existence (of mathematical entities), 242
extraction of cube root, 49, 56, 64, 65, 68, 69, 71, 73, 78
extraction of square root, 49, 54, 56, 64, 65, 67, 69, 78

Fano, Gino, 271–274, 281, 288–293, 303
Favaro, Antonio, 265
  Lezioni di statica grafica, 265
Febvre, Lucien, 2, 132
Fermat, Pierre de, 31
Ficino, Marsilio, 169, 184
  Theologia Platonica de Immortalitate Animorum (Platonic Theology), 184
fluxional equations, 213
foreignization (in translation theory), 14, 26
formalism, 235, 236, 242, 243

forms (in Plato's philosophy), 110, 115
Fourier, Jean-Baptiste Joseph, 286
fractions, history of, 43
Fraenkel, Abraham, 308, 327
Frank, Philipp, 228
Frank, Philipp and Richard von Mises
Die Differential- und Integralgleichungen der Mechanik und Physik, 228
Freig, Johannes Thomas, 190
Quaestiones geometricae et stereometricae, 190
French, Anthony Philip, 7
Frénicle de Bessy, Bernard, 31
Freudenthal, Hans, 11
Fried, Michael N., 240, 246
Friedman, Michael N., 209
Fries, Johann Jakob, 190
Bibliotheca philosophorum classicorum authorum chronologica, 190
Frobenius, Georg, xxii, 239
functional analysis, 227

Galenic medicine, 7, 9
Galilei, Galileo, 2, 198, 213
Discorsi e dimostrazione matematiche intorno a due nuove scienze attenenti alla mecanica & i movimienti locali, 7, 213
Galois, Évariste, 300
Gauss, Carl Friedrich, xx, 141, 144, 147, 150, 282–286, 289, 290, 294, 295, 300–302, 304
Disquisitiones Arithmeticae, 140, 148, 156
Werke, 286
Gaussian method, 140, 141, 151
Geertz, Clifford, 14, 15
generality, 55, 56, 62–65, 71, 74–76, 135, 159, 271–273
geometrical algebra, 2, 10–11, 13
Geometrie der Lage, 251, 252, 254, 257, 261
géométrie de situation, *see* Geometrie der Lage
geometry, 110–112, 114, 124, 127
constructions, 20
descriptive, 263, 265, 271, 289
differential, 227, 283, 289, 297, 300, 304
elementary, 271, 281, 284, 285, 288–291, 297, 299, 300, 302–304
Euclidean, 263, 283, 284, 288, 289, 291, 292, 294, 295, 299, 302
foundations of, 253, 269, 287, 303
Gaussian differential, 284, 295
higher, 281
hyperbolic, 282
modern, 261, 262
non-Euclidean, 238, 253, 255, 266, 275, 281–291, 294, 295, 297, 299–301, 303, 304
projective, 227, 252–256, 259, 261, 263, 265, 266, 268, 271–275, 284, 291
Riemannian differential, 281, 284, 300, 302, 303
synthetic, 262, 268, 271
geometry of measurement, 261
geometry of situation, *see* Geometrie der Lage
Gergonne, Joseph Diez, 258, 259
Annales des Mathématiques Pures et Appliquées, 259
Gödel, Kurt, 309
Goldstein, Catherine, 31, 151
Goldstine, Hermane Heine, 231
History of the Calculus of Variations, 231
gou-gu, 332, 345
Gould, Stephen J., 350
Grandi, Guido, 235
Grattan-Guinness, Ivor, xxiv, 6, 246, 351
Gray, Jeremy, 265
Greek mathematics, 7, 11, 168
Guicciardini, Niccolò, 282, 297
Isaac Newton and Natural Philosophy, 308
Guthrie, William Keith Chambers, 106, 107
A History of Greek Philosophy, 106

Hadot, Pierre, 107
Hall, Alfred Rupert, 27, 28
Halsted, George Bruce, 45, 274
Hamilton, William Rowan, 228
Hamilton–Jacobi theory, 227, 228, 232
Hankel, Hermann, xx, 6, 7, 135, 141, 160, 351
Zur Geschichte der Mathematik in Alterthum und Mittelalter, 135
Hardy, Godfrey Harold, 7, 238, 240, 244, 247
harmonics, 109, 111
Hartshorne, Charles, 315, 318
Harvard University, 229, 315, 321
HDSJ, *see* Liu Hui, Haidao suanjing
Hebrew University, 308, 327
Einstein Institute, 309
Heine, Heinrich Eduard, 310
Helmholtz, Hermann von, 284, 291, 296
Henrici, Olaus, 253
heritage, mathematical, 237, 246, 247
Hermann, Jacob, 17, 198, 200, 203, 215
Hero of Alexandria, 121–125, 127
Belopoeica, 123
Metrica, 123
Pneumatica, 123
Herophilus, 108
Hilbert, David, xxiii, 269, 290, 291, 299, 300, 303
Grundlagen der Geometrie, 291
Hirsch, Eric Donald

significance and meaning, 203
Historia Augusta, 185
Hogendijk, Jan Pieter, 152, 153, 158, 160
Hölder, Otto Ludwig, 239
Holgate, Thomas, 274
homogeneity, geometrical, 205, 206, 208, 214, 216, 217
Hooke, Robert, 26, 28
Houël, Guillaume-Jules, 295
Huet, Pierre-Daniel, 3
Huffman, Carl, 125
Huygens, Christiaan, 27, 198, 213, 215, 217
Horologium oscillatorium, 213, 219
hypergeometric series, 245

Iamblichus, 113
indeterminate analysis, 134, 135, 138, 141, 156
indeterminate form, 91
infinite extension principle, 235, 244
infinite series, 224, 235, 245
convergent, 234, 237–239, 247
divergent, 234, 235, 237, 238, 240, 243, 244, 247
infinitesimals, 2, 91, 95, 101, 208, 216, 308–310, 322, 324, 325, 327, 328
infinity, 243
Institute for Advanced Study, Princeton, 309
intellectual history, 24, 25
intentions, 6, 14, 25, 48, 69
interpretation
ancient texts of procedures, 78
literal, 56
of mathematical problems, 47, 56
of numerical values, 54
of rules, 47, 55–57, 59, 62
terminology, 49
texts of procedures, 44, 75
invariance (covariance), 224, 227–229, 231–233, 247
iteration, 74, 76, 78

Jacobi, Carl Gustav Jakob, 227, 228
James, William, 315
Jardine, Nicholas, 2, 30, 84, 86, 87, 91, 313, 327, 350
Jiuzhang Suanshu (The Nine Chapters on Mathematical Procedures, Nine Chapters on the Art of Mathematics), 46, 49, 58–66, 68, 70, 73, 75, 328, 339, 341
Johns Hopkins University, 315
Jones, Alexander, 111
Journal des savants, 134
Julien, Stanislas, 45
justice, 123, 125

Kant, Immanuel, 209
Katz, Victor J., 92, 98, 256
A History of Mathematics: An Introduction, 256
Keisler, Howard Jerome, 324
Keyser, Cassius Jackson, 321
Klein, Felix, 252, 254, 262, 263, 265, 266, 268–271, 273–275, 288, 290, 291, 299, 301, 303
Vorlesungen über die Entwicklung der Mathematik im 19. Jahrhundert, 267
Vergleichende Betrachtungen über neuere geometrische Forschungen, 263
Evanston Lectures (*The Evanston Colloquium: Lectures on Mathematics*), 274
Klein, Jacob, 11
Kline, Morris, xxi, 7, 238, 240, 295
Klügel, Georg Simon, 284
Knobloch, Eberhard, 237
Knopp, Konrad, 241
Knuth, Donald, xviii, 9, 45, 73–75, 79, 160
The Art of Computer Programming, 73, 74
Koselleck, Reinhart, xx, 158
Begriffsgeschichte (Conceptual History), 158
Koslow, Arnold, 345–348
Kouprianov, Alexei, 109
Kronecker, Leopold, 311, 312
Kuhn, Thomas, 6, 200
Kummer, Ernst, 310
kuttaka method, 135, 141, 148, 150

L'Hôpital, Guillaume F.A. de, 21, 313, 326
La Hire, Philippe de, 21
Ladd-Franklin, Christine, 315
Lagrange, Joseph-Louis, 197, 286, 313, 326
Lakatos, Imre, xxii, 310, 327
Lambert, Johann Heinrich, 283, 284, 286, 291, 294
Theorie der Parallellinien, 286
land distribution, 109, 123, 125
Landwehr, Achim, xx, 133, 156–158
Laplace, Pierre-Simon de, 197, 286
Le Clerc, Jean, 3–5
Leah, P. J., 244, 245
Leedham-Green, Charles, 31
Legendre, Adrien-Marie, 283, 284, 286
Leibniz, Gottfried Wilhelm, xxi, 2, 15, 89, 197–205, 208, 210, 213, 215, 216, 218, 231, 233, 235, 240, 242, 308, 309, 313, 325–327
Tentamen de motuum coelestium causis, 200
lex homogeneorum, 216, 217
calculus, 199
Leonelli, Sabina, 33
Li Huang, 329, 341, 342
Haidao suanjing xicao tushuo (*Detailed*

solutions [of the problems] of the Sea Island Mathematical Manual with figures and explanations), 329, 341
Li Jimin, 342, 343, 349, 350
"Jiuzhang suanshu" jiqi Liu Hui zhu yanjiu (*Study of the Nine chapters on the art of mathematics and the commentary by Liu Hui*), 342
Zhongguo shuxue jianshi (*A concise history of Chinese mathematics*), 342
Li Liu, 329
Haidao suanjing weibi (*Main points of the Sea Island Mathematical Manual*), 329
Li Yan, 329, 336, 341, 342
Zhongguo shuxue dagang (*Outline of the History of Mathematics in China*), 341
Li Yan and Du Shiran
A Concise History of Ancient Chinese Mathematics, 330
Libbrecht, Ulrich, 153–155
Lih Kowei, 342, 349, 350
Linacre, Thomas, 189
Littlewood, John Edensor, 7, 240
Liu Hui, 59, 60, 328–329, 332, 338, 341, 342
Haidao suanjing (*Sea Island Mathematical Manual*), 329, 331, 332, 334, 337, 349
Lloyd, Geoffrey E.R., xx, 28, 29
Lobachevskii, Nikolai Ivanovich, xxiii, 282–285, 290–304
Geometrische Untersuchungen zur Theorie der Parallellinien, 299, 302
Pangeometry, 294, 297, 298
Lotze, Rudolf Hermann, 301
Louis XI, 175
Lovejoy, Arthur O., 24
Lowenthal, David, 7
Lun, Anthony W.-C., 330

Maclaurin, Colin, 21
Mādhava, xix, 92, 98
Mādhava–Leibniz series, 92, 98
Maffei, Raffaele, 184–189
Commentaria Urbana, 184
Mahoney, Michael, 11
Marinus, 186–188
Maronne, Sébastien, 31
Martzloff, Jean-Claude, 47
Mathematical Association of America, 229
mathematical knowledge, organization of, 78
mathematical practitioners, 18, 33
Matthiessen, Heinrich Friedrich Ludwig, xx, 133, 135–161
Grundzüge der antiken und modernen Algebra der litteralen Gleichungen, 137
measurement, 253, 254, 256, 258, 259, 261, 263, 266, 271–274
mechanics, 109, 122, 123, 127
Mémoires de l'Academie Royale des Sciences, 219
meteorology, 113
Metzger, Hélène, 6, 29, 30
Meyer, Franz, 270
Mikami, Yoshio, xviii, 45, 58–78, 155, 335, 336, 341
The Development of Mathematics in China and Japan, 45, 58
Mittag-Leffler, Magnus Gustaf (Gösta), 312
Möbius, August Ferdinand, 267, 268
Mohrmann, Hans, 270
Monge, Gaspard, 265
Montucla, Jean-Étienne, 167, 168
Moore, Charles Napoleon, 234
Morrison, Robert, 51
music, 112

Nabonnand, Philippe, 260
national character, 85
Needham, Joseph, xviii, 29, 45, 68–78, 344–348
Science and Civilisation in China, 45, 344, 348
Netz, Reviel, 108
Neugebauer, Otto, 9–11, 16, 267
Newton, Isaac, xix, 15, 16, 21, 27, 31, 89, 196–197, 212, 213, 217, 219, 308, 327
Arithmetica Universalis, 24
Opticks, 197
Philosophiae Naturalis Principia Mathematica, 7, 17, 31, 197, 213, 219, 220
Newtonian mathematics, 197
Nicomachus of Gerasa, 111, 112
Introduction to Arithmetic, 112
Manual of Harmonics, 112
Nieuwentijt, Bernard, 31
Niḷā, 98
Nine Chapters, *see* Jiuzhang Suanshu
Northwestern University, 274
Nouvelles Annales de Mathématiques: journal des candidats aux écoles polytechnique et normale, 134, 263

O'Meara, Dominic, 113
Oakeshott, Michael, 246
Oberwolfach, Mathematisches Forschungsinstitut, 309
objectivity, 236
Ó Mathúna, Diarmuid, 203
Oreskes, Naomi, xxiv, 350, 351
Ostrogradskii, Mikhail Vasilyevich, 301
out–in complementary principle, 331, 335, 336

Padmanabha Rao, Anantapur, 90, 91
Pallavicino, Sforza, 3
Pappus, 20, 22
Mathematical Collection, 21
Paracelsian alchemy, 7
paradigm, 63, 66, 68–71, 76
parallel postulate, 284–286, 288–296, 299, 302
Pasch, Moritz, xxiii, 291
Peacock, George, 237
Peirce, Benjamin, 315
Peirce, Charles Sanders, xxiv, 308, 315–324, 327, 349
The Collected Papers, 315–320
Petau, Denys, 3
Philoponus, John, 113, 185
Pieri, Mario, 299
Pierre de la Ramée, *see* Ramus, Petrus
Plücker, Julius, 263, 267, 268
Plato, 7, 106, 111, 113, 115, 125, 168–172, 174–176, 191
Epinomis, 111
Meno, 126
Phaedrus, 169
Philebus, 169
Republic, 110, 111, 114, 169, 185
Sophist, 169
Theaetetus, 119
Timaeus, 111, 113, 119, 120
Academy, 107, 168, 171, 173, 175, 176, 184–188, 192
Platonism, 169, 171, 174, 191, 192
Plimpton, George Arthur, 315
Plotinus, 179, 191
pluritemporality, xx, 133, 149, 157–159, 161
Plutarch of Athens, 191
Plutarch of Chaeronea, 186
Life of Marcellus, 176
Moralia, 186
Parallel Lives, 186
Poincaré, Henri, 219, 227, 300
point-set theory, 312
polar coordinates, 211, 226, 232, 233
Polyaenus of Lampsacus, 114
Poncelet, Jean-Victor, xxii, 251–253, 255–259, 261, 263, 265–267, 269, 271–273, 275
Applications d'analyse et de géométrie qui ont servi de principal fondement au Traité des propriétés projectives des figures, 258
Traité des propriétés projectives des figures, 257, 271
Popper, Karl, 109
Porphyry, 185
Prime Mover
Aristotle, 121
Ptolemy, 121
Princeton University, 309
procedure, shu, 61
Proclus, xxi, 113, 168, 171, 174, 179, 182, 184–192, 351
Commentary on the First Book of Euclid's Elements, 113, 179, 181, 183, 189, 190
proportion theory, 208, 213
proportionality, 48, 124, 126, 177
Proust, Christine, 10
Ptolemy, Claudius, 108, 115, 117–121, 127
Almagest, 115, 117, 118, 120, 121
Geography, 117, 186
On the Kritêrion and Hêgemonikon, 117
Putnam, Hilary Whitehall, xxiv, 245, 313, 321, 323, 324, 327
Pythagoras's theorem, 50, 55
Pythagoreans, 85, 106, 173

Qian Baocong, 341
Qin Jiushao (Tsin Keu chaou), 141
Mathematical Writings in Nine Chapters (Nine Sections of the Art of Numbers), 139
quadratures, 230
quantum physics, 227

Ramus, Petrus, xx, 168–180, 189–193
Dialecticae Institutiones, 169
Prooemium Mathematicum, 168
Scholae Mathematicae, 168, 176, 177, 190
Scholae Metaphysicae, 170, 172
Scholae Physicae, 170, 172
rapport anharmonique, *see* cross ratio
Rashed, Roshdi, 153, 160
reading (of mathematical texts), history of, 43–46
receptionism, 30–32
reconstruction, 132, 133, 135, 147, 153, 155, 157, 159–161
actor's category, 153
historical, 132
mathematical, 132, 133, 138, 145, 146, 149–155, 158, 160, 161
historiographically sensitive, 153, 160
rational, 132
revolutions in mathematics, 8, 16, 351
Reye, Theodor, 261, 262
Die Geometrie der Lage, 261, 262, 274
Reyneau, Charles-René, 21
Riemann, Georg Friedrich Bernhard, xxiii, 284, 291, 293, 295, 296, 299, 300, 302, 304
spherical geometry, 290
Riemannian metric, 289
rigor, 236, 237
Robinson, Abraham, 308–310, 313, 324–327

Non-standard Analysis, 326
Roche, John James, 208, 214
Rorty, Richard, 7
Rossi, Paolo, 14, 15, 26, 33, 199
Rostock, University of, 136
Rowe, David, 11, 265
Royal College of Aeronautics, Cranfield, 309
Royce, Josiah, 315, 321
rules of false position, 60

Saccheri, Giovanni Girolamo, 283, 284, 286, 291, 293
Euclides ab omni naevo vindicatus, 286
Sagan, Hans, 233
Saito, Ken, 132
Śaṅkara, 93–96, 98
Scaliger, Joseph Justus, 3
Schlömilch, Oskar Xavier, 136
Schoenflies, Arthur Moritz, 271–274, 312
Schreckenfuchs, Erasmus Oswald, 188
Schwartz, Hermann A., 311
Schweikart, Ferdinand Karl, 284, 286, 291
Sedley, David, 114
Segre, Corrado, 288, 291
Seki Kōwa, 336
Seki Takakazu, *see* Seki Kōwa
series, summation of, 96–98, 239
set theory, 312
Severus, 174
Shang Gao, 345, 346, 348
Shen Qinpei, 329
Zhongcha tushuo (*Figures and explanations for the double-difference [method]*), 329
Siku quanshu (*Complete Library of the Four Branches of Literature*), 329
simplicity in mathematics, 21, 22
skepticism, 169
Skinner, Quentin, 24, 26
Sluse, René-François de, 21
Smith, David Eugene, 43, 155
History of Mathematics, 43
Socrates, 110, 114, 169
Sommerfeld, Arnold, 227, 228
Song Jingchang, 157
soul, 118, 121, 124
soul (in Plato's philosophy), 110, 120
Speiser, David, 203
St. Petersburg Academy, 234
Stäckel, Paul, 286, 289, 290
Staudt, Karl Georg Gustav von, xxii
Steiner, Jakob, xxii, 251, 253, 255, 256, 258, 259, 261, 265–269, 271, 272
Systematische Entwickelung der Abhängigkeit geometrischer Gestalten von einander, 267
stereometry, 111
Stieltjes, Thomas Joannes, 155
Stöffler, Johannes, 187, 188
Stoics, 122
Stolz, Otto, 262
Suan shu shu (*A book on numbers and computations*), 338, 339
Suanfa tongzong (Complete Treatise), 46
Suidas, 168
sum (according to Euler), 241, 244
summability, 224, 234, 239, 240, 242, 245, 247
Sun tsze, *see* Sunzi
Sunzi (Master Sun), 64–66, 68, 138, 141, 150
Mathematical Classic, 64–66, 68, 138, 150
Sunzi Suanjing, *see* Sunzi (Master Sun)
Syrianus, 171, 173, 174, 185–187
Syrjämäki, Sami, 29
systems of linear equations, 62, 65, 68, 70, 76, 143
Szabó, Árpád , 11

taccheda ([having] that [as] divisor), 89
Talon, Omer, 169
tasks and criteria of quality control (according to Bos), 19–24, 204
Taurinus, Franz Adolph, 284, 286, 291, 294
technical terminology, 49, 57, 67
tensor analysis, 228
Terquem, Olry, 134, 263, 265
Terrier, Paul, 265
Géométrie de position (Leçons de statique graphique. Première partie, Géométrie de position), 265
texts of procedures
ancient, 78
cuneiform, 74
interpretation, 75
textual components, 44, 79
The Gnomon of the Zhou, 50, 52, 54
theology, 115, 117, 118, 121, 127
Theon of Alexandria, 179–183, 190
Theon of Smyrna, 111–113
On Mathematics Useful for the Understanding of Plato, 111
Ting-fa (fixed divisor), 70, 72, 75
Ting-fa$_1$, 72
Ting-fa$_2$, 72
Torricelli, Evangelista, 213
tranquility, 119, 122, 123, 125
transfinite numbers, 311–313, 328
transfinite set theory, 312, 313, 321
translation, 14, 32, 34, 134–135, 140, 141, 154, 218, 220
domestication, 14, 16, 218, 220
foreignization, 14
modernizing, 17

practice of, 53
translational transformation, 211
trigonometry
elementary, 92, 93, 289
hyperbolic, 286, 290, 291, 293–296, 303
spherical, 295, 302
Truesdell, Clifford Ambrose, 197, 213, 219, 220
types of text, 77, 79

Unguru, Sabetai, 2, 11–13, 84, 132
universal mathematics, 158, 161
University of Berlin, 310–312
University of Halle, 310
University of Kiel, 136
University of London, 308
University of Paris, 168, 169, 174
University of Toronto, 309

Vacca, Giovanni, 155
van Hee, Louis, 155, 329, 330
van Schooten, Frans, 31
variational calculus, *see* calculus of variations
Varignon, Pierre, 17, 197, 198, 200, 204, 207, 213, 215, 217, 233
Veblen, Oswald, xxii, 252
Venuti, Lawrence, xxi, 14, 16, 201
Viète, François, 208, 214
virtue, 121
void, 123, 125, 127
von Humboldt, Alexander, 137, 313
Kosmos, 137
von Mises, Richard, 228
von Schäwen, Paul, 144, 146
von Staudt, Karl Georg Christian, 251, 253–256, 259–263, 266, 268–275
Geometrie der Lage, 254, 257, 259, 261, 265, 270

Waerden, Bartel Leendert Van der, 11
Wallin, Nils-Bertil, 89
Wallis, John, 21, 284, 286, 299
Wang Ling, xviii, 45, 68–78, 344–346, 348
Weber, Heinrich, 291
Weierstrass, Karl Theodor Wilhelm, 2
Weil, André, 11, 13, 231
Weiss, Paul, 315, 318
Wellstein, Josef, 291
Whiteside, Derek Thomas, 16, 17, 28
Whittaker, Edmund, 227
Wiles, Andrew, 7
Wirkungsgeschichte, 30
Wolff, Christian, 21
Wu Wenjun, 74, 75, 330, 331, 334, 337
Wu Wentsun, *see* Wu Wenjun
Wylie, Alexander, xx, 55, 134, 135, 138–150, 157

Xenocrates, 173, 175
Xiahou Yang suanjing (*Xiahou Yang's Mathematical Classic*), 341
Xu Yibao, 342

Yale University, 309
Yang Hui, 332, 334
Suanfa tongbian benmo (*Alpha and Omega of Variations on Methods of Computation*), 332, 336
Xugu zhaiqi suanfa (*Continuation of ancient mathematical methods for elucidating the strange properties of numbers*), 334
Yang Shou-Jing, 336
Yi Xing (Yih-Hing, Yih King, I-Hing), 139, 145, 147, 150, 152, 154–157
Yongle dadian (*Yongle encyclopaedia*), 329, 339
Young, John Wesley, xxii, 252
Yushkevich, Adolph-Andre Pavlovich, xxii, 240

Zacharias, Max, 271, 281, 290, 291, 295, 299, 303, 304
Zamberti, Bartolomeo, 181–183
ZBSJ, *see* Zhoubi suanjing
Zentralausschuss für Innere Mission (Central Commission for Home Mission), 134
Zeuthen, Hieronymus G., 10
Zhao Shuang, 334, 348, 350
Zhmud, Leonid, 109
Zhoubi suanjing (*Mathematical classic of the Zhou gnomon*), 332, 334, 341, 345, 348, 350

For EU product safety concerns, contact us at Calle de José Abascal, 56–1°,
28003 Madrid, Spain or eugpsr@cambridge.org.

www.ingramcontent.com/pod-product-compliance
Lightning Source LLC
LaVergne TN
LVHW101456200926
848483LV00001B/140

* 9 7 8 1 1 0 8 8 3 4 9 6 4 *